STRATEGIES FOR CENTRAL AND EASTERN EUROPE

Other books written by Andrzej K. Koźmiński

Catching Up? Organizational and Management Change in the Ex-Socialist Bloc

Winning: Continuous Improvement Theory in High Performance Organizations
  co-authored with Donald P. Cushman and Krzysztof Obloj

Organizational Communication and Management
  co-authored with Donald P. Cushman

Other books written by George S. Yip

Asian Advantage: Key Strategies for Winning in the Asia-Pacific Region

Total Global Strategy: Managing for World-wide Competitive Advantage

Barriers to Entry: A Corporate Strategy Perspective

Strategies for Central and Eastern Europe

Edited by

Andrzej K. Koźmiński and George S. Yip

STRATEGIES FOR CENTRAL AND EASTERN EUROPE

Copyright © 2000 by Andrzej K. Koźmiński and George S. Yip

All rights reserved. No part of this book may be used or reproduced in any manner whatsoever without written permission except in the case of brief quotations embodied in critical articles or reviews. For information, address:

St. Martin's Press, Scholarly and Reference Division,
175 Fifth Avenue, New York, N.Y. 10010

First published in the United States of America in 2000

This book is printed on paper suitable for recycling and made from fully managed and sustained forest sources.

Printed in Great Britain

ISBN 0–312–23550–X

Library of Congress Cataloging-in-Publication Data

Strategies for Central and Eastern Europe / edited by Andrzej K. Koźmiński and George S. Yip
    p. cm
    Includes biographical reference and index.
    ISBN 0–312–23550–X
    1. Europe, Eastern—Economic conditions—1989– . 2. Europe, Eastern—Commerce. I. Koźmiński, Andrzej K. II. Yip, George S.

HC244 .S7628 2000
338.947—dc21   00–034491

*We dedicate this book to all those
whose sacrifices helped to free
the people of Central and Eastern Europe*

Contents

List of Tables and Figures		viii
Preface		x
About the Authors		xii
1	**Evaluating the Central and Eastern European Opportunity: Eastern Promise** Andrzej K. Koźmiński, George S. Yip and Anna M. Dempster	1
2	**Hungary: Goulash Capitalism** János Vecsenyi and Gábor Szigeti	33
3	**Poland: From Solidarity to Solid Economy** Adrian Szumski	57
4	**The Czech Republic: A Window of Opportunity** Helena Hruzová and Zdenek Soucek	86
5	**Slovakia: Seeking Alliances** Ján Morovič	119
6	**Slovenia: Small is Successful** Nenad Filipovič	150
7	**The Baltic States: Trading Hub** Robertas Jucevičius	177
8	**Romania: Resources for the Region** Cristian M. Băleanu	209
9	**Bulgaria: Building on the Black Sea** Bojil Dobrev and Stefan Dobrev	240
10	**Ukraine: Europe's Frontier** Bohdan Hawrylyshyn and Pavlo Sheremeta	264
11	**Conclusions: The Sun Rises in the East** Andrzej K. Koźmiński and George S. Yip	284
Appendix Globalization Measures		304
Index		315

List of Tables and Figures

Figure
1.0 Map of Central and Eastern European Region   5

Table
1.1 Key Economic Statistics of the C&EE Region   2
1.2 Dependence on Trade   7
1.3 Mutual Dependence of C&EE Economies   8
1.4 Statistical Data for CEFTA Countries   11
1.5 Foreign Direct Investment Flow (US$m)   14
1.6 Average Monthly Wage (US$)   19
1.7 Transparency International   22

2.1 Macro-Economic Indicators for Hungary   37
2.2 Annual Flow of Foreign Direct Investment in Hungary   37
2.3 Structure of Foreign Direct Investment in Hungary   38
2.4. Largest Multinational Companies in Hungary   39
2.5. Largest Foreign Investments in Hungary   40
2.6 Largest Hungarian Exporting Companies   41
2.7 Largest Hungarian Firms   42
2.8 Foreign Investors' Motivations in Hungarian Investments   45

3.1 Macro-Economic Indicators for Poland   60
3.2 Largest Investors in Poland   63
3.3 Foreign Investing Countries in Poland   64
3.4 Foreign Direct Investments – Branches of Investments   65

4.1 Macro-Economic Indicators for Czech Republic   88
4.2 Commodity Structure of Czech External Trade   91
4.3 Largest Czech Companies   91
4.4 Largest Exporting Domestic Companies in the CR   92
4.5 Largest Foreign Investments in 1990–97   93
4.6 Scale of the Largest Foreign Investments in the CR   94
4.7 Top Ten Largest Joint Ventures in the CR   95
4.8 Top Ten Largest Foreign Firms in the CR   98

5.1 Macro-Economic Indicators for Slovakia   122
5.2 Largest Domestic Companies Active in Slovakia   126
5.3 Development of FDI in Slovakia   127
5.4 Foreign Direct Investment into Slovakia by Country   128
5.5 Main Foreign Companies investing in Slovakia   128
5.6 Distribution of FDI in Slovakia by Sector   129
5.7 Largest Investors in Slovakia   130
5.8 Wage–Education Structure in Slovakia   133

6.1 Macro-Economic Indicators of Slovenia   152
6.2 Largest Slovenian Companies   156

Table

6.3	Largest Slovenian Exporters	156
6.4	Top Countries of Origin of FDI in Slovenia	158
6.5	Largest Foreign Investors in Slovenia	159
6.6	Patterns of Household Consumption	161
7.1	GDP Levels of the Baltic States	179
7.2	Largest Lithuanian Compnaies	181
7.3	Largest Latvian Companies	182
7.4	Largest Estonian Companies	182
7.5	Largest Foreign Investors in Lithuania	190
7.6	Largest Foreign Investors in Estonia	192
7.7	Detailed Statistics for the Baltic States	205
8.1	Macro-Economic Indicators for Romania	211
8.2	Privatization by Sector	212
8.3	Largest Source Countries of Foreign Capital	213
8.4	Foreign Investment in Romania by Sector of Activity	214
8.5	Largest Investors in Romania	215
8.6	Largest Romanian companies	217
8.7	Gross Monthly Base Salaries for Management	222
8.8	Major Joint Ventures in Romania by Sector	232
9.1	Macro-Economic Indicators for Bulgaria	242
9.2	Main Foreign Investors in Bulgaria	244
9.3	Leading Foreign Companies in Bulgaria	245
9.4	Leading Bulgarian Companies	246
10.1	Macro-Economic Indicators for Ukraine	266
10.2	Largest Ukrainian Companies	267
11.1	Summary of Market Globalization Drivers	288
11.2	Summary of Cost Globalization Drivers	290
11.3	Summary of Government Globalization Drivers	293
11.4	Summary of Competitive Globalization Drivers	295
11.5	Summary of Recommended MNC Overall Strategies	296
11.6	Summary of Recommended MNC Activity Location Strategies	298
11.7	Summary of Recommended MNC Organization Approaches	300
A1	GDP (% change)	310
A2	Industrial Production (% change)	310
A3	Budget Balance (% of GDP)	311
A4	Unemployment (%)	311
A5	Inflation (%)	312
A6	Foreign Debt (US$bn)	312
A7	Foreign Exchange Reserves (US$bn)	313
A8	Exchange Rate (US$)	313
A9	Trade Balance (US$bn)	314
A10	Current Account Balance (US$bn)	314

Preface

We are writing this Preface on the day that the world is commemorating the tenth anniversary of the Fall of the Berlin Wall. That opening and the long series of preceding events made this book possible. The Fall of the Berlin Wall was the culmination of the long resistance of the people of Central and Eastern Europe against communism. This ever-present resistance was marked by landmark events such as peasant rebellions in post World-War II Poland, Ukraine and the Baltics; workers' riots in Berlin in 1954, in Poznan (Poland) in 1956, and in Gdansk (Poland) in 1970; the Hungarian uprising of 1956; the invasion of Czechoslovakia in 1968; and martial law in Poland in 1981. Also, some positive changes marked the disintegration of the Soviet bloc; 'Polish October' of 1956, 'Prague Spring' of 1968, the foundation of 'Solidarity' in Poland in 1980, *perestroika* in the USSR and finally the first democratic election in Poland in 1989. These events opened the way to the demolition of the Berlin Wall and the collapse of communism in the whole region, opening a new era of democracy and open markets. In this new era, foreign multinational companies play a very significant role.

This book is written to help managers in multinational companies develop strategies for Central and Eastern Europe as a region and for individual countries. The book is intended for both those interested in entering the region and those already there. Our research clearly demonstrates that in spite of the common communist past, which was really common only to a certain degree (different 'models of socialism'), different countries of the region developed different transition patterns. Foreign entrants need guidance in such a diversified and heterogeneous business environment. This book is intended to provide such guidance.

This book is the culmination of a three year effort by a team of sixteen contributors spread across eleven European countries. All of them are both highly respected academics in their respective countries as well as business practitioners. We have met as a group three times in Warsaw at the Leon Koźmiński Academy of Entrepreneurship and Management, individually in different parts of the world, and electronically frequently. We wish to thank all our contributors for, first, joining the team and, second, for all the efforts they have put in.

We particularly wish to thank Anna Dempster and Adrian Szumski, who took on many tasks in addition to their contributions to specific chapters. Anna Dempster edited and rewrote every chapter to ensure consistency to the common framework and to provide a common voice and tone. Adrian Szumski took on the brunt of the administrative effort of coordinating the meetings and communications with the contributors. We also thank a number of support staff, especially Dawn Perry and Jenny de Rivaz in Cambridge as well as Anna Putro and Jola Prokop in Warsaw.

Our project was partially financed by the Leon Koźmiński Academy of Entrepreneurship and Management in Warsaw as a part of its international research drive. Funding from LKAEM made our research possible. We thank the Academy for it.

Special thanks are due to Stephen Rutt and Gloria Hart of the publisher for their support of this project.

Lastly, we thank our wives, Alicja and Moira, for coping with our absences on trips or while working in our studies and offices.

Andrzej Koźmiński, *Warsaw* *November 1999*
George S. Yip, *Cambridge and London*

About the Authors

Cristian M. Băleanu, *Deputy Director, The International Management Foundation; Director General, FIMAN Development Services; and Visiting Professor, University of Bucharest.* MSEE Bucharest Politechnic Institute; PhD Sociology, University of Bucharest. Author of *Management of Continuous Improvement*, and co-author of *Project Management Manual*. Articles in *Strategic Change, Information Processing Letters, Computers and Artificial Intelligence, Romanian Management Journal*. Previous experience in research and development with several major Romanian companies. (cristi@fiman.ro)

Anna M. Dempster, *Researcher in Global Strategy and Marketing, Cambridge University's Judge Institute of Management Studies.* MPhil. in Historical Studies, Thesis on Audience Perception, Cambridge University. Currently completing doctoral degree in management at Cambridge University, England. (a.dempster@jims.cam.ac.uk)

Bojil Dobrev, *Academic Director International University Sofia, Professor in Informatics at International University, Vice Chairman of the Bulgarian Information Industry Association.* MSc Technical University, Sofia. Training and Research at IBM, CEPIA, Tokyo University, universities in Germany. Author of forty publications, surveys and methodologies in English, German and Bulgarian languages. Participant and consultant at international projects. Previous management positions in information technologies research and development institutes and companies. (intunv@ttm.bg)

Stefan Dobrev, at present *East-West Business Analyst, GMD Techno Park, Bonn.* BA Economics, University of Oxford. Experience in sales operations in the United States and in marketing strategy planning in Germany. Currently Associate in Strategic Consulting with LEK Consulting, London (s_dobrev@lec.com).

Nenad Filipović, *Deputy Director and Professor of General Management, International Executive Development Center, Slovenia.* MSc University of Zagreb. Author of *Dictionary of Management Terminology*, articles in international journals and more than twenty case studies. Board member of Central and East European Management Development Association (CEEMAN). Board director of European companies. Twelve years of management experience in executive positions with international companies. (nenad@iedc-brdo.sl)

About the Authors

Bohdan Hawrylyshyn, *Chairman, International Center for Policy Studies (Kyiv), International Management Institute (IMI-Kyiv), Foundation Vidrodgenia (Geneva)*, President of Ternopil Academy of National Economy. For 29 years, between 1968 and 1986 as Director, International Management Institute (IMI-Geneva), MAA Mechanical Engineering, University of Toronto, PhD Economics, University of Geneva. After the independence of Ukraine he served as advisor for Leonid Kravchuk, the first President of Ukraine, and devoted the bulk of his time and energy to institution building in the country. (hruzova@vse.cz)

Helena Hruzová, *Senior Lecturer, Department of Business Management, UEP and Visiting Professor in MBA Program, Prague International Business School*. Visiting Professor in DMS Program (Prague) initiated by Brunel University, UK (1993–95). Research Degree from the University of Economics, Prague, Stanford Executive Program and Central and East European Teachers Program at Massachusetts Institute of Technology. A planner in a construction company. International experience (for example, team member of Interfaculty Group of the Community of European Management Schools (CEMS) and active in international research programs). Researcher for enterprises, cooperator with European and North American Universities, cooperator with industrial and telecommunication enterprise and public sectors. (hruzova@vse.cz)

Robertas Jucevičius, *Director of School of International Business and Management, Kaunas University of Technology*, and head of qualification commission for research, academic and doctoral studies in management and business administration. Diploma of Engineer in Chemical Technology, Kaunas Politechnical Institute; PhD economics, Vilnius University; habilitated doctor's degree, Vytautas Magnus University and Kaunas University of Technology. Former Professor at Vytantas Magnus University and Fulbright Scholar, Purdue University. Research on strategic management practice and business intelligence. Heading the group of researchers and business executives developing Lithuanian Industrial Policy. Has made more than eighty publications locally and internationally. (robertas_jucevicius@fc.vdu.lt)

Andrzej K. Koźmiński, *founder and Rector of Leon Koźmiński Academy of Entrepreneurship and Management in Warsaw, President of International Business School S.A.* PhD, Warsaw School of Economics. Author of 17 books and over 250 articles in all major languages, including *Catching Up? Organizational and Management Change in the Ex-Socialist Block, Winning: Continuous Improvement Theory in High Performance Organizations, Organizational Communication and Management*. Articles in *International Studies of Management and Organization, Communist and*

Post-Communist Studies, Communist Economies, Journal of Management Development, Revue d'Etudes Comparatives, Est–Ouest and other international journals. Elected member of Polish Academy of Sciences, International Academy of Management, Académie Européene, Academie des Sciences Commerciales. Board member of EFMD (European Foundation for Management Development), Vice President of CEEMAN (Central and East European Management Development Association). Member of Prime Minister's Poland Council for Social and Economic Strategy. Visiting Professor at several American and European Universities including UCLA, Université d'Orleans, Donau Universitat. Board director of companies operating in Poland. (kozmin@it.pl)

Ján Morovič, *Agricultural Minister's Advisor for Integration and International Cooperation, General Executive Secretary for Central and Eastern Europe of GDLA (Global Distance Learning Association).* Founder and President of the City University Bratislava (1990–98), Member of Parliament, Federal Assembly of the Czech and Slovak Federal Republic (1990–92), Research Scholar, International Institute for Applied Systems Analysis (1980–85). MSc equivalent, Informatics, Institute of Automation and Regulation, School of Electrical Engineering, Slovak Technical University, Bratislava. PhD, Applied Cybernetics – Computer Science, Slovak Technical University, Bratislava. (da0027@ba.entry.sk)

Pavlo Sheremeta, *Dean, Kyiv Mohyla Business School.* MBA, Emory University, Graduate Management Program Certificate, International Management Center in Budapest and Katz Graduate School of Business, University of Pittsburgh; Diploma in Economics, Lviv State University. Formerly MBA Program Director, International Management Institute IMI-Kyiv; Project Manager, Open Society Institute, Budapest; and assistant to member of the Ukrainian Parliament. Teaches international business and human resource management at several business schools in Ukraine. (sheremet@pavlo.relc.com)

Zdeněk Souček, *Professor and Director, Institute of Consultancy and Economic Expertises, Prague Economic University. Managing Director, Management Focus International Consulting Group Ltd and Strategia Ltd*; DSc Prague School of Economics. Author of four extensive books on industrial dynamics, strategic management and management of industrial firms; co-author of many other books and university text-books; of several hundred articles and monographs on management and application of mathematical methods, published in many countries. Former deputy director of large industrial organisations in Czechoslovakia and Poland.

About the Authors

Gábor Szigeti, *Freelance research fellow and consultant in corporate management.* BA, sociology, Budapest University. Author of several case studies on organization change. Co-author of articles in Management Science in Hungary on companies in transition.

Adrian Szumski, *Assistant Professor, (International Management and Organisational Theory), Leon Koźmiński Academy of Entrepreneurship and Management*, MA Warsaw University, co-author of Organisational Behaviour. PhD candidate, disputation planned for summer 2000. Previous faculty positions at Warsaw University. Experience in consulting and management in small business. Internal consultant for automotive chemistry distributor and computer distributor and production company. (adrian@wspiz.edu.pl)

János Vecsenyi, *Director of Home of Learning at Budapest Bank Hungary, a GE Capital Affiliate, Adjunct Professor at Budapest University of Economic Sciences (Management).* MS Budapest Technical University; PhD University of Economic Sciences. Author of *Entrepreneurial Organizations and Strategies* (1999), *Decision Analysis with Managers* (1989), *Total Quality Management* (1979). Articles in the fields of total quality management, innovation, decision making, strategic management, and entrepreneurship in *Management Studies* (Hungary), *Risk Management, Applied Systems Analysis*, and *Applied Psychology*. Member of Management Science Committee at Hungarian Academy of Sciences. Board director of Hungarian companies. (vecsenyij@bbrt.hu)

George S. Yip, *Beckwith Professor (Marketing and Strategy), Cambridge University's Judge Institute of Management Studies; and Co-Chairman, Perseus Consulting.* MA Cambridge; MBA Cranfield and Harvard; DBA Harvard. Author of *Barriers to Entry, Total Global Strategy*, and *Asian Advantage*. Articles in *California Management Review, Columbia Journal of World Business, Harvard Business Review, International Marketing Review, Sloan Management Review*, and *Strategic Management Journal*. Fellow of Academy of International Business. Board director of European companies. Previous faculty positions at Georgetown, Harvard, Oxford, Stanford, and UCLA. Management experience with Price Waterhouse and Unilever. (g.yip@jims.cam.ac.uk)

1 Evaluating the Central and Eastern European Opportunity – Eastern Promise

ANDRZEJ KOŹMIŃSKI, GEORGE S. YIP and
ANNA MEDOVA DEMPSTER

CENTRAL and Eastern Europe has undergone some of the most dramatic and momentous changes in recent history. The fall of the Berlin Wall in November 1989, the storming of the Russian White House in August 1991, and the declaration by the USSR Supreme Soviet in December 1991 that 'the Soviet Union no longer exists' came after years of suppressed opposition to the *status quo*. Goulash communism in Hungary, the Velvet Revolution in Czechoslovakia, Solidarity, and the trade union movement in Poland, and *perestroika* in Russia – groups and single individuals – all played a role in the tide of events that swept the region. As the Iron Curtain crumbled and communism retreated, a massive, previously off-limits market opened to the international business community.

The response has been impressive. The region, as a whole, has attracted a great deal of attention from foreign investors and considerable amounts of both direct and portfolio investment have been made. Taking advantage of new opportunities, multinational and other companies have chosen to establish sales, production, and increasingly research and development activities in previously uncharted territory. Conscious decisions have been made by many governments to actively promote foreign investment. Although the development of individual countries has varied dramatically, the fact that so much has been achieved in such a short time is testimony to the adaptability and perseverance of people in the region. That Central and Eastern Europe continues to experience growth despite major setbacks, like the Russian financial crisis of 1998, is evidence that there is much Eastern promise for the future.

Central and Eastern Europe (C&EE) offers many opportunities to become an integral part of a global company's value chain. As the region increases in international importance, managers of multinational companies (MNCs) will need to develop effective strategies to take both the region as a whole and individual countries into account. Table 1.1 shows the key statistics for countries described in the individual country chapters. This book is written to help managers to develop these strategies: both for those interested in entering the region and those already there. As markets in C&EE expand and competition increases, success will depend on integrated strategies that take account of global, regional and local variables.

Table 1.1 Key Economic Statistics of the C&EE Region

Country	Population 1999 (m)	Nominal GDP 1999 (US$bn)	Avg. GDP % Change 1995–1999	Nominal GDP per capita	GDP PPP* 1999 (US$bn)	GDP Per Capita PPP* 1999, (US$)
Hungary	10.5	48.9	3.3	4657.1	119.1	11346.0
Poland	38.7	154.6	5.8	3994.8	344.8	8910.0
Czech Rep.	10.3	53.5	1.4	5194.2	137.8	13374.0
Slovak Rep.	5.4	18.6	5.3	3444.4	56.3	10418.0
Slovenia	2.0	19.6	4.0	9800.0	31.3	15633.0
Latvia (98)	2.5	6.4	3.7	2560.0	10.3	4136.0
Lithuania (98)	3.7	10.7	4.7	2891.9	16.4	4425.0
Estonia	1.5	5.3	5.1	3533.3	8.2	5455.8
Romania	22.5	32.6	-1.0	1448.8	132.1	5872.0
Bulgaria	8.2	12.0	-1.6	1463.4	42.8	5218.0
Ukraine	49.7	30.8	-5.5	619.7	167.8	3376.0

Source: Business Central Europe Magazine, The Economist Group, http://www.bcemag.com/
Historical data collected from WIIW, EBRD, Reuters, National Statistics.

Note: Indicators for Latvia and Lithuania are for 1998

* Purchasing power parity

Developing a Successful Regional Strategy[1]

To succeed in the C&EE region, a company needs to work on four levels:

1. Overall global strategy.
2. Regional strategy.
3. Country strategy.
4. Country operations.

Overall global strategy. Before deciding whether and how to do business in Central and Eastern Europe (or in any other region of the world), a company needs to have a clear global strategy. Key elements of this strategy include the core business strategy, the competitive objectives for the business, and the extent to which the business will be operated as a single integrated business or as a looser collection of geographically independent units.

Regional strategy. Next, a company needs to decide on the overall role of C&EE within the global strategy. Should C&EE be primarily a source of growth, or profit, or both? Should C&EE be primarily a source of supply or a locus of markets? Or should different countries play different roles? In which countries in C&EE should the company do business?

Country strategy. Having selected the countries in which to be involved, the company should develop a country strategy that includes the mode of entry, partner selection, the usual elements of a business strategy (including what activities to conduct and what parts of the value chain to locate in the country), and how activities in the country will relate to those in the rest of the region and the world.

Country operations. Lastly, the company has to be concerned with implementation at the operational level. Here the company has to deal with detailed matters such as how to adapt to local culture and business practices, how to develop the right kinds of contacts, how to find customers, and how to cope with a country's written and unwritten regulations.

Defining Central and Eastern Europe

How does one define the region of Central and Eastern Europe? What relationship does a given definition have with the rest of Europe and the world? This is a multifaceted question. While taking geography into account, a region is also defined by less tangible characteristics. A multinational company must consider all these characteristics to develop a successful long-term strategy. These include:

- Geography.
- Culture/religion.
- History.
- Language.

- Way of doing business.
- Form of government.
- Institutional arrangements.
- Cross-investment.
- Intra-regional trade.
- Trade policies and agreements.
- Economic performance and prospects.
- Infrastructure (physical and institutional).

For travel and administrative reasons alone, the geographic coherence of a region provides the necessary links between new and existing operations. From this point of view the C&EE region has a number of natural advantages. Geographically, the 'New Europe' of the East is only a stone's throw away from the West. For example, much of Central Europe is just across the border from Germany, the current industrial powerhouse of Europe. Not many people realize that the Polish border is only 80 km from Berlin, and the Czech border is 100 km from Nuremberg, while the Slovak and Hungarian borders are within 50 km of Vienna. Figure 1.0 provides a map of the region of Central and Eastern Europe, with the countries studied in this book shaded darker.

In terms of social make-up – culture, history, religion, language and ethnic groups – the C&EE region is diverse and varied. Many existing social divisions do not correspond to constructed political and geographic boundaries. For example, there is a large population of ethnic Hungarians in Romania and the former Yugoslavia, Slovaks in Hungary, Ukrainians in the Slovak Republic and Poland, and Russians in the Ukraine and the Baltic Republics and Romanians in the Ukraine. Throughout the region large minority groups, of equal importance, co-exist within a shared geographical area. It is not surprising that much recent history has been one of conflict and territorial dispute. And yet it is also this diversity and adaptability that gives the region as a whole positive advantages to develop a multicultural approach to the business of the future, and adopt a variety of new management ideas and production techniques that are rapidly being introduced by foreign companies.

The Logic of Transition[2]

Transition to a market economy is a multi-phase process. These phases can, and often do, overlap. For example, institutional changes can and should be initiated in parallel with inflation control. Early 'marketization' usually occurs as the political phases gain momentum. The general sequence, however, seems to be valid in the case of most transition economies.

Economic differences among countries are correlated with the phase which they have reached in the transition process, from planned to market economy. Different countries in the region are still at very different phases. Six distinct phases can be identified:

Figure 1.0 Map of central and eastern European region

Note: Dark shaded area = countries and economies studied

- *Political change.* In this phase, the communist political monopoly is abolished, political freedoms are secured, and ethnic conflicts are dealt with through a democratic process.
- *Early marketization.* In this phase, price controls are abolished, the main legal barriers to private entrepreneurship are lifted, and internal convertibility of the currency is introduced (including the freedom to buy and sell foreign currency).

- *Inflation control.* In this phase, government subsidies for ailing state-owned enterprises are eliminated, budget deficits reduced and tight monetary policies are implemented to control inflation.
- *Market institutions building.* In this phase, a massive privatization drive is combined with the introduction (or reactivation) of a legal framework to regulate markets and help build effective market structures. In order to stimulate the privatization drive, banking reforms, government administration reforms, financial markets, and stock exchange start-ups have to accompany this phase. Commercialization of social services (health care, education, retirement) also occur.
- *Anti-recession policies.* In this phase, entrepreneurial activities as well as internal demands and exports are stimulated by fiscal and monetary policies (such as interest rates), free trade policies and foreign capital inflow.
- *Sustainable growth policies.* In this phase, a solid institutional framework is combined with government policies and mature business strategies that provide for sustainable growth of mature market economies.

The sequence and timing of phases is crucial for successful transition. The more advanced phases cannot be initiated without successful completion of earlier ones. For example, privatization without completion of the political change leads to massive appropriation of national wealth by the representatives of the old communist party 'nomenklatura', the stripping of assets and massive illegal capital outflow.[3] Capital markets can only really take-off when inflation control ends. Government structures inherited from communism make anti-recession policies costly and inefficient. The examples can go on. Once a phase is completed the next one must be initiated without delay. Experience shows that delays in institutional changes in particular are likely to hinder successful transition. Delays and the lack of proper sequencing can easily lead to dangerous loops in the process. For example, lack of energetic and efficient anti-recession policies can compromise economic and social equilibrium and lead to a return of inflation and even political disorder.

It is clear that different countries of Central and Eastern Europe (C&EE) have developed at different rates and consequently find themselves at different phases of the transition process. The Czech Republic, Hungary, Poland and Slovenia have reached the last phases, anti-recession and sustainable growth. In contrast, several post-Yugoslav and post-Soviet countries are still dealing with the first phase, political reform. Belarus and Russia are experiencing a setback in their transition process because of the unexpected financial crisis of 1998 and have even returned to some elements of the former system such as price controls. Clearly, as a region, Central and Eastern Europe is heterogeneous and highly diverse.

Of the countries in the C&EE region, the Czech Republic, Poland, Hungary, Slovenia and Estonia are considered on the 'fast track' to full European Union (EU) membership. Historically, these countries have many links with Western Europe. They have common cultural elements, their industrial base was relatively developed before World War II and politically, experienced communism more than a quarter of a century later than did Russia. For these reasons, the recent transformation process to a market economy has been much faster here than in the rest of the region. The relatively stable and risk free environment of Poland, Hungary and the Czech Republic was recognized by the international community and West European countries when they became members of NATO in April, 1999. Slovenia, as the most advanced of the former Yugoslav republics, shares much with Austria, its neighbor and traditional ally and this has similarly aided its transition process, the development of its economy, market institutions and assimilation of Western business practices.

Further from the challenge, Romania and Bulgaria are attempting to control inflation and build up effective market institutions in order to create a more stable investment environment for foreign companies. The Ukraine, still in a relatively early phase of the transition process, having suffered setbacks (due in part to the Russian financial crisis), is also attempting to restructure its institutions and develop a market place along Western European lines. Similarly to all the countries included in this book, the Ukraine has consciously turned its gaze westward in an attempt to attract foreign direct investment into the region. Of the three Baltic Republics, Estonia is the most 'Westernized' and is also set to join the European Union in the first wave of countries from C&EE. Its two neighbors, Lithuania and Latvia, have set in place most of the key institutions for market building though they have been more guarded in their adoption of the transition process. They retain a large ethnic Russian population and links with Russian in terms of intra-regional trade. The degree to which countries depend on trade varies throughout the region. Table 1.2 shows how much each country depends on imports and exports.

Table 1.2 Dependence on Trade

	Exports	Imports
Hungary	44.7	49.3
Poland	20.2	29.3
Czech Rep.	46.5	50.8
Slovak Rep.	50.2	60.9
Slovenia	47.9	53.2
Latvia	n.a.	n.a.
Lithuania	34.9	54.7
Estonia	n.a.	n.a.
Romania	22.5	32.1
Bulgaria	36.7	42.7
Ukraine	26.5	31.0

Source: CEEBIC, The Central And Eastern Europe Business Information Center Market Access and Compliance Division of the US Department of Commerce.

http://www.itaiep.doc.gov/eebic/cefta/data.htm

The Role of Russia

The role of Russia plays an important part in the development of all the Central and Eastern European countries. However, Russia has not been included in this study as it is so large and diverse a region in itself, and the recent changes there have been so dramatic, that it warrants a book on its own.

From an international business point of view, these countries all share three important characteristics that are dictated by their past alliance within the Soviet Union:

- A recent communist past within the Soviet Union.
- The perception by the West as being part of the former communist bloc.
- The present process of transition from authoritarian states and centrally planned economies, to democratic states and market economies.

For these reasons, the investment climate of the C&EE region is, to a degree, tied to developments in Russia. The Russian financial crisis, which rocked the world in 1998, also contributed to diversifying the reaction of foreign investors, their optimism and their confidence in the whole C&EE region. It has highlighted the differences between those countries that are still linked to the Russian economy and those that are largely independent of it. Table 1.3 shows the extent to which countries in the region depend on imports and exports from Russia and thus the degree to which they are affected by events in Russia. The most economically developed countries, such as Poland, Hungary and the Czech Republic, have escaped relatively unscathed from the Russian crisis. Overall they have proved to be more stable than their giant, volatile neighbor. Other countries in the C&EE region that have retained more links with their past ally, have been more vulnerable to shifts within the Russian marketplace and the loss of investor confidence. Finally, countries in the region that have not been included in this book are those that have not yet begun to compete successfully internationally, are still plagued by internal problems, or have chosen not to turn their gaze westward. They remain relatively impenetrable to the international business community and include much of the former Yugoslavia, Albania, Moldova, and Belarus.

Table 1.3 Mutual Dependence of C&EE Economies

	Imports into Russia		Exports from Russia	
	% of all	% of GDP	% of all	% of GDP
Hungary	5.0	2.4	9.6	5.0
Poland	6.5	1.6	6.3	2.0
Czech Rep.	3.3	1.4	5.0	2.4
Slovak Rep.	3.7	1.7	15.6	8.2
Romania	3.0	0.7	12.2	3.9
Bulgaria	8.0	4.3	28.1	14.9
Average in Region	4.9	1.6	9.0	3.6

Source: EBRD data: Transition Report, London, 1996.

Role of Integrating Factors for Business in Central and Eastern Europe

In spite of the diversity found in the region and many differences mentioned above (past, present and future), a number of important factors integrate the region from a business point of view. These include past factors such as a shared history and past trade blocs, shared cultural, similarities in language and, attitudes to business. However, integrating factors also include recent developments such as changes in the form of government, legal and institutional arrangements, cross-investment and the entrance of multinational companies. Together, these developments have changed the business landscape of C&EE and made intra-regional trade, as well as links world-wide, simpler to establish and support.

Inter-regional trade

CMEA (Council for Mutual Economic Assistance), known as COMECON, was created by Josef Stalin in 1949 as a response to the Marshall Plan for Europe. As an instrument of economic integration of the 'socialist bloc', it was used to promote and finance the military build-up of the Warsaw Pact. Within the framework of COMECON, countries were assigned a particular 'specialization' intended to capitalize on economies of scale and serve the 'common socialist market'. Under this scheme manufacturers located in one country supplied all the other countries. Russia supplied the whole region with oil, natural gas, ore, cotton and other raw materials, as well as aircraft (military and civil) and most of the armaments. Poland specialized in shipbuilding and coal mining. Bulgaria specialized in computers, Lithuania in consumer electronics, Hungary in pharmaceuticals, and so on. However this plan was never as effective as was hoped because countries did not trust each other and wanted to develop their own capabilities. The simultaneous development of the automobile industry in the USSR, Poland and Czech Republic is a good example of such tendencies. Nevertheless, COMECON trade was of key importance for the whole region. In 1975, it accounted for 55.7 per cent of imports and 61.5 per cent of exports of six countries (USSR, Bulgaria, the Czechoslovak Republic, Hungary, Poland and Romania). Respective figures for 1989 were 48.8 per cent and 50.3 per cent.[4] Within the USSR these economic ties were even stronger. Many of these interdependencies still remain valid. For example, several Eastern European countries still import most of their natural gas and considerable quantities of oil from Russia and recent figures of intra-regional trade remain impressive.

Several factors contribute to the future prospects of intra-regional trade:

- The success of regionally established, large brand names such as the Czech Republic's Skoda automobiles and Pilsner Urquell beer, the former East Germany's motorcycles, Poland's Pollena cosmetics and Hungary's Tungsram light bulbs.

- Established commercial contacts used by the same people who traded with each other under communism.
- The cost advantage of C&EE producers over Western ones combined with the high price sensitivity of C&EE buyers. For example, this is the primary reason why Russian, Ukrainian and Belorussian tractors and automobiles, such as Minsk, Tavria, Ziguli, or Vladimiretz, still find buyers in Central Europe.
- The superior understanding of local tastes by producers from neighboring countries. For example, Polish clothing and furniture manufacturers specialize and successfully compete in Commonwealth of Independent States (CIS) markets.
- Similar technical standards adopted throughout the whole Soviet bloc, for example, electrical turbines of Russian design.
- Economies of scale and location advantages that result from serving the whole region from one location or developing synergies between geographically neighboring markets which exhibit similar characteristics.
- A pattern of 'small cross-border trade' carried out by hundreds of thousands of individuals buying small quantities of goods in one country and selling them in another, making a small profit by evading taxes, tariffs and administrative costs, that legal entities have to pay. For example, in Poland the amount of cross-border trade, due to both German and CIS citizens coming into the country for cheaper goods, was estimated at $4bn in 1997.

Some MNCs have eagerly taken advantage of these integrating factors. For example, Volkswagen acquired the local Skoda brand name. Asea Brown Boveri (ABB) is capitalizing on the similarity of technical standards in design of power generating equipment. Benckiser (detergents and cosmetics) is serving the whole region from two locations, Poland and the Czech Republic.

CEFTA (Central European Free Trade Association) was established by the Czech Republic, Hungary, Poland, Slovak Republic and joined later by Slovenia, Romania and Bulgaria in order to promote trade between countries considered the most advanced in the transition process in the region. Statistical data for the CEFTA countries is illustrated in Table 1.4. The free trade agreement between the four countries was created to gradually establish a free trade zone by the year 2001 through reducing and eliminating tariffs. The objectives of CEFTA are to promote the development of mutual economic relations between members through the expansion of trade, provide fair conditions of competition, and expand world trade by removing tariff and non-tariff barriers. The creation of CEFTA was an attempt to increase trade among Central European nations after the collapse of old co-operation programs. The combined CEFTA market has a value of $316 bn and covers 97 million people. CEFTA members reduced their taxes on industrial goods by up to 80 per cent, and by up to 50 per cent on agricultural goods. Agriculture is the most difficult area of co-operation between members. Another problem on the horizon is how CEFTA will survive access to the EU.

Table 1.4 Statistical Data for CEFTA Countries (1997)

	FDI (US$ m)	Exports (US$ bn)	Imports (US$ bn)	GDP US$ (per capita)	Population (millions)
Hungary	2.1	19.6	21.4	4462	10.1
Poland	3.0	27.2	38.5	3512	38.7
Czech Republic	1.3	1.4	27.1	5050	10.3
Slovak Rep.	0.1	8.8	10.3	3624	5.4
Slovenia	321.0	8.4	9.2	9101	22.5
Romania	1224.0	8.4	10.4	1549	22.5
Bulgaria	497.0	4.9	4.5	1227	8.3
Total	2085.5	78.7	121.4	28 525	117.8

Source: CEEBIC The Central And Eastern Europe Business Information Center Market Access and Compliance Division of the US Department of Commerce, http://www.itaiep.doc.gov/eebic/cefta/data.htm

Shared culture

In spite of the many differences, a shared culture in the region is an important integrating factor. A communist heritage and the process of transition can be considered strong bases for common characteristics. These include:

- A Western consumption model and Western life style are the essence of aspirations of Central and Eastern European consumers, facilitating standardized approaches to marketing;
- People are willing to work hard in order to make money and to enjoy Western consumption patterns.
- Local managers (particularly younger ones) are willing and able to accept western management philosophies and techniques (facilitated by good educational backgrounds and the unprecedented explosion in Western management education).
- An aggressive and entrepreneurial middle class (particularly numerous and influential in the more advanced countries) is rapidly emerging.
- Similar problems in the business environment exist throughout the region (although to different degrees in different countries). These include:
  - The public's sensitivity to a 'national interest' being put in jeopardy by foreign businesses and multinational companies.
  - The important role of informal relations in business, a lack of accepted standards of business ethics, corruption, and criminal activity.
  - A tradition of strong hierarchies in business translating into multi-level structures and authoritarian management practices.
  - Highly suspicious, confrontational and militant trade unions.

The existence of common characteristics, however, even if they are negative, can facilitate the entrance of foreign MNCs. Because many of the problems that can hamper MNC involvement are common throughout the region, regional strategies can be developed to deal with them effectively.

Languages

English is the dominant foreign language, with German in second place. Overall, the knowledge of English, and Business English, is on the rise and this trend is certain to continue. This is due to the popularity of English among the younger generation who perceive it as a 'ticket to a brighter future' and to the proliferation of American-style management education. In spite of this trend, the number of people who have absolute working knowledge of Business English is still limited, while translators and interpreters often do not know business terminology. This is why local language skills are highly valued by MNCs. For political reasons, Russian cannot be considered a regional business language (outside of the CIS) and using Russian in non-Russian speaking countries is a mistake leading to emotional, negative reactions. Knowledge of local languages, especially within such a diverse region, is highly valued by local business partners and employees, and facilitates business relationships. The knowledge of local history, culture and art, which are important distinguishing factors, is also appreciated.

Ways of doing business[5]

Four identifiable ways of doing business prevail in Central and Eastern Europe. The first type is represented by large, state-owned companies, recently privatized or retaining a dominant government share. These companies combine bureaucratic organizational cultures with the central importance of informal relations and still involving representatives of the old communist 'nomenklatura'. The second mode is represented by small and medium-sized, mostly family owned, private enterprises. These are often aggressive and entrepreneurial in spirit but suffer from poor managerial skills and a limited knowledge of Western business practices. The business ethics of these companies are often dubious due to the still remaining (particularly in the countries less advanced in transition) links to the 'second' (black) economy. The third mode is represented by large private enterprises founded after 1989. Especially in the most advanced countries only the most efficient of these have survived. These companies present a good knowledge of the local consumers and regional markets. In some cases they make use of already established networks and valuable business contacts. Most of these companies are still managed by their owners and founders, and they remain undercapitalized by Western standards. The process of mergers, acquisitions, and alliance formation has not yet started among local companies. Some of them, for example in the computer and software industry, are valuable business partners for leading multinationals. They can act as agents, licensed dealers, maintenance and training providers, and so on. The fourth mode is represented by subsidiaries of western MNCs or joint ventures practising standard Western-style management. But subsidiaries located in Central and Eastern Europe, in most cases, enjoy less autonomy than subsidiaries located in Western Europe.

Forms of government

All central European and Baltic post-communist countries can be considered as well-established democracies, with a free press, and independent courts. Some post-Soviet Asian republics such as Khazakstan (not covered by this book), and Belarus (also not covered) experience authoritarian regimes to different degrees. In all post-communist countries, however, law enforcement and corruption, particularly at the local government level, is a real and ongoing problem.

Institutional arrangements

Overall, local populations and the new political elite welcomed the disintegration of the Soviet Union, the Warsaw pact and COMECON in the 1990s. Because of this history, and being tied into a strict and often unprofitable trading bloc, institutional arrangements leading to regional integration are still regarded with some suspicion. Some bilateral arrangements are emerging like the Czecho–Slovak, Polish–Ukrainian, Russian-Ukrainian, and so on. The trade agreement, CEFTA, is also gradually gaining in importance.

Cross-investment

There is relatively little cross-investment between C&EE countries. This is due mainly to the low capitalization of local enterprises, the low degree of consolidation of industries in the region, and a strong local market orientation. Russian oil and gas producers such as Gazprom or Lukoil are the notable exceptions. They try to control distribution networks and refineries in many C&EE countries by importing their products.

Multinational companies

Some leading multinationals invest in many countries in the region and have begun to capitalize on cross-country synergies. Examples of such companies are:

- Asea Brown Boveri in power generating equipment and the electro-technical industry.
- Benckiser, Colgate-Palmolive, Procter & Gamble, and Henkel in detergents and cosmetics.
- Daewoo in automotive industry and electronics.
- ING, Citibank, and GE Capital in banking and financial services.
- Neste and Shell in oil.
- McDonald's in restaurants.
- Coca-Cola, Pepsico, and Nestlé in beverages and foods.
- Arthur Andersen, Deloitte Touche, and McKinsey in auditing and consulting.

Cross-country synergies result from regional coordination, pooling of resources (such as qualified personnel), developing regional economies of scale (in production), and distribution and marketing. Securing business from other multinational companies, operating in several countries in the region, developing region-specific capabilities, and restructuring existing companies are also important in capitalizing on cross-country synergies. Table 1.5 shows the amount of foreign direct investment made into the different countries of C&EE and how that grew from 1990 to 1998.

Table 1.5 Foreign Direct Investment Flow (US $ millions)

	1990	1991	1992	1993	1994	1995	1996	1997	1998
Hungary	300	1500	1500	2300	1100	4500	2000	2100	1150
Poland	89	100	300	600	500	1100	2800	3000	7500
Czech Rep.	n.a.	n.a.	100	600	700	2500	1400	1300	2540
Slovak Rep.	18	82	130	200	200	200	200	50	350
Slovenia	4	41	113	111	128	176	186	321	301
Latvia	n.a.	n.a.	43	51	155	244	376	515	200
Lithuania	n.a.	n.a.	25	30	31	72	152	328	950
Estonia	n.a.	n.a.	n.a.	157	215	199	111	130	565
Romania	18	37	73	97	341	417	263	1224	884
Bulgaria	n.a.	56	42	40	105	82	100	497	270
Ukraine	n.a.	n.a.	200	200	100	400	500	600	700

Source: Business Central Europe Magazine, The Economist Group. http://www.bcemag.com/bcedb/stat_main.htm
Historical data collected from WIIW, EBRD, Reuters, National Statistics.

Economic performance and prospects

The Russian financial crisis of 1998 combined with economic turmoil in Asia enhanced the split of the economies of the region into two groups:

- Economies which suffered only minor effects from the Russian crisis due to the minor role of imports and exports from Russia.
- Economies of Russia, and countries closely linked to it, where the crisis had a powerful impact on economic activity.

The prospects of the first group of economies depend mainly upon their own ability to continue market reforms and retain a high level of economic activity in Western Europe (particularly in Germany), as well as continue in the process of EU enlargement. The prospects of the second group of economies depend mainly upon the economic recovery of Russia and institutional and political developments there.

Infrastructure

Hub cities in the region, such as Prague, Budapest and Warsaw, are easily accessible by air from all the business capitals of the world. Hotels and telecommunications have considerably improved. Modern office space is available but in many cases is more expensive than in the United States or Western Europe. Road and railway transportation, as well as border controls and warehouse space, can be problematic. Trucks carrying merchandise from Western Europe to Russia can easily lose several days waiting for customs controls at the borders of Poland, the Ukraine or Belarus. Corruption at the border crossings is often quoted as an additional cost, inconvenience and source of uncertainty.

Diagnosing Globalization Drivers in Central and Eastern Europe

Four sets of industry globalization drivers are used to diagnose the current and potential situation of each country's economy. In this section we describe each driver and provide examples of how they apply in Central and Eastern Europe. Each country chapter evaluates these drivers qualitatively. The concluding chapter provides a quantitative rating of each driver to allow for cross-country comparisons. In the Appendix of this book there is a technical definition of each driver.

Market globalization drivers

Market globalization drivers (common customer needs, global/regional customers, global/regional channels, transferable marketing and lead countries) have an effect on whether MNCs should participate in C&EE markets, the types of services and products they should offer and the marketing approaches they should take.

Common customer needs

Common customer needs represent the extent to which customers in different countries have the same needs in the product or service category that defines an industry. Many factors affect whether customer needs are similar between countries. This includes similarities in culture, economic development, climate, physical environment, as well as the stage of the product life cycle. Under communism, Western products were often considered a 'forbidden fruit' and by the same token became highly attractive to local consumers. 'Western appeal' can still be used as a marketing tool, but a variety of other, specific marketing approaches, should be developed. Income levels and income distribution of consumers also play a decisive role.

The most developed countries of the C&EE region have a rapidly growing middle class with increasingly West European consumption patterns. However, even in these countries some global products, such as ketchup, mayonnaise, tea, coffee and other foodstuffs, have to be adapted to local tastes. Successful MNCs make use of established local brands in foodstuffs such as beer, detergent, sweets and cosmetics. Another reason to maximize the use of cheaper local products is the high price sensitivity that is a common feature of the new middle class consumer. Countries that are less advanced in the transition process have seen a slower emergence of the middle-class consumer. With a lower average level of disposable income there is a preference for cheaper, lower quality local goods. Only a small group of 'super-rich' consumers develop 'conspicuous consumption' patterns and can afford even the most expensive, global, luxury products. Only this small group demonstrates low price sensitivity.

Global/regional customers and channels.

Global/regional customers or channels of distribution buy on a centralized or co-ordinated basis for decentralized use, or at the very least select vendors centrally. For this to be a significant factor, a country needs to have the global/regional MNC headquarters located there. In this respect, the massive entry of global retailers, such as Auchan, Carrefour, Leclerc, Metro, Billa and Makro, is an important development. Capital cities, such as Budapest, Prague and Warsaw, compete to become the regional headquarters of MNCs operating in the region.

Transferable marketing

The nature of purchasing decisions in a country may be such that marketing elements, such as brand names and advertising, require little local adaptation. In other words, global brand names and advertising are readily transferable. Language is an important factor influencing marketing transferability. Brand names in English are generally acceptable throughout the region. Content labels, such as ingredients, are often printed in several local languages. There are practically no forms of regional mass media. Each country has its own system, which includes global TV channels (such as CNN) or European channels (such as Canal Plus or BBC), available by satellite or cable (both widespread in the region). Television advertising can be used throughout the region. Western European (mainly German) commercials are often dubbed and aired in C&EE countries. However, using foreign commercials may not be a successful strategy if, for example, dialogues are not translated properly or the setting or content does not makes sense in the context of local traditions and customs. In some instances such an approach can be unsuccessful. This is why leading international advertising agencies, operating in the region, develop advertising campaigns specifically designed for the larger domestic markets such as Poland and the Czech Republic. Such campaigns take into account and make use of local literature, folklore and history.

Lead countries

Because of the presence of innovative competitors and demanding customers, innovation in products or markets can be concentrated in either one or several countries at once. Lead markets can be found only in the most advanced economies of the world. And even the most advanced countries of C&EE have a lot of catching up to do, in order to become global lead countries.

Cost globalization drivers

Cost globalization drivers affect where MNCs should locate their activities. They include: global/regional economies of scale, sourcing efficiencies, favorable logistics, good infrastructure, favorable country costs and the role of technology.

Global/regional scale economies

Global/regional scale economies or scope economies apply when single-country markets are not large enough to allow competitors to achieve optimum scale. This concerns the extent to which a country has markets that can contribute sales volume to MNCs that need to achieve global or regional scale economies. Alternatively, MNCs must consider if the local market is large enough to support a minimum efficient scale plant? If not, they must consider if there are sufficient exports to support such a plant? In Central and Eastern Europe, particularly at the beginning of the transition process, the size and dynamics of markets was difficult to assess and predict. Even large countries such as Russia and the Ukraine can be characterized as relatively small markets in terms of high value-added goods. Production location, in politically and financially unstable regions, was considered risky. It is at this point that regional scale economies are needed. The most common solution is to locate production in countries with relatively large and growing domestic markets, that can provide a more stable environment. A base in such a relatively stable market can be used as a platform for penetrating neighboring markets. Of the Central European countries, the Czech Republic, Hungary and Poland meet these requirements best.

Sourcing efficiencies

A country may be able to provide critical factors of production in efficient volumes. In this respect Central and Eastern Europe provides plenty of opportunities.

- Well skilled and relatively inexpensive labor (technical workers) are in abundance in all countries of the region. MNCs operating in industries such as automotive, textile (clothing), mechanical, electrical, and electronic can take advantage of this opportunity.
- All post-communist countries have relatively well developed 'linkage industries' (such as metal, paper, chemicals, etc.) Several enterprises operating in these industries are being privatized, modernized and often taken over by foreign MNCs.
- Much of the C&EE, Russia and other post-Soviet countries is an abundant

source of raw materials such as oil, gas, and timber. Poland is one of the leading producers of copper. Coal, metals, precious stones and minerals are also found in the region.

Favorable logistics

A favorable ratio of sales volume to transportation cost enhances the ability of MNCs to concentrate on production. In this respect the more advanced countries have a considerable competitive advantage. Poland, the Czech Republic and Hungary can be considered as a 'bridge' between Western and Eastern Europe. Enterprises located in these countries have developed close ties to Western European and particularly German markets. Slovenia, located between Austria, Italy and post-Yugoslav countries, can be considered as a gateway to the Balkans, while the Baltic countries may be treated as a bridge between Scandinavian and CIS markets.

Good infrastructure

The quality of a country's infrastructure, its roads, power, communications and so on, affect the cost and effectiveness of MNC operations. For the purposes of this book, infrastructure differs from logistics. Infrastructure concerns primarily internal conditions, while logistics primarily involves shipments to and from the outside world. For example, Slovenia probably has the best infrastructure in the region, but because of the Balkan war and post-Yugoslav crisis it is not connected to any economically viable region south of the border except Croatia, which is still involved in the conflict. Hungary and the Czech Republic have better infrastructures than Poland, but Poland has a larger domestic market and is centrally located in the region. In order to control inflation, which is still a real threat, post-communist countries have to maintain very tight budget discipline. Therefore, government financing of radical infrastructure improvement programs remains rather unlikely, while private financing is insufficient.

Favorable country costs

Countries vary in their production costs, which take into account not only labor wage costs but also overhead costs, taxes, customs duties and other transaction costs, such as exchange rates. Wage rates in countries that are more advanced in the transition process (Czech Republic, Hungary, Poland, Slovenia and Estonia), are rising steadily, although they are still low by Western European standards.

The increases of wages are to some degree compensated by increased productivity. The integration of the more advanced countries into the EU will accelerate the process of wage increases. In spite of this, they will still retain a considerable labor cost advantage compared to much of Western Europe for the next 10–15 years. Less advanced countries, further from full EU membership, will almost certainly maintain low wages for much longer. Table 1.6 shows the

Table 1.6 Average Monthly Wage (US $)

	1990	1991	1992	1993	1994	1995	1996	1997	1998
Hungary	212.8	239.8	282.2	295.2	316.8	309.5	307.0	310.5	325.1
Poland	n.a.	n.a.	213.0	215.7	231.3	285.5	323.0	352.7	384.5
Czech Rep	182.6	128.5	164.3	199.6	239.5	307.8	356.4	333.4	362.8
Slovak Rep.	178.7	127.7	160.6	175.3	196.4	241.9	266.0	285.7	283.9
Slovenia	900.2	609.9	627.9	666.1	734.6	945.0	953.9	901.2	943.7
Latvia	n.a.	n.a.	n.a.	n.a.	119.0	161.0	179.0	228.6	225.9
Lithuania	n.a.	n.a.	n.a.	n.a.	91.0	163.3	193.5	202.3	288.1
Estonia	n.a.	n.a.	n.a.	n.a.	130.7	186.2	234.3	256.9	293.1
Romania	138.6	97.6	82.6	103.1	109.8	138.3	138.4	121.8	153.0
Bulgaria	157.5	55.0	87.7	116.9	91.4	113.1	75.5	82.1	118.2
Ukraine	n.a.	n.a.	50.0	35.4	48.3	54.7	75.3	86.0	61.1

Source: Business Central Europe Magazine, The Economist Group, http://www.bcemag.com/ Historical data collected from WIIW, EBRD, Reuters, National Statistics.

level of wages, and how they have changed from 1990 to 1998 in C&EE countries. Although wages have grown, they remain lower than many Western European countries, adding to favorable country costs for MNCs.

Technology role

Countries vary in the extent to which they can be used as a base for developing technology. The Soviet Union had world-class science in many domains (often linked to the military). In recent years, however, leading scientific labs and institutes have deteriorated considerably due to the lack of government funding, and many scientists have emigrated. The use of local scientific resources by MNCs is limited by economic and political instability. Only gradually are MNCs learning how to make use of valuable scientific resources such as research institutes and world class scientists that can be found throughout the region. The number of research and development facilities supported by MNCs in Central and Eastern Europe is gradually increasing.

Government globalization drivers

Government globalization drivers affect whether, and the way in which, MNCs can participate in a country. These include favorable trade policies, favorable investment rules, participation in trade blocks, absence of government intervention, absence of state-owned competitors, legal protection, compatible technical standards, and common marketing regulations. An unfortunate effect of the transition process is the high level of corruption, particularly in countries that are still dealing with earlier phases of transition or experiencing setbacks and 'loops' in the process.

Favorable trade policies

Countries negotiating accession to the EU are gradually becoming a part of the EU free trade zone. This does not exclude some degree of protectionism practised towards other parties. These countries, however, are under constant scrutiny from international bodies such as the IMF, World Bank and NATO. Consequently, their economies are becoming more and more open. Governments of these post-communist countries are under pressure to protect their own industries and ailing enterprises, which have suddenly become exposed to international competition. Because of their technological backwardness, over-employment, lack of managerial skills and slow, bureaucratic decision-making processes, state owned enterprises are often and clearly unfit to compete in the voracious global marketplace. The less advanced countries often give in to pressures to support the ailing industrial dinosaurs. At the same time, the international business community and international organizations such as WTO and IMF are encouraging them to open their markets. Governments that are uncomfortably wedged between two opposing camps are, all too often, tempted to change legal regulations and trade policies. Trying to accommodate open marketplace ideals and protectionist forces at the same time, they can aggravate the situation further and cause instability.

Favorable investment rules

Rules on foreign direct investment vary greatly and continue to change, mostly towards greater openness. The positive impact of Foreign Direct Investment (FDI), on those countries that have attracted the most FDI, is clearly visible in their economic development. A much more cautious attitude is taken towards foreign portfolio investment. It is widely believed that massive inflows and almost instantaneous outflows of foreign speculative capital considerably aggravated economic and financial difficulties experienced by the Czech Republic in 1997 and Russia in 1998. This is partly because most countries in the region attempted to impose restrictions on massive international transfers of money.

Participation in trade blocs

Participation in regional trade blocs and free trade zones help to open up economies to MNC activity. The CEFTA (Central European Free Trade Association) grouping including the Czech Republic, the Slovak Republic, Hungary, Poland, Slovenia and Romania is such a trade block. Because of strong pressures to protect local producers, CEFTA was unable to deal effectively with the issues of 'sensitive goods' (in particular foods and textiles). Its impact on trade liberalization was, at best, moderate when compared to the impact of EU and international organizations such as the WTO or IMF. The CIS grouping of Russia and several post-Soviet states is another regional trade bloc. Its workings, however, are almost completely paralyzed by political, financial and macroeconomic instability.

Absence of government intervention

Most governments are sensitive to foreign dominance of, or even participation in, key industries and hence intervene. In the more advanced transition countries, the scope of such intervention is very limited. It covers, to different degrees, infrastructure, the defense industry, and the energy sector. Telecommunications and media are also open to foreign participation as well as the financial services sector. Post-communist governments have not yet learned how to push MNCs to locate higher value-added activities, such as research and development, design and marketing, in their countries, but they are beginning to learn. Governments of the less advanced countries are much more inclined to intervene in cases of foreign capital participation. This is mainly due to political and legal instability.

Absence of state owned competitor

State-owned companies that are crying out for protection in the region are mainly concentrated in low value-added industries such as steel, bulk chemicals, mining, heavy machinery and shipbuilding. Governments try to protect these producers, at least temporarily, in order to avoid unemployment and confrontation with militant trade unions. As a result, MNCs operating in these industries and markets are put at a disadvantage. In higher value-added industries, governments usually attempt privatization through foreign direct investment (FDI). In such cases protectionist measures become negotiable for MNCs.

Legal protection

A major concern in Central and Eastern Europe is poor protection of contracts, trademarks, and intellectual property. A combination of an underdeveloped legal system, lax law enforcement, and general corruption can make these problems particularly severe in some countries. Software, music, film and video games producers are especially susceptible to pirating activities. The more advanced countries have made concerted, legislative efforts to provide better legal protection for MNCs operations. As accession to the EU approaches, law enforcement measures will probably follow. The Transparency International global corruption watch group, in its 1998 report, ranks some countries of the region (Russia and Ukraine) close to the bottom. On the other hand, the Czech Republic, Poland, Hungary and Slovak Republic are in the middle of an eighty-country ranking list, on a similar level to Italy, Spain and other southern European countries. Scandinavian countries, Canada and New Zealand are ranked at the top of the list as the least corrupt countries.[6] Table 1.7 on the following page gives the ranking of C&EE and several other countries for comparison.

Table 1.7 Transparency International

Country Rank	Country	1998 CPI Score	Standard Deviation	Surveys used
1	Denmark	10.0	0.7	9
11	United Kingdom	8.7	0.5	10
15	Germany	7.9	0.4	10
16	Hong Kong	7.8	1.1	12
17	United States	7.5	0.9	8
21	France	6.7	0.6	9
25	Japan	5.8	1.6	11
26	Estonia	5.7	0.5	3
28	Belgium	5.4	1.4	9
33	Hungary	5.0	1.2	9
36	Greece	4.9	1.7	9
37	Czech Republic	4.8	0.8	9
39	Italy	4.6	0.8	10
	Poland	4.6	1.6	8
47	Belarus	3.9	1.9	3
	Slovak Republic	3.9	1.6	5
61	Romania	3.0	1.5	3
	Yugoslavia	3.0	1.5	3
66	Bulgaria	2.9	2.3	4
	India	2.9	0.6	12
69	Bolivia	2.8	1.2	4
	Ukraine	2.8	1.6	6
71	Latvia	2.7	1.9	3
	Pakistan	2.7	1.4	3
76	Russia	2.4	0.9	10

Source: Transparency International, http://www.gwdg.de/~uwvw/CPI1998.html ;

Notes: 1998 CPI Score – relates to perceptions of the degree of corruption as seen by businesspeople, risk analysts and the general public and ranges between 10 (highly clean) and 0 (highly corrupt).
Surveys Used – refers to the number of surveys that assessed a country's performance. Twelve surveys were used and at least three surveys were required for a country to be included into the 1998 CPI.
Standard Deviation – indicates differences in the values of the sources: the greater the standard deviation, the greater the differences of perceptions of a country among the sources.

Compatible technical standards

Differences in technical standards among countries affect the extent to which products can be standardized. Government restrictions in terms of technical standards can make or break efforts at product standardization and be used to restrict imports. Four clusters of such standards can be identified in Central and Eastern Europe:

- Post-Soviet technical standards resulting from Soviet technological domination in the region.
- European technical standards imposed by the EU.

- Standards resulting from technology imports from non-European countries such as the United States or Japan.
- Local standards.

Countries negotiating accession to the EU will have predominantly European or compatible with European standards. Other countries will be probably be slower in replacing post-Soviet and local standards.

In addition to differences in technical standards, MNCs need to be concerned with the ability of each country to maintain international technical standards in, for example, maintenance, operating procedures and safety issues. The Chernobyl nuclear disaster, for example, resulted from the near total neglect of such standards, including emergency procedures. The same behavioral patterns may still be found throughout the region, but countries aspiring to EU membership are being forced to meet European standards.

Common marketing regulation

The marketing environment of individual countries affects the extent to which uniform global marketing approaches can be used. In this respect all countries of Central and Eastern European countries fall in line with generally accepted Western standards. Limitations on tobacco and alcohol advertising, the use of nudity and language, exist to different degrees in different countries.

Competitive globalization drivers

Competitive globalization drivers all spur MNCs to participate in a country and also affect the kinds of strategies they should adopt. These include: the global/regional strategic importance of a country, globalized domestic competitors, the presence of foreign competitors and the interdependence of countries.

Global/regional strategic importance

The global/regional strategic importance of a country affects what role each individual country plays in the portfolio of an MNC. Global/regional strategic importance differs by industry and is defined in terms of a country being a:

- Large source of revenues or profits;
- Significant market of global competitors;
- Major source of industry innovation;
- Home market of global customers;
- Home market of global competitors.

In terms of being a large source of revenues or profits, or being a significant market for global competitors, countries with the largest and fastest growing

GNPs, as well as the largest populations, may be key in becoming of global and regional strategic importance. Some Central and Eastern European markets are becoming battlegrounds for MNCs. Poland plays such a role in the automotive industry, post-Soviet countries in the tobacco industry, Hungary and Poland in the media, and so on.

Internationalized domestic competitors

Local competitors with significant foreign activities usually pose a greater challenge for foreign MNCs. In Central and Eastern Europe this challenge does not exist due to the low degree of internationalization of local enterprises, usually undercapitalized and lacking international strategic thrust. The giant Russian gas and oil companies, Gazprom or LUKoil, may be the only notable exception. These companies attempt to dominate the energy supply of many countries in C&EE by acquiring local distributors, retail networks (gas stations) and processing plants (refineries). This strategy has proved successful to different degrees in different countries, since at times it is perceived as a threat to the new-found national sovereignty of previously Soviet countries.

Presence of foreign competitors

Many major MNCs are already present in most key industries in Central and Eastern Europe. This creates greater challenges for them but also signals good future prospects in the region. MNCs have entered in three waves, each one specific to a country. The first wave (1989–93) was mainly directed towards Hungary, the second (1992–96) towards the Czech Republic and the third (1996 onwards) towards Poland. Up until the financial crisis of 1998 all major MNCs maintained their presence in Russia and are likely to do so in the future, although not to the same degree.

Interdependence via MNC value chains

MNCs create strategic interdependence between a country and others by sharing key activities, such as factories, or other parts of the value chain. Many Central and East European countries are already part of such interdependent networks built up mainly by German and other European MNCs. Poland is an example of this in the automotive industry and clothing, the Czech Republic and Hungary in the mechanical industry.

Developing Global/Regional Strategies

The types of global/regional strategies that MNCs have developed can be grouped into five major categories or global/regional strategy levers: market participation, products and services, location of activities, marketing, and competitive moves. This section explores each of these strategies and provides some examples. Each country chapter gives a few countrywide generalizations and some industry specific strategies. *The reader should use this framework and these analyses and examples as a guide to develop their own best strategies, remembering also the need to continually monitor change if necessary.*

Global/regional market participation strategies

MNCs need to choose the country-markets in which to conduct business as well as the nature and level of their activity, particularly in terms of target market share. Managers need to select country-markets not just on the basis of stand-alone attractiveness but also on the basis of how participation in a particular country will contribute to globalization benefits and the global/regional competitive position of the business. Furthermore, different countries can play different strategic roles as markets.

For example, locations in Russia and in other CIS countries are usually selected for the sake of the CIS market itself and locations in Central European countries are often considered as platforms for further expansion both eastward and westward. Expansion to the east capitalizes on past business contacts, connections and brand name awareness established in COMECON times. Expansion to the west capitalizes on the gradually developing economic links between present and possible future EU countries, on high labor costs in the EU (particularly in Germany), and on geographic proximity to key Western European markets. ABB is an example of a company using such a strategic approach in Central and Eastern Europe.

Participating in markets outside the home country acts as a lever for both internationalization (the geographic expansion of activities) and globalization (the global integration strategy). But in the internationalization mode managers select countries based on stand-alone attractiveness. On these grounds, Hungary was considered the most attractive location in the earlier transition phases, the Czech Republic in the mid 1990s and Poland in the second half of the decade. In contrast, when used as a global strategy lever, market participation involves selecting countries on the basis of their (future) potential contribution to globalization benefits, and to the global competitive position of the business. The same considerations also apply to determining the level at which to participate (primarily the target market share) and to determining the nature of participation (building a plant, setting up a joint venture and so on).

For Central and Eastern Europe, companies need to determine first the overall role that the region as a whole should play in the global portfolio and second, the role of individual countries in the regional portfolio. Typically these roles depend on a combination of the country–market characteristics and of the company's history and position in that country–market. For example, IKEA has a long history of outsourcing from several countries at the same time, even during communism. In the 1990s, IKEA started to build its sales network following rising incomes and an emerging middle class in these countries. Similarly, Fiat has a long tradition of licensing agreements in Russia, Poland and the former Yugoslavia. The company decided to use Poland as its mass production site for its subcompact models, the Cinquecento and Seicento. Korean companies, which were completely new to the region, chose to locate in almost all C&EE countries, assuring a massive European presence in the automotive industry, electronics and financial services.

Global/regional product and service strategies

MNCs need to decide which products and services to offer in each country and whether and how much to adapt them to local markets. The conventional wisdom has been to adapt as much as possible to local tastes and needs. But global thinking now recognizes the costs of adaptation and product proliferation and emphasizes the potential benefits from maximizing standardization instead. Furthermore, regardless of cost, over-adaptation can reduce the universal appeal of MNCs offerings. Central and Eastern Europeans come to McDonald's, Kentucky Fried Chicken or Burger King precisely because they are Western and not local. For the same reason, consumers buying clothes look for distinctively Western styles and brand names. Even local producers try to advertise their products as 'American Style' or 'Taste of America' or 'big success on German market' and so on. The stripped-off, basic models of automobiles, presumably adapted to the price sensitivity of local buyers, were often ignored by more affluent consumers.

Conversely, insufficient adaptation can lead to product failure. For example, certain household appliances were too big for small kitchens. Women who were not accustomed to conditioning after washing their hair could not understand the advantages of a 'conditioner and shampoo in one'. Tea, sold in tea bags (as opposed to loose), was condemned as too pale and mild tasting. Producers of frozen foods and concentrates had to add local dishes such as 'perogi' to their product portfolio in Russia and 'knedliki' in Poland, the Czech Republic and Slovak Republic. The challenge for MNCs is to find the right balance between standardization and adaptation. Each country chapter provides many examples of what companies have done to reach this balance.

Global/regional activity location strategies

MNCs have the opportunity and challenge to optimally configure and locate their entire value added chain – research, development, procurement, raw material processing, intermediate production/subassembly, final production/assembly, marketing, selling, distribution, customer service, management and support activities. A global approach to activity location means deploying one integrated but globally dispersed value chain or network that serves the entire worldwide business, rather than separate country value chains or one home-based value chain. Central and Eastern European countries play an increasing role in the value chains of European, American and Asian (mainly Korean) companies as they seek new market opportunities and lower production costs.

The role of most C&EE countries is continually evolving. At the start of the transition they were targeted by MNCs as 'emerging markets,' primarily for export and as sources of cheap labor. Gradually, as transition progressed, new elements of the value chains were deployed. Distribution centers and sales networks were established in the countries where revenues were higher and grew faster. The marketing function followed naturally. Assembly and production facilities were established in countries which were considered more stable and consistent in the transition drive and had good prospects for market growth and exports. In some cases regional headquarters were established and even transferred to Central and Eastern Europe from neighboring Western countries such as Austria or Germany. Recently, some MNCs have been establishing research and development and design centers, taking advantage of existing scientific, engineering and creative talents available at a much lower cost than in Western Europe. Each country chapter discusses the distinct location strategies of MNCs and the likely role that the country will play in the overall regional and global value chains of MNCs.

Global/regional marketing strategies

MNCs need to decide on the extent to which they adapt their marketing to each country for each element of the marketing mix: positioning, brand names, packaging, labeling, advertising, promotion, distribution and sales techniques, sales representatives and service personnel. They also need to identify aspects of a successful marketing mix needed specific to each country. As discussed under Market Globalization Drivers, C&EE countries have distinctive cultural and behavioral aspects that require market adaptation. As with all product adaptation MNCs need to strike the right balance. Over-adaptation can be as inappropriate as under-adaptation.

In most product categories Central and East Europeans prefer Western brand names and many local firms try to use Western sounding names. For example, in computing and software two successful local companies are Computerland (Polish) and Graphisoft (Hungarian). However a 'local touch' is also appreciated, particularly in mass consumption products such as foods: for example, 'Grandma Mayonnaise' by Hellmans or Dosia detergent powder by Benckiser. If any generalization can be made, it is that more strategic elements of the marketing mix, such as brand names, packaging, appearance, and advertising themes, should tend towards being either global or regional. With increasing travel within and outside the region, the important role of small cross-border trade and increasing exposure to global and regional media, cross-country spill-overs become more and more common. For example, between the eastern parts of

Poland, Belorusssia, Lithuania, and Ukraine, and the Southern parts of Poland, the Czech Republic, the Slovak Republic, Hungary and Romania, television viewers can tune into each other's programs and advertising. Moreover shopping abroad, or buying small quantities of goods abroad for resale, is still quite common. This is due to the considerable differences in price and market saturation in different countries of the region with the less advanced countries still experiencing shortages. Therefore companies need to maintain regional uniformity to avoid contradictions and lost opportunities for recognition and purchase. On the other hand more tactical elements of the marketing mix – selling, promotion and distribution – usually need to be adapted to local conditions. For example, Henkel is using local literary and historical characters to promote its detergents. A similar approach is taken in other promotional campaigns of consumer products throughout the region.

Global/regional competitive move strategies

MNCs should make competitive moves in individual countries part of a successful global competitive strategy. So for individual countries, it is important to consider the extent to which an MNC includes the country in their global/regional competitive moves, as opposed to making competitive moves in the country independent of other countries. The competitive globalization drivers of a country, the presence of foreign competitors and interdependence of countries, affect whether and how they make such moves. Oil and gas deposits in Russia and Azerbaijan and metals in Kazakhstan cannot leave indifferent any major player active in the industry. The same applies to the rapidly growing regional telecommunications market. When GE acquired Tungsram, the Hungarian lighting sources manufacturer, the world leader in the lighting sources industry, Philips, responded by acquiring Polam Pila, the Polish producer. In spite of the effects of the 1998 financial crisis on the Russian economy, most global players still maintain their presence in the country, at least to keep an eye on competitors.

Overall Global Organization and Management Issues

Organization factors affect both what the nature of global strategy should be and the effectiveness of its implementation. The nature of organization – its structure, management processes, people and culture – significantly affect the ability to implement global strategy. The more centralized, integrated, and uniform the organization elements are, the easier it is to formulate and implement global strategy. The appropriate degree of integration also depends on company history. On the other hand, local characteristics, such as different economic and institutional environments, culture and work style, require MNCs to tailor their organization specifically to each country. Local characteristics often demand significant modification of global/regional organizational approaches.

Organization structure

Three critical issues arise in terms of organization structure in Central and Eastern Europe: the need for local partners, the need for regional headquarters and the degree of autonomy of subsidiaries.

Choose local partners carefully

In most C&EE countries, a critical issue in organization structure is whether the foreign company can enter alone or if there is need for a local partner. Local partners can provide the legal right to operate at all, as well as access to production facilities, access to local markets and customers, government contacts and local expertise. Some kind of partner is almost always necessary since market structures and market institutions in most countries are not yet fully developed and still riddled with peculiarities that are not easily understood by foreigners. Partners with direct and indirect access to authorities and well-developed market contacts are particularly important. In all cases of business partnerships, an MNC should maintain management control enabling them to implement global/regional strategy. State owned enterprises or recently privatized enterprises carried out by mass government privatization (voucher) schemes (such as implemented in Russia, in the Czech Republic, and on a smaller scale, in other countries) make particularly difficult partners. They remain highly politicized with the influence of old communist 'nomenklatura'. This can jeopardize the stability of management teams as well as the consistency of company strategies.

Private enterprises are generally too small to be equal partners to major MNCs. However, some MNCs use local partners as distributors and dealers to penetrate local markets faster and to minimize risk. This approach is popular among companies that require extensive distribution networks such as automotive companies (Daimler Benz, Renault, Peugeot), software and computer companies (Microsoft, Oracle, Dell) and consumer electronics companies (Philips, Sony). After the initial stage of market penetration, MNCs often choose to build up their own networks, as for example, in the case of Daimler Benz in Poland.

Regional headquarters

Due to the heterogeneity of the region and the fact that different countries find themselves at different phases of the transition process, the role of regional headquarters is mostly limited to issues related to material flows, such as production planning, logistics and sales organization.

Give more autonomy

MNCs need to decide how much autonomy their operations in C&EE should have. While this decision depends upon the subsidiary's level of experience, managers and their role, it can also depend on country characteristics. In particular, countries with less business experience, a weaker work ethic or a higher degree of corruption, generally need tighter control. At the beginning of the transition process all the countries of Central and Eastern Europe were perceived by MNCs as requiring a relatively high degree of control. Increasingly, however, some countries have made much more progress than others and some Central and Eastern European markets have emerged as important sources of revenues and competitive battlefields for MNCs. This new situation calls for much more autonomy in areas such as product development, marketing strategies, human resource management and capacity development.

Management processes

Management processes comprise such activities as planning and budgeting, and making the business run smoothly. These processes include strategic planning, budgeting, cross-country coordination, motivation, performance review and compensation, human resource management (including career planning, training and skills development, employment terms) and information systems.

Use global management processes

MNCs need to decide whether and how to adapt global management processes for each country and which ones to use. A key determinant here is the level of experience that a country has with modern management techniques. The degree to which accounting standards, and the legal system of a country are compatible with international standards also plays a role. (In this respect, those countries invited to negotiate accession to the EU are much more ready for standardization of management processes.) MNCs usually find themselves in the position of having to 'push the envelope' in changing and upgrading practice. The country chapters provide examples of what individual companies have done so far.

Participate in global processes

A second issue concerns the extent to which MNCs can expect subsidiaries and partners in a country to participate in global management processes. Participation of executives based in Central and Eastern Europe (including local nationals) in global forums is constantly rising as the countries of the region become integrated into the global strategies of MNCs. The degree of participation differs, however, and is much higher in the case of the more advanced countries in the region.

Human resources

'People issues' can be summarized in two categories: using and developing locally capable managers and staff, and using non-local managers such as expatriates.

Can use local managers

A shortage of local management talent was a significant constraint in Central and Eastern Europe, particularly in the early years of transition. This obstacle has been largely overcome due to the unprecedented explosion of Western management education in the region.[7] Hundreds of executive MBA programs (some of them established jointly with high profile American and European business schools) have been set up throughout the region. Undergraduate and graduate management training programs have been developed by local universities, in most cases with the help of Western schools. These programs have become a favorite choice of high school graduates. Some MNCs operating in the region (such as ABB) have made an effort to upgrade the management skills of their local personnel by sending them to management training offered locally or by leading Western business schools and setting up their own management training centers in the region. Most of these programs are offered in English, in which most aspiring young managers are fluent.

Can use non-local managers

In many instances, however, the presence of expatriate managers is needed to maintain coherence and control over local operations. Expatriates are often resented by local managers because of their financially privileged position and lack of sensitivity to 'local ways'. Such attitudes can be detrimental to the entire local operation, particularly if expatriates do not represent the highest level of competencies and skills. In order to avoid a situation that may cause tensions, highly competent people have to be selected and provided with intercultural training. For most MNCs operating in the region there is scope for improvement in this area.

Culture

Culture comprises the values and rules that guide behavior in a corporation. Local, regional and national cultures clearly affect how managers behave and how they fit into the corporate culture. National cultures differ throughout the region and any hypothesis that there is one distinct 'Central and Eastern European culture' is almost impossible to maintain, because of the variety of historical forces, cultures and foreign influences that have shaped C&EE. The one thing all these countries have in common is their shared communist past and as Soviet bloc countries, which has been experienced to different degrees in different countries, and is a much stronger influence in the CIS countries.

MNCs need to take all of the many cultural differences into account, but at the same time, strive to go beyond them. As Percy Barnevik, the former CEO of Asea Brown Boveri observed, 'We must be sensitive to national cultures but not paralyzed by them.' Ultimately, MNCs need to decide how far they can go in instilling their global corporate culture and how much they need to adapt to local culture.

Bibliography

CEEBIC: The Central And Eastern Europe Business Information Center Market Access and Compliance Division of the US Department of Commerce. http://www.itaiep.doc.gov/eebic/cefta/data.htm

Intrilligator, M.D. (1994), 'Privatization in Russia has led to Criminalization' in *The Australian Economic Review*, 2nd qtr, pp. 4–14.

Koźmiński, A.K. (1993), *Catching Up? Organizational and Management Change in the Ex-Socialist Bloc* (Albany, NY: SUNY Press).

Koźmiński A.K. (1996), 'Management Education in Central and Eastern Europe', in *International Encyclopedia of Business & Management, vol. 3* (London: Thomson Business Press), pp. 2775–80.

Lavigne, M. (1995), *The Economics of Transition* (New York: St. Martin's Press).

Lee, M., Letiche, H., Cranshaw, R. and Thomas, *Management Education in the New Europe* (London: International Thomson Business Press).

Rzeczpospolita (1998), no. 291 (5151) 12-13 XII: 5. Transparency International 1998 Corruption Perception Index. http://www.gwdg.de/~uwvw/CPI1998.html Transparency International.

Notes

1 The methodology for this book was first developed in George S. Yip, *Asian Advantage*, and applied to the Asian–Pacific region.
2 The methodology for such an analysis of the transition process is first developed by A. Koźmiński in *Catching Up? Organizational and Management Change in the Ex-Socialist Bloc* (Albany, NY: SUNY Press, 1993).
3 Intrilligator M.D. 1994. 'Privatization in Russia has Led to Criminalization', in *The Australian Economic Review*, qtr 2, pp. 4–14.
4 Lavigne, M. (1995), *The Economics of Transition*, (New York: St. Martin's Press).
5 A. Kozminski in *Catching Up? Organizational and Management Change in the Ex-Socialist Bloc* (Albany, NY: SUNY Press, 1993).
6 *Rzeczpospolita* (1998), no. 291 (5151) 12–13 XII: 5.
7 Lee, M., Letiche, H., Craweshaw, R. and Thomas (1996); Koźmiński (1996).

2 Hungary – Goulash Capitalism

JÁNOS VECSENYI and GÁBOR SZIGETI

Overview

HUNGARY was among the first countries in the Central and Eastern Europe to attempt reforms that would bring it ever closer to political democracy and a Western-style market economy. Within the region Hungary attracted a large proportion of the initial foreign investment. Consequently, Hungary had a head start in developing regulatory and financial institutions and, although there is still room for improvement, in many areas it has reached EU standards. Many multinational companies and small and medium enterprises have chosen Hungary as their regional headquarters. These foreign companies have invested not only in the physical operations but also in training local employees and acquiring research and development (R&D) facilities in strategies that look towards the future.

Hungary has a substantial domestic market of over ten million inhabitants and consumer behavior patterns that are very similar to Western Europe, making marketing techniques readily transferable. Although not a source of cheap labor, Hungary has a highly educated labor force and there is a lot of potential capacity for R&D facilities. The Hungarian government is stable and business friendly and provides a positive environment for foreign multinational companies (MNCs).

History

The origins of the Hungarian people lie in the Central Asian steppes. Descended from a small group of Finno-Ugric peoples at least 4000 years ago, the Magyar and Onogur people split off to seek new territory. Continuing the move westward, the consolidated Magyar tribes conquered the Carpathian Basin between 895 and 896 AD. Missionaries sent by the Pope and the military force of the Christian king eventually converted the pagan Magyars to Christianity. The Church also helped to establish Hungary, along with emerging Bohemia, Poland and Russia, as an important feudal state. For nearly 250 years, Hungary held its own under increasingly powerful kings who were receptive to economic and cultural influences from both the Byzantine and Holy Roman Empire.

Throughout its long history Hungary has had to struggle to survive as an independent state and has suffered a number of major setbacks in its development. In 1241, Hungary suffered the Mongol invasion of the Carpathian

Basin which devastated the countryside, destroyed 20–40 per cent of the settlements and marked the end of an era of over 700 years, the longest period of Hungarian self-rule. Over the next 200 years, the throne of Hungary was fought over and occupied by a variety of rulers from Bohemia, Bavaria, Luxembourg, Austria, Poland and Naples. Nonetheless, Hungary survived and by the end of the fifteenth century had become one of the richest and most powerful Renaissance centers in Europe.

In 1526, Hungary was left leaderless when the king and most of the ruling class nobility were slaughtered in a battle against the Ottoman Turks. The Turks occupied and ruled most of the country for the next 150 years, destroying much of the Christian population, its culture and institutions. The Western slice of what remained of Hungary elected Ferdinand Habsburg as King of Hungary in the hope that his powerful family could stop further Ottoman advances. The Habsburgs, who saw Hungary largely as a buffer for their own lands and interests, maintained their rule until 1918.

When the Ottoman Turks were defeated by the Holy Alliance in 1699, the Habsburgs were able to reunite the Hungarian lands. Inspired by the American and French Revolutions, the first half of the nineteenth century became a period of Hungarian nationalistic fervor. The War of Independence of 1848, though unsuccessful, paved the way for eventual Habsburg concessions resulting in the Compromise of 1867 and the creation of the Austro-Hungarian Dual Monarchy. Spurred by economic development and the rise of internationally reputed musicians, scientists, and writers, the celebration of the Magyar Millennium in 1896 marked a zenith in Hungarian culture. In 1849, the Scottish engineer Adam Clark built a suspension bridge across the Danube between Buda and Pest and Budapest was born. Tungsram, the Hungarian light bulb company was founded in 1896 at the height of these exciting and intoxicating times.

The effect of the national catastrophes of the twentieth century is still immediate in the minds of most Hungarians. Thrust into the First World War as part of the Austro-Hungarian Empire, Hungary found itself on the wrong side of victory. In 1920, at Versailles, the Treaty of Trianon carved away two-thirds of Greater Hungary. The loss of territory and people was also economically devastating, since it effectively removed 89 per cent of iron production, 84 per cent of forests and nearly half the food processing industry. A massive 62 per cent of the railroad network was lost with the treaty with added clauses that strictly limited the building of new railway lines in the future.

Inter-war Hungary became economically stagnant, bitterly right wing and ripe to be drawn into the orbit of Mussolini and Hitler who promised, and temporarily delivered, the restoration of Greater Hungary. Once more Hungary was on the wrong side of victory. Hitler's decision to spare Vienna by sacrificing Budapest as the last Nazi stronghold resulted in the near total destruction of the city. The Soviet Army first liberated and then occupied Hungary for the next 45 years. Despite a failed revolutionary attempted in 1956 and moderate political and economic reforms in 1957 and 1968, two full generations of Hungarians knew only a socialist, centrally planned and controlled system.

The seeds of change, planted by János Kadar (that is, 'goulash capitalism') through market reforms of 1957 and 1968, gradually began to take root and grow. On 16 June 1989, a demonstration held in Hero's Square, signaled that a 'silent revolution', aimed at reforming the political system without the use of armed forces, was gaining momentum. The demonstration was exactly 33 years after the revolution of 1956. And on 23 October 1989, the communist 'Peoples' Republic' became the Republic of Hungary. The change in name had fundamental political and social implications. Hungary once again established itself as an independent nation state. Further political changes in March 1990 and the 'reopening' of the Budapest Stock Exchange in June 1990 had a direct influence on the business community.

In June 1991, NATO leaders agreed that Hungary, Poland and the Czech Republic would be invited to join the alliance. Several days later, the summit of the European Union in Amsterdam accepted a list of five nations including Hungary to participate in membership negotiations. In November 1997, the majority of Hungarians voted for NATO membership in a referendum. These have set Hungary back on track to the Euro-Atlantic world, where many Hungarians have always felt they belong. As the land of 'goulash communism', Hungary was one of the first countries of Central and Eastern Europe to establish East-West joint ventures, privatize its industry, and be granted associate membership in the EU. Hungarians were also Central and Eastern Europe's pioneers in adopting democratic and pro-market reforms.

Economic reforms

In the 40 years between 1948–88, many efforts were made to reform the socialist system. The first of these was the 'New Economic Mechanism', implemented in 1968, under which direct state intervention was replaced by economic regulators. It introduced price policy and price control, financial policy of state-budgetary revenues and outlays, enterprise profits through income and other taxes, regulation of personal income (wages), credit policy, foreign trade policy and investment policy. The managers of state-owned companies were afforded greater autonomy in organizing production and framing their strategies and investment plans. The New Economic Mechanism eliminated the centrally planned economic system, but did not create a real, free market. By the late 1970s, with growing concern for the country's deteriorating trade position, the reform movement gained new momentum. Reforms were aimed at improving the efficiency of Hungarian industry in order to increase exports, especially to the West. The passing of The Small Businesses Act of 1982 was the first step to allowing private business to operate in the country again.

It was in the period 1988–9 that fundamental economic laws were accepted, ironically by the communist parliament. In order 'to develop private economy' the Corporation (Business Associations) Law created six different types of enterprise ownership including unlimited partnership, deposit partnership, limited company, business union, joint enterprise and two of the most popular forms in Hungary, the limited liability company and the joint stock company. The Law was aimed at enabling easy, unhindered flow and reallocation of capital in the economy. A Foreign Investment Law was issued in 1989 to facilitate direct investments into Hungary and to protect foreign investors from nationalization or expropriation of investment. To encourage technology transfer, tax benefits were given to foreign investors.

To decrease the proportion of state-owned enterprises, the Business Transformation Law offered new forms of ownership of companies. The new laws allowed the transformation of state-owned companies into business associations, which declared simultaneously the principal of general (legal) succession. The law also provided for the transformation of a company into another type of firm, which also covered mergers and splitting. Supporting the privatization of large state-owned enterprises, the State Property Agency was formed in early 1990. A new accounting system was introduced to develop mutually acceptable accounting and bookkeeping practices. The basic principles of the bookkeeping methods and the contents of the balance sheet follow the rules of the Generally Accepted Accounting Principles of Western countries. Important new institutions, such as the bankruptcy law, a competitive banking system, personal and value added tax systems and a bond and stock market, were formed to reduce the role of government.

Small businesses were created in increasing numbers from 1982. By 1998 the total number of small businesses was about 200 000, not to mention the 6–700 000 self-employed individuals representing 6–7 per cent of the country's population. Ownership structure changed dramatically with private ownership replacing state ownership in all sectors. The main macro-economic indictators describing the changes in the Hungarian economy are illustrated in Table 2.1.

Foreign investment

Hungary was among the first countries in Central and Eastern Europe to attract foreign direct investment. By the time privatization was virtually completed an annual US$ 1.5bn to $ 2bn in direct foreign working capital was expected to flow into the country. By the end of 1998 the total value of foreign direct investment (FDI) amounted to US$ 17.6bn, of which over 90 per cent was invested by large multinational companies. Although the import of foreign capital became possible with the law on joint ventures, only the subsequent political transformation created appropriate internal and external conditions for foreign investment. Hungary took the lead in creating these conditions, therefore it is no accident that Hungary was the first in the region to see an influx of foreign capital at this early stage. Table 2.2 illustrates the flow of FDI into Hungary over the past eight years. Hungary has attracted 37 per cent of all direct foreign capital investment into Central and Eastern Europe since 1989.[1]

Table 2.1 Macro-Economic Indicators for Hungary

	1990	1991	1992	1993	1994	1995	1996	1997	1998
Nominal GDP ($bn)	33.1	33.4	37.3	38.6	41.5	44.6	45.1	45.6	47.6
GDP per capita PPP ($1000)	n.a.	n.a.	n.a.	n.a.	8.4	8.9	9.3	10.0	10.7
GDP (% change)	−3.5	−11.9	−3.1	−0.6	2.9	1.5	1.3	4.4	5.1
Industrial production (% change)	−10.2	−16.6	−9.7	4.0	9.6	4.6	3.4	11.1	12.6
Budget balance (% of GDP)	0.4	−2.9	−6.8	−5.5	−8.1	−5.5	−1.9	−4.0	−5.4
Unemployment (%)	1.9	7.4	12.3	12.1	10.4	10.4	11.4	11.0	9.6
Average monthly wage ($)	212.8	239.8	282.2	295.2	316.8	309.5	307.0	306.7	316.0
Inflation (%)	28.9	35.0	23.0	22.5	18.8	28.2	23.6	18.3	14.3
Exports ($bn)	9.5	9.3	10.0	8.1	10.7	12.9	13.1	19.1	23.0
Imports ($bn)	8.6	9.1	10.1	11.3	14.6	15.4	16.2	21.2	25.7
Trade Balance ($bn)	0.5	0.2	0.0	−3.2	−3.9	−2.5	−3.1	−2.1	−2.7
Current–account balance ($bn)	0.4	0.3	0.3	−3.5	−3.9	−2.5	−1.7	−1.0	−2.3
Foreign direct investment flow ($m)	300.0	1651.0	1487.0	2294.0	1684.0	4945.0	2828.0	1436.0	1971.0
Foreign exchange reserves ($bn)	1.1	3.9	4.3	6.7	6.8	12.0	9.7	8.4	9.4
Foreign debt ($bn)	21.3	22.7	21.4	24.6	28.5	31.7	27.6	23.7	26.7
Discount rate (%)	20.0	26.0	20.0	22.2	29.0	27.5	21.8	19.5	17.0
Exchange rate ($)	63.2	74.8	79.0	92.0	105.1	125.7	152.6	186.8	214.5
Population (m)	10.4	10.3	10.3	10.3	10.2	10.2	10.2	10.1	10.1

Source: Business Central Europe Magazine, The Economist Group, http://www.bcemag.com/

Table 2.2 Annual Flow of Foreign Direct Investment in Hungary (US$m)

Year	Cash	Contribution in kind	Total
1990	311	589	900
1991	1459	155	1614
1992	1471	170	1641
1993	2339	142	2481
1994	1147	173	1320
1995	4453	117	4570
1996	1983	57	2040
1997	1676	12	1688
1998[e]	n.a.	n.a.	1760

Source: Hungarian Ministry of Industry, Trade and Tourism
Note: [e] = expected

By 1990, foreign capital was already invested in one-fifth of all Hungarian companies. By the end of 1995, from among companies conducting double-entry accounting, partly or fully foreign owned companies had 47 per cent of the registered capital, 33 per cent of the employees, 45 per cent of the net sales, 61 per cent of investment and 66 per cent of exports. Certain sub-sectors were almost completely in foreign ownership, such as the manufacturing of tobacco products, where the ratio of foreign working capital to registered capital was already 97 per cent; electric machine production (84 per cent); and the production of non-metal mineral products (69 per cent). This ratio exceeded 50 per cent in the manufacturing of rubber and plastic products and in the manufacturing of public road vehicles and in the paper industry. It was also close to that proportion in telecommunications technology and in the manufacturing of office machines and computers.[2] Table 2.3 describes how foreign direct investment is distributed throughout different sectors of the economy.

Table 2.3 Structure of Foreign Direct Investment in Hungary

Sector	Total Capital Base (Hungarian + foreign)		Foreign Capital (as proportion of total)	
	HUF (bn)	%	HUF (bn)	%
Agriculture	26	1	16	1
Mining	15	1	11	1
Manufacturing	773	39	560	43
Food and beverages	201	10	158	12
Textiles and clothing	31	2	26	2
Wood, paper, printing	450	2	38	3
Chemical industry	206	10	109	8
Non mineral product	50	3	41	3
Metallurgy	48	2	34	3
Machinery	185	9	148	11
Other manufacturing	7	0.5	6	0.4
Electric engineering, gas, water	358	181	172	13
Construction	55	3	46	4
Trade, maintenance, services	203	10	156	12

Note: Exchange rate in 1995, 125.7 HUF = US$1.00.

Green field investment represents 43 per cent of foreign investments.[3] The ratio of foreign ownership increases with experience in the country. By 1993, in the firms with foreign ownership 70 per cent represented majority ownership and 44 per cent had 100 per cent ownership.[4] In certain sectors of foreign trade (such as the market for vegetable oil and fat) foreign working capital was in an almost exclusive position by 1993, and it also had majority ownership in almost all other product groups. In 1995 (according to customs statistics), 55 per cent of total exports and 61 per cent of total imports were effected by partly or fully foreign owned companies.

MNC investments have been broad-based and range through manufacturing industry, utilities (electric engineering, gas and water supply), food and beverages, the machinery industry and trade, storage, post and telecom industries have all attracted especially significant foreign investment. Table 2.4 lists the largest MNCs present in Hungary. Those investors already present in Hungary continue to invest in projects to raise the added value content of their operation and to integrate Hungarian affiliates into their global network. The dynamics in the development of foreign-owned companies has grown by 36 to 38 per cent in the last three years, while their export has increased by 40 to 42 per cent. Table 2.4 shows the largest foreign investors.

Table 2.4 Largest Multinational Companies in Hungary

Rank	Company	Nationality	Main Business	Revenues 1996 ($m)
1	Hungarian Telecom	Germany, USA	Telecommunication	2536
2	Opel Hungary	USA	Automobile	1040
3	RWE-EVS	Germany	Electricity	652
4	Panrusgáz	Russia	Petroleum	469
5	Metro Holding Hungary	Germany	Wholesale trading	412
6	Electricité de France	France	Electricity	389
7	GE Lighting Tungsram	USA	Lighting engineering	335
8	Unilever Magyarország	Netherlands, UK	Consumer goods	329
9	Bayernwerk	Germany	Electricity	310
10	IBM Strorage Products	USA	Computers	309
11	Magyar Suzuki	Japan	Automobile	306
12	Audi Hungaria	Germany	Automobile	292
13	Shell Hungary	Netherlands, UK	Petroleum	283
14	Julius Meinl	Austria	Supermarket	254
15	Westel 900	USA	Mobile telecoms	198
16	Alcoa-Köfém	USA	Aluminium	193
17	OMV-Hungária	Austria	Petroleum	187
18	Lehel Zanussi	Sweden	Electrical goods	178
19	Plus	Germany	Supermarket	172
20	IR3 Video (Philips)	Netherlands	Electronics	171

Source: Figyelö TOP 200, 1996.

A unique target area of the capital inflows is privatization. Hungary is practically the only country in Central and Eastern Europe that has not discriminated against foreign investors throughout the privatization process. In 1997, privatization revenue exceeded US$ 7bn, or make up almost 40 per cent of all FDI. The telecommunications sector received some 25 per cent of major foreign investment, followed by the electricity, gas and water suppliers, with a 15 per cent share, and food processing, also with a 15 per cent share. Financial institutions and banks represent 10 per cent of the total FDI. The largest investors in the energy sector are RWE Energie, Bayernwerk and Isar-Amperwerka of Germany, Electricité de France and Gaz de France and Tractebel of Belgium. Together these companies invested more than US$ 1.25bn. In the food industry, soft-drink manufacturing absorbed US$ 400m, sugar processing US$ 600m, the confectionery industry US$ 200m and the tobacco industry US$ 190m. More than US$ 230m was poured into the retail sector, mostly due to the development of large hypermarket chains. The largest investors in banking and finance are the Italian BCI, with US$ 370m and Dutch banks ABN-Amro (US$ 240m) and ING Group (US$ 110m). The companies that have made the largest foreign investments, and their countries of origin are illustrated in Table 2.5.

Table 2.5 Largest Foreign Investments in Hungary

Company	Country	Sector	Value of investment (US$m)
Ameritech & Deutsche Telekom	(US–German)	Telecoms	22 230
General Electric Lighting	(US)	Lighting	726
RWE-EVS	(German)	Energy	628
Volkswagen-Audi	(German)	Automotive	550
Eridania- Béghin-Say	(Italian, French)	Sugar industry	540
General Motors	(US, German)	Automotive	500
BCI		Banking	370
Aegon Insurance Group	(Netherlands)	Insurance	366
PTT TelecomDenmark	(Denmark)	Telecoms	340
USWEST Int.	(US)	Telecoms	330

Note: as of January 1998.
Source: Napi Gazdaság. 23 September, 1998.

In many industries, particularly in food and trade services, although large state-owned companies have been dismantled another type of concentrated market has emerged in the last five years. By the mid-1990s, market penetration had led to two or three large companies dominating in each sector. Most of these companies are multinationals and to a lesser degree state-owned or semi-state Hungarian enterprises. For example, the sugar industry has three foreign firms: Béghin-say, Agrana, Eastern Sugar – and one Hungarian firm. The tobacco industry has four foreign companies – Philip Morris, BAT, Rentsma and Reynolds. The car industry has Opel, Suzuki, Audi and Ford.

According to a recent research study based on more than one hundred foreign investors, three major categories of firms could be identified: assembly export oriented; domestic-based export-oriented; and non-export domestic-based firms.[5] Export-oriented firms were those where the export ratio was higher than 50 per cent. Table 2.6 shows the top exporting firms in Hungary.

A recent study for the American Chamber of Commerce by McKinsey revealed the concerns of foreign MNCs in Hungary.[6] Firstly, the MNCs' representatives would like to have an ongoing forum of conversation with the

Table 2.6 Largest Hungarian Exporting Companies

Name of Company	Export Sales Revenue (HUF m)
IBM Kft.	283 000
Audi Hungária Motor Kft	188 735
Philips group	154 361
Opel Magyarország	121 832
MOL Hunagraian Oil and Gas Co.	82 106
GE Lighting Tungsram	n.a.
Magyar SuzukiRt	57 918
Hungarian Intellectual Property Agency	54 246
Malev Hunarian Airlines	51 320
Alcoa-Köfém	45 373

Note: Exchange Rate in 1997, 186.8 HUF = US$1.00.
Source: Figyelö Top 200, September 1998.

government on business related issues. The survey showed that the conditions provided by Hungary and the needs of the investors met on the highest level were political stability, development of infrastructure, and working conditions. Secondly, the majority of the concerns the investors expressed were legal and administrative obstacles, taxation and the starting of working hours. Potential concerns were energy prices, and the market size. MNCs were less interested in closeness to market, favorable labour costs and existence of universities and research institutes. The list of MNCs' requests included the following: help in the interpretation of laws; clear and evident laws and policies, clear and more punctual tax and customs policies, simpler legal frameworks, longer lasting economic regulations, more consistent government commitments, faster court decisions, and acceleration of the industrial park program.

Domestic companies

The traditional large state owned companies, that represented the majority of firms in 1990, have gradually all but disappeared. Most of them either went bankrupt like MOM and Medicor while others were privatized and became a part of MNCs. Tungsram, one of the best European light bulb makers, was acquired and merged with General Electric Lighting as early as 1989. Taurus Rubber Works, famous for its truck tires and rubber products was privatized and later integrated into Michelin.

However, a few large local firms survived and became quite successful. The largest ten are listed in Table 2.7. MOL, the Hungarian Petroleum company, Dunaferr ironworks, Hungarian Railways, are on the top ten list. Pannonplast, a most successful Central-European company producing plastic products, and Richter pharmaceutical company were able to grow without a Western strategic partner. Most of these companies are very active in the international arena. MOL the petroleum company, BorsodChem the chemical company, the pharmaceutical companies Richter Gedeon and EGIS, Zwack Unicum the fine spirit company, among others have developed international reputations. MOL is heavily investing in gasoline stations primarily in neighboring countries, while Pannonplast is expanding its operation internationally. Ikarus, at one time Europe's largest bus maker totalling 13 000 buses per year in the mid-1980s, made its way back into the Russian market at the end of the 1990s by establishing assembly lines in Russia.

Table 2.7 Largest Hungarian Firms

Name of company	Sector	Sales Revenue (HUF m)
MOL Hungarian Oil and Gas	Petroleum	603 624
Dunaferr Group	Metallurgy	205 912
Hungarian Railway	Railway	133 380
Tiszai Vegyikombinát	Plastics	90 425
Hungarian Post	Postal Services	62 634
Hungarian Intellectual Property	Foreign trade	54 246
BorsofChem	Plastics	53 588
Richter Gedeon Chemical	Pharmaceutical	52 568
Hungaropharma	Pharmaceutical wholesale	51 396
Paks Nuclear Power Station	Energy	49 442

Source: Figyelö TOP 200 of 1997.
Note: Exchange Rate in 1997, 186.8 HUF = US$1.00.

A number of new firms successfully grew and survived the first stage of development. Fotex, the large retail conglomerate, is successfully competing with the large multinational supermarket chains.[7] However it is software development, in general, which is the major driving force behind fast growing ventures in Hungary.[8] Recognita developed world leading optical recognition software. Geometria with its geographic information systems is one of the top ten in Europe.

The software company Graphisoft is perhaps the best known example of fast growing, Hungarian ventures.[9] Graphisoft is a developer of solutions for building design, marketing, and maintenance, with balanced worldwide sales and cutting-edge technology. ArchiCAD®, its flagship product, is an integrated, object-oriented, three-dimensional Computer Aided Design (CAD) solution for architecture and building industry. Available on Windows 95/NT and Macintosh operating systems, ArchiCAD® is sold in 80 countries and 22 languages through its independent marketing partners and wholly owned subsidiaries. Graphisoft was founded in 1982 by Gábor Bojár and his friends and is one of the first private partnerships in Hungary. In 1998, it was among the top three architecture, engineering and construction software companies in the world with close to 200 employees in its Budapest headquarters, R&D and subsidiaries in Munich, San Francisco, Tokyo, Hong Kong, London, Madrid, and Dubai. With sales of more than US$ 22m in 1997, Graphisoft has maintained 25–30 per cent net income and close to 40 per cent average growth since 1993. On June 8, 1998 Graphisoft NV, the Netherlands based holding company for Graphisoft Group, listed its shares on Neuer Markt, the high growth technology segment of the Frankfurt Stock Exchange.

Overall Globalization Drivers for Hungary

Most of Hungary's globalization drivers favor the participation of foreign MNCs. In addition, Hungary aspires to become increasingly attractive as a gateway and hub for emerging economies of the former Soviet Union. The growing purchasing power of ten million inhabitants, the advanced economic reforms and the industrial traditions of a skilled labor force have made Hungary a natural target for MNCs seeking new markets and production facilities. While progress is apparent nearly everywhere, not all sectors of the Hungarian economy are developing at the same pace. Some segments, such as the food industry, face huge competition while other markets are remarkably under-exploited. Over the last decade, many MNCs have chosen Hungary as the location for their first investment into Central Europe. One attraction for larger firms is the desire to establish a foothold in a potential growth market. But the interest in Hungary is not limited to large multinational corporations. Many foreign owned, small-and-medium-sized firms have also begun operations here and some of these investors began to make profits shortly after their initial outlay. Others have faced a longer start-up period. Almost all were attracted by the large number of 'market niche' opportunities, which many investors consider to be more plentiful in Hungary than in Western Europe.[10]

Market globalization drivers

Hungarian customers readily accept Western brands. This is partly because Western brands were already familiar to Hungarians who could travel abroad in the 1980s. As a consequence of extensive foreign investment more local access was created for Western brands and firms in the 1990s. Companies such as Coca-Cola, PepsiCo, McDonald's, Procter and Gamble, Unilever, and their brands like Coke, Pepsi, Big Mac, Vidal Sassoon, Pampers, Eskimo ice cream, and Lipton tea are all familiar and accepted by Hungarians. Hungarian consumption patterns are becoming increasingly similar to those of industrialized Western nations. This looks to become the trend for the future as Hungarian youngsters' favorite gathering places are often fast food restaurant chains such as McDonald's, Burger King and Pizza Hut.

Although some marketing adaptation is required due to language and cultural differences, Hungary offers MNCs a relatively high degree of marketing transferability, especially with regard to brand names. Western satellite and cable television programs and their commercials further increase consumers' awareness of the latest trends and products.

Cost globalization drivers

Hungary is a source of relatively cheap labour for production and expansion in Europe. However, wages have a minor role in the investment decisions of MNCs since they can be described as low only when compared to those of Western European countries. Hungary's relative comparative labor cost advantage has been more or less eroded because Hungarian wages have gradually increased within multinational companies. The further opening of Eastern European countries, such as the countries of the former Soviet Union, with even cheaper labour costs has created an even less favorable position in terms of cost competition.

The new National Basic Educational Plan, launched in 1998, re-establishes the possibility of a very high standard of education of a new generation of Hungarians. With the 'school computerization program' Hungarian students are entering the information age on the information superhighway.

The availability and distribution of skilled workers is also a potential area of considerable competitive advantage for Hungary. Although machine engineering has historically employed the most skilled apprentices, other sectors including communications, have been gaining skilled workers at an ever increasing pace.

The abundance of well-trained employees and the potential for Research and Development facilities has not been overlooked by foreign MNCs. As Jack Welch, CEO of General Electric pointed out. 'Success sometimes takes time. Hungary (where GE bought Tungsram in 1989) was very difficult. The problem wasn't the people themselves – it was that they didn't understand the market economy. It was that simple. We had an eager, thirsty crowd who we weren't able to move along fast. Today, though, we are a world-class lighting company in Hungary – with at least half of the advanced technology done there. Patient capital pays off.'[11] GE's acquisition of Tungsram is considered a strategic investment, as production in Hungary raised GE Lighting's market share in Europe from 5 to 20 per cent.

There is however a tight labor market in some segments of the labour force and this serves as a driving force for the implementation of modern business practices in Hungary. Other driving forces include changing attitudes, convertible expertise and the increasing availability and use of consultants familiar with a market economy business environment. On the negative side, however, bureaucracy remains entrenched, resistant and reactive to change.

Unlike many of its neighbors, from the 1980s onwards Hungary has had reasonably extensive experience in dealing with Western business practices. During the early phases of the transition, company managers were allowed to experiment with Western management techniques and introduce some Western technology. Hungary became the Soviet Union's conduit for hi-tech goods, such as computers, and pharmaceuticals. During the last twenty years Hungary has gained experience in technology transfer.

A fairly developed road and railway system, the waterway on the Danube, the main water-road of Europe, and the expanding international airport of Budapest provide relatively good transportation conditions. Situated on the crossroads from West to East and from North to South of Europe, transportation infrastructure is a key to attracting foreign investors. Rapid development in telecommunication occurred when MNCs started to invest into this sector.

Government globalization drivers

Overall, the Hungarian government is politically stable and business friendly. Favorable foreign trade and investment policies have been pursued by past governments and most foreign MNCs consider the Hungarian economy to be highly attractive. The factors identified by foreigners attractive to them as investors are mapped out in Table 2.8.

Table 2.8 Foreign Investors' Motivations in Hungarian Investments

Motivation Factors	Strengths of Implication				
	insignificant	slight	average	high	significant
Entry to Hungarian market					×
Access to East European market			×		
Close to West European market		×			
Cheap labour force				×	
Developed infrastructure		×			
Cheap raw materials	×				
Governmental benefits (e.g. tax)		×			
Development of the Hungarian banking system		×			
Size of privatization offering				×	
Political stability				×	
Other*				×	

Note: *Personal contacts in small businesses.
Source: Árva László: Külföldi müködötöke, hazai beszállítói kapcsolatok. Foreign working capital, domestic subcontractor relations. *Közgazdasági szemle.* November 1997.

Absence of state-owned competitors

The most influential factor in attracting foreign investors was the privatization of Hungarian industry, trade and utilities. As in most Central European countries, privatization started slowly, but the pace has picked up considerably. During the period from 1995 through to early 1998, the government succeeded in selling the controlling stakes of most of the country's electric companies (with the exception of the central transmission grid, nuclear plant and one coal fired plant). This includes all six gas distributors, significant stakes in the national oil and gas company (MOL), the state telephone company (MATÁV) and many other large companies. In 1995 alone, privatization revenues amounted to US$ 3.7bn. Hungary's privatization strategy differed significantly from neighboring countries in that companies were sold primarily for cash to principally strategic investors, who were also committed to inject capital and bring new management, technology and know-how. In contrast, in some neighboring countries, companies were given away via coupon schemes and tended to be controlled by financial investors or state controlled banks. The 'Hungarian privatization model' is widely considered to be a success story and is being emulated, to a certain degree by neighbors such as Romania. As a consequence foreign MNCs implemented the 'best buy' market purchase investment strategy in Hungary.[12]

Foreign ownership is encouraged not only in the manufacturing industry and trade but also in the banking industry. The Hungarian government aims to become the Central Eastern European financial center. The stake held by foreign shareholders in the combined capital of the Hungarian banking sector is 61 per cent, which amounts to over HUF3.4bn (Hungarian Florints). However, the high degree of foreign presence, unmatched by Central and Eastern Europe standards has not resulted in a concentration of the banking sector. No foreign shareholder holds a stake of more than eight per cent. In the privatization of Hungarian banks, German strategic investors such as Bayerische Hypoteken und Wechselbank, Westdeutsche Landesbank, Commerzbank, DG Bank, Dresdner Bank and Deutsche Bank, played the most important roles and their combined market share is still below 20 per cent. In addition to German banks, Austrian (Erste Bank, BankAustria-Creditanstalt, Raiffeisen Zentralbank, Österreichischen, Volksbanken), Dutch (ABN Amro, ING, Rabobank), Italian (Banca Commerciale Italiana, Banca San Paolo di Torino), French (BNP, Crédit Lyonais, Cetelem) and Belgian (Kredietbank) banks have appeared on the Hungarian banking market. American banks (Citibank and GE Capital) and a Russian bank (Gazprom Bank) are also present. Asian banks also conduct a considerable amount of business here, including the South Korean Daewoo and Hanwha, the Malaysian IC Bank and the Japanese Long-Term Credit Bank of Japan, as well as Nomura.

Favorable trade policies

Trade with Central and Eastern European countries is on the rise, but more significant trade relations are being established with EU member states. In fact, a remarkable development in trade with EU countries has already taken place, furthering Hungary along the road towards European integration. Hungary began early in developing trade relations with the nations of the Organization for Economic Cooperation and Development (OECD) and became a member in

March 1996. Other international groups to which Hungary belongs include the World Trade Organization (WTO), and the Central European Free Trade Association (CEFTA). Hungary made an important step towards European integration with the conclusion of the Europe Agreement in 1991, which partly deals with the breakdown of trade barriers. Other topics covered include the harmonization of EU and Hungarian laws, agriculture, industry, environment, the financial sector and anti-trust policy.

Favorable foreign direct investment rules

On 1 January 1996, new foreign exchange regulations were established, making the florint convertible with other currencies. Nonetheless, according to Article 8 of the Statutes of the IMF, Hungarian and foreign legal entities are obliged to keep receipts, invoices and other specified documents from Hungarian banks and the Hungarian National Bank for a period of five years. The new rules cover considering capital transactions include:

- Hungarian companies and individuals are allowed to acquire ownership in foreign companies without the preliminary permission of the Hungarian National Bank if certain requirements are met.
- There is no need for a foreign exchange licence to obtain a foreign loan with a duration exceeding one year, however the Hungarian National Bank needs to be informed within eight days after conclusion of the loan contract.
- The value of inheritances of foreigners are convertible into hard currency and may be freely transferred abroad.
- Hungarian residents are allowed to buy or build real estate abroad without any permission, however the Hungarian National Bank needs to be informed.

In 1988, the Foreign Investment Act established a clear legal basis for investments from abroad. Since then, the development of economic legislation has made the Act largely superfluous. Still, the Act expresses Hungary's commitment to providing a stable legal environment for foreign investors. The law forbids expropriation, except in the case of acute national concern. Should this happen, foreign owners will be compensated immediately, based on the real value of their investment. Foreign owners are guaranteed the ability to take profits, or income from liquidation of their business, out of the country in their home currency.

Role of regional trade blocs

Hungary is a member of CEFTA, OECD, and the World Trade Organization and is in the process of entry talks with the European Union. As part of the Central European Finance and Trade Association (CEFTA) agreement, free movement of goods among Poland, the Czech Republic, Slovakia, Slovenia and Hungary creates conditions which will smooth the transition into the European Union.

Reliable legal protection

Hungary has historically accepted the protection of trademarks and intellectual property rights. Hungary is represented in almost all international organizations on international copyright protection.

Hungary has also concluded bilateral treaties on investment protection with many countries. These treaties provide a more detailed description of the general legal guarantees enjoyed by foreign investors. In the absence of a tax treaty, they might also provide protection from double taxation. Further investment and trade protection is also available from foreign government funds and guarantees. There are even commercial insurance plans offering protection against the risks involved in foreign investment and trade in Hungary.

Compatible technical standards

In technical areas, Hungary's standardization has a history of more than a hundred years. International standards are part of the Hungarian standard system. The country subscribed to ISO 9000 from as early as 1990. On top of that, in 1996 the government established the National Quality Award based on the model of European Quality Award. In 1996 eighteen, and in 1997 seventeen, firms applied for the Award. Among those awarded were Hungarian firms such as the manufacturer Herend Porcelain, Taurus Agricultural Tires and MNCs like Westel 900 GSM and Opel Automobiles.

Competitive globalization drivers

Hungary's global strategic importance lies in the size of its domestic market, and its potential to become a strong regional hub for MNCs. Because of current political stability, a strong technological background and the historical presence of MNCs, Hungary is in the process of developing as a regional center of financial services, information technology and pharmaceutical research.[13] The potential of research and development in Hungary is illustrated by Sanofi, the French pharmaceutical company, which invested in the Hungarian pharmaceutical company Chinoin not only to get access to its market but also to access its research capacity. Other examples include Nokia and Ericsson, the mobile telephone giants, who have established one of their software development centers in Hungary.

With entry into the European Union, Hungary will become more important to export-oriented MNCs. For example, Suzuki, the 97 per cent Japanese-owned public limited company with two major shareholders, Suzuki Motor Corporation and the Itochu Trading House, is intending to use Hungary as a kind of European center for the brand with the Hungarian plant as a base. With possible future expansion of the plant, there is even a chance that the technical development of some special spare parts for the European market, will be sited here. The company also demonstrates the benefit of becoming part of MNCs' interdependent value chains. Of the value of a Suzuki Swift assembled here, 33 per cent comes from Japanese imports, 29 per cent from Hungarian suppliers, 26 per cent from value added by the plant, and 12 per cent from spare parts bought from suppliers in the European Union. In 1997, about 4 000 cars were exported, and the domestic market will receive nearly 15 000.

Suzuki, however, is not alone in Hungary. Other automobile makers such as Opel of General Motors, Ford and Audi are competing on the Hungarian market with their locally produced cars. MNCs located in Hungary compete with each other for Hungarian customers. This is the case with Philips and IBM in computers, Coca-Cola Amatil (the Australian Coke bottling in Hungary) and PepsiCo in the soft drink market, the big accounting firms (such as Ernst and Young, PriceWaterhouseCoopers, Deloitte and Touche) in their services. In the retail market the German Metro and Tengelman compete with Tesco from the UK and Cora from France.

Overall Global Strategies For Hungary

MNCs from Europe, the United States, and South Asia have all pursued different strategies in the Hungarian market. Out of the top forty multinationals more than thirty are competing on the Hungarian market with each other and with local and other foreign non-multinational firms. In all industries, competition is increasingly intense and strategies pursued by these companies reflect the growing global pattern of competition. In this new world large local firms play a decreasing role. MNCs increased their participation through privatization of local firms, joint ventures and green field operations. A growing number of small and medium sized local companies are trying to fill niches.

This section concentrates on the strategies of six MNCs with successful operations in Hungary – Coca-Cola Amatil, Unilever, General Electric Tungsram, General Electric Capital, Budapest Bank, United Technologies Automobile, and Westel 900 GSM. Except for GE Capital, Budapest Bank, and Westel 900 GSM, all the other companies are almost 100 per cent owned by MNCs.

Market participation strategies

By European standards, Hungary is large enough to be an attractive market in its own right. It now has the added advantage of being a potential gateway to Albania, Bulgaria, Croatia and Romania as well as other countries of the former Soviet Union. Out of the six focus companies, GE Tungsram, United Technologies Automobile and Unilever are export oriented. Meanwhile Coca-Cola Amatil, GE Capital-Budapest Bank, and Westel 900 GSM focus primarily on the domestic market.

Although Unilever had a subsidiary in Hungary before World War II, it was nationalized only after the war. Unilever has been in the Hungarian market in imported products since the 1970s. The company began directly investing in Hungary in 1991. Unilever Hungary divides its activities into two major segments: production and marketing/sales. Unilever has four manufacturing facilities in the country. The newly built Eskimo plant in the west of the country, produces ice cream. In the Budapest factory, Rama and Hera produce margarine and related products. In the north east, plants produce detergents such as OMO, Biopon, and Domestos. A plant in the south produces quick frozen vegetables under the Iglo brand. Unilever competes, among others, with Procter & Gamble, Henkel, and Benckiser.

The Australian company Coca-Cola Amatil (CCA), which has 100 per cent ownership of Coca-Cola Amatil Hungary, entered the Hungarian market in 1968, as a distributor to a Hungarian bottler. Prior to 1991, the primary focus for CCA was bottling and manufacturing. Since entering the Hungarian market, CCA expanded facilities, diversified as well as globalized its bottling operations. The growth of international competition played an important part in the investment decision. The threat from Pepsi, who already had bottling operations in Eastern Europe, was becoming very strong. Acquiring the bottling operations in Hungary helped CCA to gain competitive advantage over Pepsi by taking existing resources away.

General Electric was another company that acquired a large Hungarian lighting bulb producer Tungsram, even before privatization began in 1988. Considered as the flagship of MNCs operating in the country, GE Tungsram became the learning ground for the turnaround of a Hungarian state-owned conglomerate.

United Technologies Automobile Hungary (UTAH) was founded in 1992 as a green field investment for an automobile cable assembly company. UTAH is owned 100 per cent by United Technologies, the US based multinational company. The primary focus of its production is supplying major automakers with pre-assembled electric cables all over Europe.

Westel 900 Mobile Telecommunications Company was founded in 1993 as a joint venture by Hungarian Telecom (44 per cent) US West International (42 per cent) Westel Radio Telephone (9 per cent), International Finance Corporation (5 per cent). Westel 900 is the largest GSM service provider in the Hungarian market. Westel 900 GSM was the first Hungarian mobile supplier to obtain the ISO 9001 quality certificate and won the National Quality Award.

Budapest Bank was one of the first Hungarian banks to be privatized in 1995. General Electric Capital Services became a strategic shareholder with 27 per cent ownership, while the European Bank for Reconstruction and Development (EBRD) supported the deal with 34 per cent ownership. By this investment, Budapest Bank gained an opportunity to become a premier financial service company on the Hungarian market. Other banks privatized by Dutch ABN-Amro, or ING, or German banks like Westdeutsche Bank, have created fierce competition in the financial market.

Product and service strategies

Most products and services need only moderate adaptation for Hungary. MNCs, on the whole, try to leverage their strong global products in the Hungarian market, with some tailoring to meet local needs. At the same time, local brands can also be included in the multinational product mix, or at least into the domestic product mix.

Coca-Cola Amatil produces and markets only global brands. GE Tungsram carries the internationally accepted original Tungsram brands as well as GE brands. Unilever also incorporated some local brands into the firm's portfolio for primarily domestic marketing. *Amodent* was a well-accepted local brand. With product development and marketing the brand is extremely successful. Other than that, the new version of Magnum ice cream filled with fruit was developed and field-tested in Hungary. Based on the successful innovation the product is ready for worldwide introduction.[14] UTAH as a subcontractor designs its products according to car makers' request.

Only a few domestic Hungarian products have become successful players in the international arena so far. Among them, Rubic's Cube, by far the most well known Hungarian product, was produced outside the country, and the profits were not made in Hungary. Bristol-Myers Squibb acquired the Hungarian company Pharmavit, after it developed the bubbling vitamin tablets 'Plusssz' which was successful in Hungarian and CEE markets. The small Hungarian software company Graphisoft developed the three dimensional software 'Archicad' for the architecture market. Now Graphisoft has a world wide distribution network and has become a market share leader in its category.

Activity location strategies

MNCs in Hungary produce for the local market, for export, or for both. Most of them are developing an increasing number of local subcontractors and providing them with impetus to grow and develop quality delivery. Few MNCs developed regional marketing distribution centers and only a few MNCs concentrate on local R&D activity. But Nokia established a software center with 500 engineers in 1998 to develop a new generation of mobile phone switchboards.

Philips employs more than 6000 people in Hungary. Its export income almost trebled from 1996 to 1997, which made Philips one of the largest exporters in Hungary. The success is primarily attributed to video, computer monitor, and car radio production. One of the stars of the company is Philips' Monitor subsidiary, which is the European manufacturing center of the Philips group. After operating for just over a year, the firm received the ISO 9002 certificate, attesting the quality of its products and manufacturing processes.[15]

After the political transformation, Hungary's national automobile industry was revived. MNCs such as Suzuki, General Motors (Opel), Audi and Ford came to Hungary: Suzuki and GM with the assembly of Swift and Astra, respectively,

Audi with engine manufacturing, Ford with spare parts manufacturing. More than that, they helped to create a continuously growing background industry of suppliers, which ensures not only the creation of jobs but also the adoption of the most up-to-date technology.

GE Lighting Tungsram created its European distribution center in Hungary in addition to its manufacturing units. Now, it is in the process of reorganizing its R&D center. United Biscuits from UK acquired Győri Keksz in 1992 and in addition to its production, the company organized local R&D and procurement units in Hungary.

Marketing strategies

Products and services are often marketed in Hungary in the same way as in Western countries. Several dozen local and multinational advertising agencies help MNCs to develop and implement creative marketing strategies. Most companies are able to use their global brand names in the Hungarian market. On the other hand, they share local adaptation needs: translating information on product packages and enclosed documents into Hungarian. But translating is not always essential as in the case of Philips which chose not to translate its global slogan 'Let's make things better!'.

Advertising strategy plays a critical role for almost all six of our focus companies. These companies consider television as a primary advertising outlet followed by newspapers, radio and billboard posters. In their television campaigns some use global advertising, some develop local or use a combination. For example, Coca-Cola Amatil use the global Coca-Cola advertising for their winter campaign. A local film for the summer campaign at Lake Balaton, a favorite summer resort for teenagers, and a global-local combination for the Christmas commercial was applied when the marketing team realized the local needs of domestic customers.[16] However, the 'Always Coca-Cola' slogan has always appeared. When the first combination advertisement was showed in Christmas 1996, sales figures doubled in the following weeks.

Unilever uses many international brand names such as Lipton, Ponds, Lux, Sunlight, Signal, Omo, Brut, Rama, Eskimo, which are so completely accepted by Hungarians that they are considered in the same way as local brands. Unilever added about 40 per cent to its total production by incorporating well known local brands such as Biopon detergent, Amodent tooth paste, Amo, and Baba soaps, Liga margarine. Unilever spent US$10m for television advertisements in 1996.

Distribution channels for fast moving consumer goods are the major retail chains and a great variety of smaller outlets. This market is rather segmented. Most manufacturers of FMCG have a role in training local retailers for merchandizing and selling space management. Large shopping malls are in the early phase of development in Budapest. The first two Western styled malls were opened in 1996. Duna Plaza was based on an Israeli investment, and Polus Center funded by Canadian and US capital. Two additional malls were being opened in

1997 and several others in 1998. There are plans for opening more malls in other cities as well.

Competitive move strategies

Competition in the Hungarian market is increasingly intense. Major global competitors compete against other MNCs. Many MNCs also face a mixture of foreign and local competition, depending on the industry. For example Unilever competes with Procter and Gamble, Henkel and Benckiser. Coca-Cola battles with Pepsi and some local brands for market share. GE Tungsram has its own rivals such as Philips and Osram. GE Capital-Budapest Bank competes with local and foreign banks like OTP, the largest local saving bank, the Citibank, Creditanstalt, ABN-Amro, ING. Head to head competitors on the mobile telecommunication market are Westel 900 and Pannon GSM, a Scandinavian-Hungarian joint venture.

Organization And Management Approaches For Hungary

Hungary is facing a period of fundamental transformation. At the same time, never before has it been so critical to recognize the importance of human resources, and of the need to make massive investments in development and training, especially at the managerial level. Hungary's skilled workers must be managed more effectively and efficiently in order to maintain a long-term competitive advantage for the MNCs employing them.

In the past, Hungarian management has been criticized for its lack of market-oriented skills, its reliance on antiquated management practices and its lack of commitment to performance management. Western, mostly American, managers typically characterize Hungarian managers as slow decision-makers, unwilling to take risks, too inflexible and lacking a 'spirit of adventure'. In turn Hungarian managers consider American colleagues responsible for a number of difficulties and obstacles in communication, due to a lack of respect for local ways and customs, and a lack of detailed knowledge of the regional market. An employee service/customer orientation is still the exception in most former state enterprises and even joint ventures are finding it very difficult to change established work practices developed under the former socialist system. Part of the problem is that previously, the Hungarian higher educational system was reluctant to recognize business and management studies as respectable academic disciplines. Business administration was considered a sophisticated type of vocational training. Management and business subjects were associated, if considered at all, with traditional university disciplines such as economics, law or engineering.

The situation has changed in business and management education. There are almost ten different MBA programs offered at private business schools such as the International Management Center in Budapest and in major universities, such as Budapest University of Economics and Technical University of Budapest or in distance learning schools such as Open Business School. Management training schools and consulting firms offer a great variety of advanced professional knowledge and skills in marketing, customer service, financial management, organizational development and human resources management.

Most MNCs spend time and money on training their management teams and workforce. All six MNCs mentioned above have organized local and international training programs for their employees. The major focuses of training programs are sales and services, financial analysis and new methods of people management. UTAH, GE Tungsram and Budapest Bank, CCAH and Westel 900 have special executive training programs for fast-track high-potential Hungarian managers.

There is no one rule for who fills managerial positions. In two of the six MNCs (UTAH, Westel 900) there are Hungarians in the CEO position. At UTAH all the managers are Hungarians. In all six companies, local nationals occupy many top management team positions. More and more Hungarian managers have become the members of their company's international executive team. Unilever, Coca-Cola, GE Lighting now have Hungarian managers on expatriate assignments.

Hungarian workers are skilled, but their skills are not being adequately utilized. This has partly been a product of inadequately trained managers, but is also the consequence of organizational systems that exploited workers rather than energizing and empowering them. The establishing or reestablishing of relationships with trade unions will be a critical emerging factor. Their co-operation with the government is critical to any long term market restructuring. The alienation and exclusion of trade unions would provide an organized institutional focus for worker alienation.

The reality of business life in Hungary is that all the changes have produced an increasing degree of uncertainty that will only increase in the future. Consequently, organizations and their employees will have to become much more flexible. There is also a visible change in the attitudes of managers. While for the older generation work is a tool for living, for a younger generation of managers aged 25 to 35, 'live for work' is the normal attitude in business. Furthermore, the younger generation is more flexible in changing positions within the firm and between industries.

Conclusion

The Hungarian economic policy is aimed at establishing a free market economy, and initiating a fundamental restructuring of enterprise ownership through new laws and through the various privatization programs. From 1989 onwards, the importance of the private sector, as opposed to the public sector, has increased dramatically.

MNCs have generally found favorable conditions for setting up their operations in Hungary. What started as an 'easy buy' business opportunity, sometimes required blood and tears to get started, but often turned out to be a profitable and promising enterprise. MNCs learned to cope with gaps, not only in different organizational structures, but also in a very different business environment and national culture. The question of the future fate of Hungary is still an open one.

Hungary still faces a number of problems. These include continued high inflation and high public debt and a low rate of economic growth, which will have to be significantly higher than the EU average over a long period in order to raise living standards to EU levels. The question of the fate of Hungary is still very much unresolved. What are the challenges faced by a country with a European heritage, previously shackled by a controlled economy, and now facing the competition of a global business environment?

The need to reform the social security system, whose deficit is financed from the budget, is concomitant with the problem of public debt. Improvements must also be made in infrastructure, industry and technology to raise Hungary's competitiveness, and a further reform of the banking system is required. However priority needs to be given to reforming agriculture, as Hungary has a surplus in agricultural trade with the EU.

In 1998, the European Union awarded Hungary an A-ranking for political stability and its economic situation. Hungary was considered to have a well operating market economy with mid-term competitiveness in Europe, although there is still a need to fight corruption and increase the standard of living. According to EU experts, legal conditions for firms and banks should be improved, and care must be taken to retain financial balance. An A-grade was also given to Hungary regarding harmonizing with the European Union. As pioneers in adopting democratic and pro-market reforms in Central and Eastern Europe, the Hungarian government and companies have learned to utilize the educational system and workforce effectively and to everyone's advantage. They have also promoted opportunities for foreign investors and multinational companies entering the region. There are well founded hopes that this second Hungarian renaissance will continue strongly into the future.

Notes

1 Hungary holds the lead. In *No Limits. Joint Ventures in Hungary: 25 years*. 1997.
2 Hamar, J.: Success Story? *In No Limits. Joint Ventures in Hungary: 25 years*. 1997.
3 Companies starting their operation with no local partner.
4 Foreign Working Capital in Hungary, 1994. KSH (Central Statistical Office).
5 Étetö Andrea-Sass Magnolna, 'Influencing Factors on Foreign Investors' Decisions. *Közgazdásagi Szemle*, June 1997, pp. 531–46.
6 Henk J. Back report. *Népszabadság*, November 1997, p. 2.
7 'Faded Fotex'. Business Central Europe, June 1998.

8. 'The Need for a Niche'. Software industry survey, Business Central Europe, March 1998.
9. Hisrich, R., Vecsenyi, J. (1997) 'The Entry of a Hungarian Software Venture into the US Market'. In Hisrich, McDouglall, Oviatt (eds), *Cases in International Entrepreneurship*. Chicago: Irwin.
10. 'Doing Business in Hungary'. Arthur Anderson. 1998 edn.
11. 'Transfer of Best Ideas from Everyone, Everywhere'. Interview with Jack Welch, chief executive, General Electric. Financial Times, 1 October 1997.
12. Árva László, 'Foreign Working Capital, Domestic Subcontractor Relationship, Foreign Trade Balance and Technology Transfer (in Hungarian). *Közgazdasági Szemle*, November 1997, pp. 1007–18.
13. The Budapest Business and Financial Services Center plc was set up in early 1998, with the participation of leading politicians and businessmen in order to strengthen the leading role of Hungary and Budapest in the business and financial life of the region.
14. 'The Stage of High Rivalry is in Hungary'. Interview with Fergus Balfour, president of Unilever Hungary and South Center Europe. *Népszabadság*, 5 November 1997.
15. Philips in Hungary. In *No Limits*.
16. In the world-wide commercial the picture of Chain Bridge,one of Budapest's symbols, appeared.

3 Poland – from Solidarity to Solid Economy

ADRIAN SZUMSKI

Overview

POLAND'S intrinsic advantages lie in its large domestic market of nearly 40 million inhabitants and central geographic location between Western and Eastern Europe. It has demonstrated the most solid economic performance of any country in the region. Poland attracted a large proportion of the initial foreign investment made into Central and Eastern Europe and many multinational companies chose to begin their activities for the whole the region from here. Poland has in turn invested to develop its physical and financial infrastructure, and continues to compete with other countries in the region that offer cheaper production and labor costs. The Polish government introduced favorable regulations for foreigners regarding direct and indirect investments, convertible currency, a stock exchange and a liberalized market. Poland was recently rewarded with prestigious NATO membership, which will bring the country ever closer to its Western European allies.

Culturally, Poles and therefore Polish consumers, are consciously Western in their tastes and ambitions. This makes many global products and marketing techniques easily adaptable. Poland has the potential to become home to regional head offices, distribution centers and production facilities. With its highly educated and relatively inexpensive scientists, Poland is also a prime location for research and development centers. Off to such a soaring start, the key question is whether Poland will continue to compete successfully with other emerging markets and sustain its lead?

History

The name *Polska* (Poland) emerged in the tenth century as a domain of a small Slavonic tribe, the *Polanie* ('the people of the open fields'). Because of its geographic location in the central open plains of Europe, from its earliest days Polish rulers struggled to safeguard an independent Polish state. Indeed, much Polish history can be seen in terms of its complicated relations with its powerful neighbors – Prussia and Russia in the eighteenth and nineteenth centuries, Germany and the USSR in the twentieth.

Polish historians have been consumed by the story of partitions. Carved up by its powerful neighbors in the eighteenth century, the republic of Poland-Lithuania, a once powerful state, completely ceased to exist. The possibility of Poland's completely disappearance from the geographic map of Europe is an important factor in understanding the Polish psyche and the existence of a strong national consciousness. The fact that Poland survived, or was resurrected as a state, has fuelled arguments that there is something intrinsically 'Polish' about the region. Catholics might argue for the nation's 'soul', while Romantics speak of a Polish 'spirit'. Whatever the arguments, the inhabitants of the former Republic survived its fall, and with them the many tangible elements of Polish life – culture, language, religion, social and political attitudes. These elements have survived throughout the centuries to form the bridge between the often difficult past and hopeful future.

Geographically Poland belongs to and has always belonged to the East. In every other sense, its strongest links have been with the West. Poland's Western connection was forged in large measure by its loyalty to the Roman Catholic Church. For much of Polish history, the Church has stood at the center, not just of politics, but of national consciousness. The church has represented reform and innovation as well as tradition and continuity. Throughout the twentieth century many religious leaders also became national leaders and were key in the opposition towards communism. Poland's Catholicism also determined that all her elected rulers came from the West, that her cultural ties lay with the Latin world, that her closest political connection would be with the (Roman Catholic) Empire, and that her sympathies lay with the Catholic peoples of the West rather than with the Orthodox of the East.

The Western connection was strengthened by trade and politics. The main axis of Polish commerce lay westwards on the roads to Germany or on the Baltic sea lanes to Holland, France and Spain. Little has been done to divert the Poles' instinctive westward gaze. In comparison, Poland's much closer physical contact with the East has done little but sharpen existing antagonisms, The ancestral memories of the Mongol invasions, the Russian armies which marched from the East, and the lack of Soviet support at the end of the Second World War, have made resentments run deep. Although geographically tied to the East, in many ways, Poles have always been consciously Western in their ambitions.

The Transition

In the late 1970s, protests over food shortages fuelled a plethora of minor grievances directed against all manner of hardships and abuses. No reasonable observer could deny that the emergence of the free trade union movement, SOLIDARITY in 1980–81, proved conclusively that the Polish communist regime had lost the confidence of everyone except its own élite. The SOLIDARITY movement was the coming together of a huge variety of people from all walks of life and beliefs: workers activists, intellectual advisors,

Catholics, students, peasants, old-age pensioners, people of all shapes and sizes, to debate openly the hardships of the current situation and propose reform. Although the activities of SOLIDARITY were swiftly and decisively put down, the damage this widespread movement did to the government could not be undone.

The military *Coup* of 13 December 1981, led by General Jaruzelski, after which SOLIDARITY was quashed, took everyone by surprise. Throughout 1982, Poland was officially ruled by a military Council of National Salvation (WRON), popularly known as *wrona* or 'The Crow'. By 1983 the mounting foreign debt headed towards $30bn and even current interest payment had to be rescheduled. The demand for food was eased by draconian, and long overdue, price rises of up to 300 per cent, while most Poles continued to survive on the breadline. Queuing remained a way of life while the standard of living plummeted. The vast gap between the purchasing power of the average family and the goods available for sale continued to widen. Poland, a rich country blessed with great natural resources, sank into abject poverty.[1]

The first market-oriented reforms were introduced in this period. The years 1981–82 saw the reduction of bureaucratic administration, an increase in private enterprise and encouragement of worker autonomy. The survival of elements of private ownership under communism in Poland was an extremely important factor aiding the eventual transition to a free market economy. The fact that about 75 per cent of agricultural land had remained in private hands was extraordinary amongst Soviet bloc countries. Because of the strong peasant traditions and the possibility of massive rural and social upheaval, the Communist Party was discouraged from completely nationalizing all land after 1946. Some land that belonged to the Polish aristocracy before the war was nationalized, but much was divided between individual small farmers. Although private ownership of agricultural land was tolerated, it was often attacked in official propaganda. However many small private enterprises, such as bakeries and butcheries, survived throughout the period.

The official state farms (PGR) were often uneconomical and heavily subsidized by the state budget. The largely ineffective economy caused a permanent lack of food in the shops. A thriving 'grey' market, which made use of legal resources to produce and sell products and services without state supervision (that is taxes), became very dynamic. By the 1970s, private economic activity, such as small businesses and private services, was increasingly tolerated. During the 1980s, because of gradual liberalization and legalization of enterprise and business, the private sector expanded rapidly. It was at this time that small Polish traders travelled to Western Europe and brought back goods to sell in Poland. Some of the largest fortunes of the 1990s grew from such humble entrepreneurial beginnings. Although the Communist Party planned to introduce changes that would eventually lead to a market economy, they still wanted to retain political control.

Economic performance

After political and economic liberalization, GDP began to grow and inflation gradually decreased. The changes in Poland's economy are illustrated by Table 3.1. Large numbers of private businesses were started. As with any healthy competitive environment, some of them have grown and prospered while others have failed or were bought-up by competitors. But it was these first small businesses which were central to the success of transition from a controlled to a market economy.

Table 3.1 Macro-Economic Indicators for Poland

	1990	1991	1992	1993	1994	1995	1996	1997	1998
Nominal GDP ($ bn)	59.0	76.4	84.3	85.9	92.6	116.7	134.6	134.2	157.5
GDP *per capita* PPP($)	4192.0	3991.0	4346.0	4668.0	5040.0	5454.0	5876.0	6407.0	8430.0
GDP (% change)	–11.6	–7.0	2.6	3.8	5.2	7.0	6.1	6.9	4.8
Industrial production (% change)	–24.2	–8.0	2.8	6.4	12.1	9.7	8.3	10.8	4.8
Budget balance (% of GDP)	3.1	–6.7	–6.6	–3.4	–2.8	–3.6	–3.1	–3.0	–2.4
Unemployment (%)	6.3	11.8	13.6	16.4	16.0	14.9	13.2	10.5	10.4
Average monthly wage (%)	n.a.	n.a.	213.0	215.7	231.3	285.5	323.0	352.7	355.2
Inflation (%)	585.8	70.3	43.0	35.3	32.2	27.8	19.9	14.9	11.8
Exports ($bn)	15.8	14.9	13.2	14.1	17.2	22.9	24.4	25.7	28.2
Imports ($bn)	12.3	15.1	15.9	18.8	21.6	29.1	37.1	42.2	47.1
Trade balance ($bn)	3.6	–1.0	–2.7	–4.7	–4.3	–6.2	–12.7	–16.5	–18.9
Current account balance ($bn)	3.1	–2.2	–0.3	–2.3	–0.9	5.5	–1.4	–4.3	–6.9
Foreign direct investment flow ($m)	89.0	100.0	300.0	1715.0	1493.0	2511.0	4000.0	5678.0	7500.0
Foreign exchange reserves ($bn)	4.5	3.6	4.3	4.3	6.0	15.0	18.0	20.7	27.4
Foreign debt ($bn)	49.4	53.4	47.0	47.3	42.2	44.0	40.4	38.0	42.7
Discount rate (%)	48.0	36.0	32.0	29.0	28.0	25.0	22.0	24.5	18.2
Exchange rate (/$)	1.0	1.1	1.4	1.8	2.3	2.4	2.7	3.3	3.5
Population ($m)	38.2	38.3	38.4	38.5	38.6	38.6	38.6	38.6	38.7

Source: Central Business Europe. The Economist Group. http://www.bcemag.com/

By 1996, the government had successfully launched a Mass Privatization Program covering 512, mostly medium-size, state-owned enterprises. Over 90 per cent of Poles participated in this program, obtaining certificates in fifteen National Investment Funds (NFI), which controlled shares of the newly privatized companies. The investment funds have been listed on the Warsaw Stock Exchange since June 1997. Out of about 2500 private enterprises, 512 state owned companies were privatized in this program. About 6000 companies still remain under government ownership, although the government has announced plans to privatize 70 per cent of the remaining state owned enterprises by the year 2001. By 1998, almost every small and medium state owned enterprise had been privatized. There are plans to privatize some of the largest Polish companies affecting almost every industry including the oil sector, the electrical energy sector and power grid, copper mining and steelworks, and banking. Privatization will include large companies such as the telephone company Telekomunikacja Polska (TP SA), the dominant Polish insurance company (PZU), and the national airlines (LOT).

Before the recent changes, Poland was lacking even the basics of a market economy. A Polish capital market was introduced in March 1991 with the Law on Public Trading in Securities and Trust Funds. The Law provides the regulatory framework for operations in the capital market and introduced its major agents, the Securities Commission, the Securities Exchange and the Stockbroker.

The Warsaw Stock Exchange (WSE) was opened on April 16, 1991 and is a joint stock company with the State Treasury as the major shareholder. The State Securities Commission establishes all operational principles and admission requirements for joint-stock companies listed on the Exchange. The number of joint stock companies listed on WSE has increased dramatically from five (April 1991) to 170 (August 1998). The capitalization of WSE has grown from US$142m in 1991 to about US$9.6bn in 1997. Currently there are no transaction taxes on stockmarket trading. The WSE began continuous quotations for several stocks in 1999 and is planning to expand to others.

The WIG Index, the total return index of WSE is correlated closely with Wall Street indexes, connecting WSE to the global financial system. Foreign investors, in particular American investors, consider Polish investments alongside other emerging markets in their portfolios. WIG is still extremely susceptible to fluctuations in world markets, and troubles in any one of the emerging markets, in particular, can cause the WIG index to fall, regardless of the conditions in Poland. Nonetheless, the high standard of regulation and operations of the Warsaw Stock Exchange has been recognized by the international community. In October 1994, the Warsaw Stock Exchange was admitted as a full member to the International Federation of Stock Exchanges (FIBV). The new Securities Act, of 4 January 1998, facilitates the further development of Polish capital markets. Changes include further alignment with regulations of OECD and the EU, introducing securities lending and borrowing mechanisms, and definition of the underwriting rules.

The creation of new types of investment vehicles was made possible by the Act of Investment Funds adopted on 21 February 1998. From March 1999, reformed pension schemes will further increase private investment into the capital markets. In the past, the state used all pension contributions to pay current pensioners and no money was invested. The new system will allow 40 per cent of contributions to be invested by private funds, and will be compulsory for all Poles aged 18–30. Most of the new funds are joint ventures between big Western insurance companies or banks such as Citibank, and local financial institutions. In all, eight million Poles are initially expected to be covered by the new scheme. The scheme is also designed to produce savings for employers who will now pay only half of the 45 per cent pension contributions, with the other half being paid by the employees when, in the past, the entire sum was paid by employers. The reform will provide a much needed increase in the rate of saving in the economy.

The privatization of large, previously state-owned companies will result in a substantial increase in the Exchange's capitalization. The Council of the WSE is counting on a radical expansion of the number of quoted companies (to around 500), with a growing turnover (to around US$50bn) and capitalization (to US$60bn). The overall strategic aim of the founders of the Warsaw Stock Exchange has been to achieve the position of the largest and most important bourse in Central Europe.

During the transition Poland became a member of the World Trade Organization and the OECD and has applied for membership to the European Union. Poland has signed 'Europe Agreements' resulting in a move towards regulations along Western European lines and increased economic freedom. Poland is also a member of the Central European Free Trade Agreement (CEFTA) which was signed in Krakow, on 21 December 1992 by the economic ministers of Poland, Hungary, the Czech Republic, and the Slovak Republic, and later Slovenia, to promote trade between these countries. On 12 March 1999 a historic milestone was reached when Poland, along with Hungary and the Czech Republic, became the first countries in Central and Eastern Europe to join the North Atlantic Treaty Organization (NATO). Becoming a member of NATO had been one of the strategic objectives of the Polish government since 1991.

Foreign investment

After implementing radical reforms in the 1980s, Poland received almost half of the early investments made into Central and Eastern Europe. After an initial period of economic shock and adjustment, from 1992 onwards, economic indicators, such as GDP, began to increase. The best year for GDP growth was 1995 with a rate of 7 per cent, which was almost reached again in 1997 with 6.9 per cent. Inflation fell from 1991 to a predicted 8.5 per cent in 1999. Unemployment was down to 10.5 per cent of the labor force by 1997, the national average of EU countries. Foreign investors evaluated Polish economic performance warmly. Poland attracted the most FDI among all economies in C&EE countries in 1996, 1997 and 1998, with inflows estimated at US$10 bn in 1998. Many investors decided to invest primarily in Poland in 1998 and the trend is set to continue. Possibilities for investment are diverse and there are many interesting opportunities. The top investors into Poland are illustrated in Table 3.2.

Many MNCs opened activities in Poland in a variety of different sectors, from consumer to investment goods. Currently, ABB have a whole range of factories to build new electric plants and provide all the necessary services. Daewoo, Fiat,

Table 3.2 Largest Investors in Poland

Rank	Investor	EQ Loans	Plans	Origin
1	Fiat	1142	815	Italy
2	Daewoo	1011	567	Korea
3	EBRD	617	216	international
4	Pol-Amer.Ent.Fund	505	0	USA
5	Pepsico	412	380	USA
6	IPC	370	0	USA
7	ING Group	350	53	Netherlands
8	Coca-Cola Amatil	285	0	Australia
9	ABB	282	258	Switzerland
10	Philip Morris	282	90	USA
11	International Finance Corporation	277	0	international
12	Nestlé	248	0	Switzerland
13	France Telecom	232	0	France
14	Reemtsma Cigarettenfabriken GmbH	226	12	Germany
15	Saint Gobain	220	180	France
16	Commerzbank AG	210	13	Germany
17	Citibank	200	0	USA
18	SHV Makro NV	200	50	Netherlands
19	Epstein	200	0	USA
20	Thomson Consumer Electronics	185	0	France
21	British Petroleum	180	350	UK
22	Framondi	175	0	Austria
23	Pilkington	169	0	UK
24	Allied Irish Bank	162	0	Ireland
25	Lafarge	150	200	France
26	Shell	150	50	UK
27	Statoil	150	400	Norway
28	BOC Group	150	0	UK
29	Unilever	140	0	UK/Netherlands
30	Michelin	136	151	France

Source: PAIZ Polish Agency for Foreign Investment. http://www.paiz.gov.pl/, as of 1 Jan. 1997.

Note: As of 1 Jan. 1997

Mercedes and recently GM and Volvo have invested in the production of private vehicles. There are many MNCs in the financial sector like Citibank, ING, and GE Capital. Some MNCs have bought Polish companies and operate under both names, like Daewoo-FSO, Alianz-BGŻ, WZWUT Siemens, ING-Bank Śląski and others. Many MNCs, like Makro, Auchan, Géant, Billa and Tasco, have entered the retail sector.

Interestingly, Poland also receives a significant income from tourists. Despite the lack of a developed tourist infrastructure, such as hotels and organized activities, Poland has many historic, cultural and natural attractions. In 1995, the income from tourists generated US$6.6bn that grew to US$8.6bn in 1996, and about US$9bn in 1997. The estimated income in 1998 was US$10bn. These

Table 3.3 Foreign Investing Countries in Poland

Country	Equity & Loans US$m	Commitments US$m	Number of Companies
USA	2966	2670	77
Germany	1524	756	113
International	1493	188	15
Italy	1224	1200	29
Holland	952	309	32
France	900	537	42
Great Britain	509	364	21
Sweden	361	83	30
Switzerland	358	14	8
Australia	328	67	3
Austria	315	19	30
Denmark	238	23	16
South Korea	185	1225	3
Ireland	106	11	3
Spain	94	104	2
Canada	94	23	19
Finland	93	63	9
Norway	80	12	5
Belgium	47	54	14
Japan	33	13	7
South Africa	25	40	1
China	25	25	1
Turkey	23	60	2
Russia	20	0	1
Singapore	13	60	1
Liechtenstein	12	10	4
Slovenia	5	0	1
Greece	4	3	2
Luxembourg	2	0	1
Total	12028	7933	492

Source: PAIZ: Polish Agency for Foreign Investment. http://www/piaz.gov.pl/

numbers include a huge border trade from tourists, mainly from Germany and the CIS, who shop in Poland. Tourism is set to become one of the most promising sectors of the future.

Domestic companies

The type of strategy pursued by Polish companies often depends on their ownership structure. Companies funded by private capital are very aggressive and innovative. However, they often lack the capital to make the necessary investment into developing management techniques. There is also a group of privatized companies with good results, which are still developing. Then there is

Table 3.4 Foreign Direct Investments – Branches of Investments

Sector	Equity & Loans US$ million	Commitments US$ million
Industry total of this:	7482.9	6816.9
Fuel-energetic	182.8	1491.1
Metallurgical	108.3	183.7
Electro-machinery	2039.6	2538.7
Chemical	642.4	239.4
Mineral	748.0	796.2
Wood and paper	608.4	89.1
Light	301.2	220.4
Food	2535.3	1224.9
Others	316.9	33.4
Construction	607.2	251.1
Agriculture	15.0	0.0
Transportation	48.0	4.5
Telecommunication	587.6	380.7
Trade	709.6	297.8
Municipal Economy	24.8	6.7
Finance	2522.9	273.4
Insurance	29.7	2.2
Total	12 027.7	7933.3

Source: PAIZ: Polish Agency for Foreign Investment. http://www/piaz.gov.pl/ December 1996.

a group of companies that have restructured, or are in the process of doing so. These are competing aggressively and often successfully. There are still a number of state-owned companies that retain a dominant or monopoly position in sectors such as telecommunications, petroleum refining and distribution, chemicals, air transport, insurance and banking, although the government has declared almost total privatization and deregulation of the Polish economy by 2001. Liberalization, the opening of all markets to competition with few exceptions, along the lines of the EU, was announced at the end of 1998. A special attempt was made by vice Prime Minister and Minister of Finance Leszek Balcerowicz to reduce administration and bureaucracy.

Many Polish companies do not sell under their own name, but are subcontractors for large MNCs. Buying a local brand and selling a global product under that brand has often proved to be a successful strategy. The Polish clothing company Próchnik, is a subcontractor for a number of well-known brands such as Pierre Cardin and BOSS. Zelmer is a subcontractor for Siemens in the white goods market. But there is also a trend to develop Poland's own products and brands.

Polish companies are trying to enter the neighboring markets of the CIS, Czech Republic and Slovak Republic. Export to the EU is more difficult because of legal restrictions, quotas, non-tariff regulations and the competitiveness of EU markets. Hortex, a leading frozen vegetable brand, is one example of a company with overseas activities which has entered CIS markets and become a leading brand in Russia.[3] Originally, Hortex was a state owned producer of food preserves (juices, jams, frozen vegetables and fruits). Hortex had a dominant position in Polish markets and exported to other countries. Like many state-owned companies, it could not compete in an open market place and by the early 1990s was on the brink of bankruptcy. Restructuring, foreign investment (by the European Bank for Research and Development and the Bank of America) and good management, pulled Hortex out of trouble. The company introduced new products, adopted new technology, and entered CIS markets. Foreign sales rose dramatically and profits were substantial. Now Hortex is recognized by 90 per cent of Polish consumers as well as exporting successfully.[4]

Another example of a private company with overseas activities is Forte, a furniture manufacturer with global ambitions, which captured over seven per cent of the Polish market. Forte offers more then 1100 types of furniture and recently the company opened a shop in Beijing. In general, the Polish furniture industry is a thriving sector and makes up a large proportion of Polish exports to Western Europe and CIS markets. Production of furniture tripled between 1990 and 1997 and export of Polish furniture is systematically rising from US$1.3bn in 1995 to US$2bn in 1997.[5] The prognosis is that the furniture market will become increasingly profitable in the future. As the traditional export sector during communism, foreign MNCs have chosen to invest extensively in this sector. With the advantage of already existing extensive human resources, such as experienced employees and design services, MNCs have contributed the foreign capital and management skills to develop the furniture industry into a market leader.

Country Globalization Drivers

Poland's central location and large domestic market are natural country globalization drivers. The fact that Poland has also become the most successful of the transition economies and offers multinational companies a relatively stable political and economic environment, means it continues to attract many foreign investors in spite of the increasingly competitive environment.

Market globalization drivers

Most market globalization drivers strongly favor Poland's central role in the regional strategies of MNCs, and Poland has the potential to play a part in their global strategies.

Common customer needs

Polish consumers are very similar in their behavior to their Western counterparts. Poland is the sixth largest market for sales of new cars in Europe, with the second fastest growth rate on the continent.[6] The dynamic growth of the cellular phone industry has surprised even specialists. Polish mobile phone operators have commented that many companies changed their business plans because of the unexpectedly high rate of growth of customers. By the end of 1998, there were over 1.8 million users of mobile phones in Poland.[7] By 1998, banks had issued more than 2.4 million credit cards.[8] After liberalization, the import of foreign consumer goods increased dramatically. When profits from sales in the domestic market went up, many more MNCs were attracted to invest in Poland. They began to offer standard products available in Western Europe, with minor changes in packaging, and opportunities and choice increased for Polish consumers.

An effect of the political and economic changes is that Polish society has evolved into distinct classes and there is a growing divergence in the levels of incomes. The rate between the salaries of workers and managers can be as high as sixty to one. A new middle class has emerged, with preferences for high quality consumer goods, but insufficient income to fulfil its aspirations. There is also a group of people with very high incomes, who profited from the transition period and who have a very high level of purchasing power.[9]

The Polish automotive market is a particularly dynamic sector and the structure of sales of new cars can be used to illustrate the dramatic changes in Polish society. In 1998, sales of new cars reached the half million mark, positioning Poland as a European leader. Furthermore, currently this market is growing by about 30 per cent per year. In Europe, only the Italian passenger car market is growing faster than the Polish market.[10] Polish consumers are also moving from cheaper models to more expensive ones. Because of the size of the Polish market, automotive companies have built production sites in the country. Both Fiat and Daewoo have factories of small and medium sized cars. A variety of components are transported and assembled in Poland. General Motors is building a factory for Opel cars. Many companies, like Ford, Hyundai and Renault have assembly facilities. In pursuit of their customers, global suppliers like Delphi and Isuzu are also extending their activities.

Global channels

A number of large MNCs, like Benckiser, Coca-Cola, PepsiCo and Novartis, have chosen to locate their regional headquarters in Poland. These companies have chosen Poland because of its successful economic development, political stability, large domestic market and central geographic location in Europe. To support the foreign multinationals there is also a variety of service companies present in Poland. These include the big five accounting firms, consulting companies and advertising agencies, which provide a full range of services to their global clients.

Some sectors in particular have experienced outstanding growth. The mobile phone sector is one such success story. The first analogue cellular phone network was opened in 1991. In 1996 Centertel had more then 200 000 subscribers and coverage in almost all of Poland with the monopoly in mobile phones. In October 1996, two more operators of GSM standard digital cellular phones won licenses to start activities. Operators were consortiums of foreign companies with experience in telecommunications, and domestic investors with more than a 51 per cent share. After a year both GSM operators had near 800 000 subscribers and coverage over large cities, towns and main roads. At the start of 1998, there were well over 1 million subscribers of mobile phones (analogue and digital), which exceeded even the most optimistic predictions. In the first quarter of 1998, Centertel opened the digital network, DCS, in large cities and combined analogue and digital networks, which increased competition. In the first half of 1998, Centertel had some 80 000 subscribers to the DCS system. For the end of 1999, nearly 2.7 million mobile phone users existed.[11] An estimated 27 million mobile phone users are predicted by 2000.

Cost globalization drivers

In terms of cost globalization drivers, Poland is increasingly high-cost in comparison to other Central and Eastern European countries that are further behind in the transition process. Foreign companies continue to enter the Polish market since although not low-cost, Poland more than compensates by providing a cost-efficient, stable and developed environment.

Labor

Poland has a large and well-educated work force. Even under communism there were specialized studies in management (although focused on central planning) and a very good technical education existed. Poland has significantly increased the number of students reaching high levels of education. Currently 29.2 per cent of the population between 19 and 24 years of age are studying in public and private universities and schools.[12] There are more than 100 private schools and colleges which focus mainly on economic and computer sciences. There are about 20 MBA courses and many professional training institutions for working people.

Polish employees are willing to learn new skills and develop themselves and many MNCs have ranked Polish workers highly. ABB was surprised at how quickly companies bought by them in Poland become a valuable part of their global value chain. Polish staff quickly adapted global management processes and actively made use of them at work. The Benckiser's factory in Poland, which produces detergents, has on several occasions won the competition between Benckiser's factories worldwide.

Research and development

There are a number of advantages in locating research and development centers in Poland. There are many well-educated scientists, with low financial

Poland – From Solidarity to Solid Economy 69

expectations, who lack the facilities and funds to provide for their diverse research. Innovations have been made in a variety of fields including cutting optic fiber, specialized software and research into blue lasers. MNCs have recognized the potential that Poland has of becoming an important research center in Europe. Motorola has signed an agreement in Cracow with one of best technical universities to build a Technological Park and finance a research center in a special economic zone. The Swedish company L.M. Ericsson, is planning to open and invest US$100m into new research facilities. British Aerospace is co-operating with PZL Mielec to produce the training fighter, Hawk 100. Some of the production will be moved to PZL Mielec.[13] Microsoft is testing the creation of a virtual city in Raszyn, a small town near Warsaw. The technology will be provided live over the internet and every task, from shopping to administration will be done via the local computer network.[14]

Infrastructure

The telecommunications infrastructure in Poland is extensive and rapidly growing. Telecommunication ratios, like telephone lines per 100, are still low but there are large investments being made by Telekomunikacja Polska (TP SA) and other companies. However, Poland is still behind other countries in Central and Eastern Europe, with a teledensity of 20 per cent on average and 14 per cent in the countryside. The competitive environment is rapidly increasing for state-owned companies such as TP SA. In March 1999, Elektrim, Poland's largest listed industrial group became Poland's second largest telecoms company after it acquired Bresen Telecommunications Poland. The deal, Poland's largest take-over by a domestic company outside the financial services sector, highlights the growing scale and sophistication of the country's corporate activities.[15] The financial infrastructure to support such activity is good by European standards, and a number of global players, such as Citibank, ING or GE Capital, are present. In comparison, Polish banks are not prepared to compete in this environment, which is an advantage for foreign multinationals in the banking sector which are entering the region.

Government globalization drivers

The Polish government is business-friendly and extremely committed to providing the legal and political framework necessary to attract and facilitate foreign investment.

Role of regional trade blocs

Poland has been a member of WTO from July 1995, which has affected its trade policy. Poland's strategic objective is to become a member of the European Union as soon as possible. Legal regulations within Poland are evolving to conform to EU standards. Poland has signed multilateral agreements with the EU, EFTA, and CEFTA. Tariff and non-tariff barriers were imposed, after the period of liberalization of 1989–91, because of an unfavorable trade balance. Licenses were kept for cigarettes, dairy products, natural gas, petroleum and spirits. Most prices have already been liberalized, except those for heating, electricity, gas, basic medicines, rents in local authority housing, and spirits. Liberalization in the price of heating, electricity and gas is gradually being introduced and should be completed by the year 2000.

Government intervention

The government Agency for Agricultural Markets intervenes in the sale of agricultural products. The Agency provides credit guaranties, manages state reserves of food products, provides export subsidies and price support. One of the EU's conditions of Polish membership is to lower these subsidies and tariff barriers in agriculture. Compared to many other sectors, Polish agriculture is in the early stages of its development. While 25 per cent of workforce is working in agriculture, this produces just five per cent of GDP. Agricultural reforms have been slow because they are politically extremely sensitive. Most farms are too small to provide economies of scale and yet their closure would bring pauperization to rural areas. The largest weakness of Polish agriculture is the lack of a modern food-processing sector. However, from 1996 the food-processing sector began to develop and there still exists much potential for Poland to become an important international player.[16] Nestlé has invested in several food-processing companies. The privatized state company Winiary, became the number two Nestlé factory in instant food products.[17] Another successful project was Farm Frites, which farms potatoes and produces chips in Poland, for McDonald's across the C&EE countries.[18]

Poland established seventeen Special Economic Zones (SEZs) on 20 October 1994.[19] These have the right to grant investors special tax benefits for a period of over 20 years. The investor has to meet either a minimum investment or employment level within the zone in order to qualify for these benefits. The purpose of these zones is to create employment in economically depressed areas (that suffer from high unemployment), by attracting foreign investment. Many large investments are made into SEZs, because of the favorable conditions and tax breaks. However there is still room for much more investment into these areas.

Absence of state-owned competitors

Although some strategically important sectors are still controlled, many key industries have already been liberalized. Privatization of government companies, in telecommunications, insurance and banking, will be finalized by 2001. Life Insurance is already fully liberalized and MNCs like Allianz, Amplico, Commercial Union, Nationale-Nederlanden are present. Communication insurance is also open to foreign companies. In 1999, the insurance market will be fully liberalized for foreign companies; there will be no limitations on opening insurance services in Poland.

Poland has created the Anti-Monopoly Office (AMO) to prevent monopolistic behavior by state owned companies with a favorable position in the market place (more then 80 per cent of the market share). There were some cases of AMO actions against TP SA, which has a monopoly in wired telecommunication and data transmission, PZU SA which is dominant in insurance (75 per cent), and even Fiat, for raising its prices.

Favorable foreign direct investment rules

The Foreign Investment Act of 1991 provides reasonably liberal regulations for foreign investors. The law permits any level of foreign investment up to 100 per cent, with the exception of telecommunications, mining, steel, defense, transportation, energy and broadcasting. After 31 December 1998, investors from OECD and EU countries will be able to establish branches and agencies in Poland. The Law on Companies with Foreign Participation does require a permit from the Treasury Ministry for certain major capital transactions, or a lease of assets, with a state-owned enterprise. There are some limitations to some Polish sectors because of considerations of 'public security'. For example, only Polish companies can establish airports, although licenses and concessions for defense production and management of seaports and airports will be granted on the basis of national treatment for investors from OECD countries. There are also limits on the number of shares a foreign investor may own in certain sectors. Foreign ownership is limited in air transport (up to 49 per cent); fisheries (49 per cent); broadcasting (33 per cent); long-distance telecommunications (49 per cent through 2003); international telecommunications (zero per cent through 2003); and gambling (zero per cent). However, all profits are freely convertible and may be freely repatriated.

Legal protection

Poland has laws enforcing international regulation of intellectual property and prohibition of imitation. There is an organization of MNCs and Polish companies, which cooperate to prevent piracy. Polish police are trained to recognize and take appropriate action in the protection of intellectual property. The new regulations were successful enough to bring the level of piracy close to the average of European countries. New laws from October 1998 have prosecuted piracy of any intellectual property.

Compatible technical standards

Many technical standards in Poland are different from the EU and there is a need to receive a special certificate. From 1998, about 1400 different products required a 'B' mark certificate (for 'Bezpieczenstwo' or 'safety' in Polish) to be sold in Poland. Testing for the 'B' mark is performed by the Polish Certificate and Testing Center (PCBC) or one of the fifteen specialized institutes authorized and supervised by the PCBC. Companies who sell products without the certificate will be penalized and charged 100 per cent of the value of the products sold. This regulation is intended to provide minimum protection to Polish consumers. There is a strong move to unify technical standards along the lines of Western Europe in preparation for EU consolidation.

Currently, foreign companies with the EU's certificate C and ISO 9000, receive Polish approval much more quickly. If they pass EU standards, the Polish certificate 'B' will be almost automatically granted. The process is still time-consuming and costly, which can deter smaller companies. In the future, the EU will inspect Polish certification institutions after which certificates from these institutions will be honored by both the EU and Poland. The number of products requiring certification will also increase, coming into line with EU regulations.

The main obstacles to government globalization drivers are the different regulations in accounting and the frequent changes in law. To overcome legal difficulties, there are many global and local consulting companies, such as Arthur Andersen, McKinsey, and Deloitte & Touche, which can provide necessary advice at the early stage of investment. These companies employ many local personnel, with a good working knowledge of the often volatile Polish market place.

Competitive globalization drivers

Because of its large population and central geographic location, Poland is a strategically important country for most multinational companies entering Central and Eastern Europe. Located between the EU and CIS markets, Poland has the potential of becoming an important economic outpost.

Global regional strategic importance

Many MNCs, hoping to enter CIS markets, first chose Poland to locate activities, from where they can gain a better understanding of the region before expanding further. Many MNCs, such as Benckiser, Novartis and PepsiCo have chosen to control their operations for the whole region from Poland. Despite its central location, and the advantages this brings to transport between countries of the region, a lack of well-developed highways can cause delays in road transportation. A national program to extend and improve highways has been plagued by mistakes and setbacks and the government has not provided sufficient funding and guarantees necessary for the companies building the highways.

Poland is a large source of revenues for MNCs in sectors such as detergents, cars and home electronics. Because of its rate of growth of its domestic market, and the constant entry of new MNCs, the Polish market in consumer goods is perceived as extremely competitive. Even large international companies have to compete effectively if they are to retain a substantial market share. For example, Cussons which had a small market share in washing powders, bought a local brand and in one year won the biggest market share. Consequently, within one year the previously dominant companies, Unilever and Procter & Gamble lost their leading position. All MNCs producing detergents (Unilever, Procter & Gamble, Benckiser, Henkel) acknowledge that the Polish market is extremely competitive. However the potential profit margins are so promising that new entrants continue to appear regularly.

Internationalized domestic competitors

Poland is the home market of a number of companies with significant domestic and international achievements. The privately owned computer company Optimus ranks in sixth place in terms of PC computer units sold in Europe, ahead of many global competitors. Optimus partners some of the largest global players such as Microsoft, Intel, NEC and Seagate. Because of the dynamic growth of the domestic market, at about 40 per cent per annum, the company has not yet gone international. It has increased market coverage and developed, under its own brand, more sophisticated products including servers, workstations, notebooks, networks and software. Optimus began exporting to CIS countries, but revenues from exports are small compared to domestic sales. However, when the domestic market matures, it is likely that Optimus will internationalize sales.

Presence of foreign competitors

The motivation of some MNCs who aggressively invest in Poland is to keep their global market share and prevent other competitors from earning large profits in the country. There are many sectors dominated by foreign enterprises. For example, global advertising agencies have 80 per cent of all advertising funds, which is exceptional. Companies that have invested in Poland at the start of the transition have good market positions and are defending their market share. For example, Unilever dominated the market in detergents for a long time. Fiat is an excellent case of a company keeping market share gained at an early stage of the transition. New entrants are finding it very capital demanding to enter the market place, but are also finding it more stable, in terms of legal regulations, than earlier entrants.

Global Strategy Levers

Strategies of MNCs in Poland vary, and to a degree depend on the global strategy of the parent company, the potential of globalization in the sector and on the legal regulations of the sector. There is no one solution, but global companies are slowly winning in the competitive environment. However there are some independent Polish companies which compete with large MNCs and some sectors where Polish companies have remained dominant. Sectors where Polish companies are especially strong are those with low globalization potential such as the food and garment sectors. Because the Polish market is very competitive and volatile, it is increasingly likely that good global strategy will be the key to success in Poland.

Market participation strategies

In December 1997, the start of official negotiations on Poland joining the EU was announced. Becoming a member of the EU will make Poland even more attractive to foreign investors. Those companies already in Poland have early entrant advantages over newcomers. In recent years, distribution channels have been established, the financial sector has becomes more developed, service companies like consulting, auditing and advertising have more experience which make investing into Poland easier and safer.

Supermarkets like Géant, Carrefour, Makro, Cash & Carry, Rema, HIT, Billa and others are developing supermarket chains throughout Poland. Big retailers are investing in cities larger than 50 000 inhabitants. Of the 77 chains in Poland, there are nine hyper-markets, 11 supermarkets, eight discount chains, four franchising chains and seven originally Polish chains. Together they have 3328 shops, which includes 35 hypermarkets, and 178 supermarkets. Most retailers are developing their chains and building new shops.[20] The turnover of all retail sales in Poland in 1996 was about US$16.7bn and in 1997 US$18.4bn. The top fifty retailers doubled their sales between 1996 and 1997. German HIT, who has only five hyper-markets, was the leader in Poland with US$233m sales.[21]

New players, such as petrol stations, have also entered the retail market. By the end of 1997 there were about 6000 petrol stations in Poland, mainly owned by private companies (63 per cent). The former state-owned monopoly and distributor of petrol, CPN now owns 24 per cent of all petrol stations. But MNCs are rapidly expanding their networks, Shell owns 66 stations, Statoil 61 and BP owns 44. Stations owned by big global companies are better located, have better facilities, provide a full range of services and are more profitable.[22]

MNCs have increasing influence in all sectors of the economy. For example, while the telecommunication services are still dominated by government-run TP SA, several independent competitors are establishing themselves in the more remote parts of Poland. These companies are often joint ventures between global companies and Polish partners. They have a small market share and limited activity area, but they are becoming a serious competitive threat to TP SA. They have the newest technology and are developing much faster than TP SA. When the government liberalizes the telecommunications sector, there will be large opportunities for the smaller companies and MNCs already present. Companies like L.M. Ericsson, Siemens, Nokia, Motorola and Alcatel have the technology, the know-how and the equipment to compete with TP SA. Cable networks are also mainly joint ventures between MNCs and local partners, who are prepared to provide network services via their cables. Poland is still a very attractive market for telecommunication services and will be an area of tough competition in the future.

Another example of a sector dominated by MNCs is consumer goods, and includes sweets, soft drinks, computers, cars, electronics, white goods and many other products. All the major players in consumer goods are present in Poland. As most local companies with attractive resources, (such as market share, strong brand, distribution networks and so on), have been purchased by foreign MNCs, the multinationals hold the dominant market share and all serious competition is between them. There are a few Polish companies competing with MNCs, but in the long run they will be acquired or will keep only a minor market share.

Global retailers, such as Makro, IKEA, Billa, Auchan, Géant, TTW, Tesco, Carrefour, have been present for some time in Poland. With the growth of consumer markets, consumer habits are changing. Small Polish retailers, located in town centers, are competing against global retailers who are gradually winning over Polish customers as they become more and more used to doing their shopping outside city centers. Polish owned shops are creating chains of smaller shops. New entrants into retail make this sector even more competitive and the situation ever more demanding.

Cultural factors make cars a very important product for Poles. There are many global suppliers and global car producers present, like FIAT, Daewoo, GM (Opel), Volkswagen, Delphi, and Isuzu. The rate of growth is relatively stable and the trend will probably continue, with Poland as one of the largest car markets in Europe, making this sector especially interesting for global players.[23]

There are just a few Russian companies active in Poland. One is Gazprom, which has invested a large amount of capital in a gas pipeline from Russia to Western Europe. It is very important to note that Poland is dependent on Russia for its natural gas. Lukoil, another fuel player, started to build a network of petrol stations in Poland with plans to reach 300 stations by 2001. Some Russian banks own shares in Polish financial institutions. Because of the small number of companies present, their early stage of development and the lack of information on them, it is difficult to determine a trend to their activities.

There are a number of large players from Central Europe. One of them, the Slovenian company KRKA, is actively competing in the pharmaceutical market. Another active player is the Croatian company Podravka, which sells instant food and spices under the brand VEGETA. Podravka, which lost market share with the decomposition of Yugoslavia, is now successfully competing with CPC Foods and local producers. The objective of Podravka is to gain the same market share as the current leader CPC Foods.

Another attractive opportunity for multinationals is the financial sector. Most local banks and insurance companies are not prepared for serious competition. MNCs have the opportunity to capture a substantial market share in a promising sector. By 1999 the banking sector will be fully liberalized and MNCs will be able to open branches and agencies in banking and insurance. Some of them, like LG, Citibank and ING are actively investing in Poland. ING acquired one of the biggest banks in Poland with offices in the main cities. Allianz formed a joint venture with BGZ agricultural bank with networks all over the country.

Poland has very liberal regulations regarding shares of foreign capital in assets of domestic banks. There is a limitation of 20 per cent of foreign capital in the stock capital of Polish banks, but this limitation is not always subscribed to. Permission to sell shares in domestic banks is given by the National Bank of Poland (NBP), which serves as the Polish central bank. In practice a foreign investor can buy 100 per cent of the shares of Polish domestic banks, if he receives permission from the NBP. The assets of the Polish banking system amounted to US$53.5bn in 1997, which constitutes approximately five per cent of the Spanish banking system.[24] Foreign investments in the Polish banking sector are estimated at 45 per cent of all capital.

The second factor to consider when evaluating the Polish banking sector is the volume of credit of the non-financial sector. In the Anglo-Saxon banking model, the credit of the non-financial sector is about 65–72 per cent of total assets. In the Polish banking sector, from 1990–92, the opened credit of the non-financial sector was only about 14–15 per cent of all assets. During the transition process, Polish companies could not credit their development on foreign capital markets because of strict regulations and lack of government guarantees. Their capital was consumed by inflation and high tax rates. Consequently, Polish companies could not develop. When liberalization in the financial sector began in 1992, the share of credit for non-financial sector grew to 40–45 per cent by 1998.[25] A big difference exists between foreign-controlled banks (those with more then 20 per cent of assets in foreign capital) which give 40–45 per cent credit to the non-financial sector, and Polish banks which only give 18–20 per cent credit. After opening Polish capital markets to the activities of European banks, Polish banks lost out in the competition because of their lack of knowledge of credit activity. The long term implications of this are very negative for Polish banks.

Global/regional product and service strategies

In general, there is a minimal need of product adaptation for the Polish market. Slightly more adaptation is necessary in the service sector, which must be more sensitive to the preferences of Polish customers. But there is still a lot of opportunity to make use of global products and services. Differences in legal regulations, culture and tradition do exist, but these factors are not of central importance in most cases. Adaptation of products is often necessary in terms of packaging, and the language used on labels and instructions, rather than the core product itself. MNCs are not keen to give out information about unsuccessful product introductions but unofficial sources suggest that there have been a number of disappointing product launches, when MNCs were not careful to adapt products and services to the needs of local consumers.

Some products in particular needed adaptation to meet local expectations. One such example was Unilever's Lipton Tea. This is a global brand, with standardized advertising, but formulated differently all over the world. In Central and Eastern Europe, Unilever introduced a Western tasting mixture that was successful everywhere except in Poland. After some research it became clear that Poles preferred a sharper tasting tea. Unilever tested and launched an Asian mixture with a stronger taste and consumers responded quickly by accepting the new taste.

Some products have had a very short life cycle. The shampoo Head & Shoulders was initially a product produced for Western Europe. However, there was a difference in the frequency of use of shampoo in Poland which was not taken into account and the product did not sell well. After adaptation and a very skilful marketing campaign that made use of local personalities, the product won the largest market share. Despite the advertising campaign, only a few months later the product almost completely disappeared from the market. This was partly because competitors had introduced similar products and had been able to successfully copy the aggressive marketing strategy, winning back their market share.

The entry of management consulting services has been largely unsuccessful. Big consulting companies like Andersen Consulting, McKinsey, Price Waterhouse Coopers, KPMG, and Deloitte & Touche, failed to reach the same percentage in revenues from management consulting when compared to other services in the region. Foreign management consulting firms have generally performed poorly because of competition from small local firms that have established reputations, connections with local government officials, and good working knowledge of what is legally and socially acceptable.

In contrast, advertising is an example of the success of global services. Large global companies, such as Leo Burnett, Saatchi & Saatchi, and McCann-Erickson, entered the Polish market in pursuit of their global clients. In the early 1990s, MNC advertising was largely standard global campaigns adapted to the region. There were some failures, but companies gained experience and trained local staff. After only a few years advertising and market research, services became fully developed. Foreign advertising agencies work for MNCs and local clients and have more then 80 per cent of market share. On average, the advertising sector is growing at 30–40 per cent each year making it ever more competitive.

Another example of successful foreign entry can be seen in the ice-cream market. The fact that consumption of ice-cream per person in Poland is 3.3 liters, (compared with 13.6 liters in Sweden, the European leader in *per capita* consumption), means there is much potential growth in this market. Of the few thousand ice cream companies that operate in Poland, most are relatively small. There are only about twenty companies with a market share larger than one per cent. The dominant position was taken by Algida of Unilever with 15 per cent and Schöller with ten per cent.[26] Algida was initially selling only expensive, imported products. But by 1992, Algida had bought a local producer and started production. To compete, Schöller also invested in a factory. Both companies changed the situation in the market, speeding up the competition with aggressive advertising, the annual introduction of new products and the building up of a distribution network.

Activity location strategies

MNCs in Poland use a variety of different activity location strategies. Some companies like Samsung, NEC, Sumitomo have established procurement offices. This is often part of the strategy in the early stages of entry. After the situation in the country has been assessed by the MNC, more advanced activities are planned.

Working in cooperation with the oldest Jagiellonian University in Poland, Motorola located its research center in Krakow. The venture is an example of a possible future trend of using highly educated, low cost, scientists for research and development. Telecommunication companies such as L.M. Ericsson, Nokia and Siemens have activities for adapting global products to the Polish market. Many MNCs such as Unilever, P&G, Daewoo, GM, Volkswagen, Danone, Philips, and ABB have located production in Poland. They are expanding production for the local market or exporting to neighboring countries. Novartis and Benckiser are examples of companies who located their regional marketing teams, regional sales forces, distribution centers, customer services and headquarters in Poland. MNCs are establishing their presence in Poland starting from procurement through to regional distribution, marketing, sales and even production.

Global/regional marketing strategies

There are few differences between Polish consumer tastes and those in Western countries and, for this reason, Poles are highly receptive to global marketing campaigns. A large number of Poles have access to satellite television and therefore exposure to global advertising campaigns which increases the effectiveness of global marketing strategies. Of course, global marketing has its limitations and much depends on product positioning. If the product is already global, it will succeed with largely global advertising. When the advertising campaign is large, it can use a combination of both global and local advertising. The Procter & Gamble 'Always' campaign is a good example of such a marketing mix using global advertising flavored with local personalities. In this way, a global campaign is tailored for local needs. Local elements, personalities and accents are added to fit with consumer expectations and some adaptation is required by law, for example, Polish instruction must be added. Global brands are generally positioned as the best quality, premium price products even though standard or medium quality Western models are often considered in Poland as premium products because of the low average income. Other examples of recent successful global campaigns include those for L.M. Ericsson, Volkswagen Golf IV and Citroen Xsara.

There is a need to localize marketing when the global product makes use of a well-known local brand. Unilever was very successful with its product Pollena 2000, a washing powder, which had a local campaign with strong accents of Polish culture and history. On the other hand, PepsiCo ignored the strong historical traditions of a Wedel, a Polish chocolate factory it had acquired, and made changes to the taste of the product. Eventually, because of this, Wedel lost its first place in terms of market share.

Generally, prices are similar throughout the region and to avoid unauthorized import, are also similar to or lower than prices in Western Europe. However, some MNCs have decided to use the opposite approach, setting prices higher than in Western Europe in the belief that larger margins and lower sales are more profitable in the region. For example, Eastman Kodak decided to set higher prices than in Germany for photo films. The result is that unauthorized imports from Germany, with a 40 per cent margin, are almost freezing sales from the official channels in Poland. There is also much social pressure to lower prices because of the low income of consumers.

Successful marketing in Poland depends to a degree on the industry, but is also dependent on competitive pricing and good advertising strategies. Competitive pricing is a very important factor because of the tough competition and still low average income of consumers. Effective advertising and promotion is the most important factor of success. Competitive marketing strategies can often be very costly and it is difficult to have competitive pricing and good promotion. But without this it is extremely hard to be successful. MNCs are fighting to obtain the largest market share possible before the situation changes to become more stable.

Global competitive moves

Because of its large population, Poland is perceived as a key market in the region. Competition in Poland is very high and the situation changes rapidly. For example, in 1995 Cussons only had a small market share but by 1996 had become the market leader in detergents ahead of Unilever, P&G, Henkel and Benckiser. Its success lay in the fact that Cussons bought local brand IXI, which had a small market share but excellent local recognition. Cussons added an aggressive marketing factor and quickly won the largest market share in detergents. The 'winner takes all' syndrome is very visible among competitors, especially in consumer goods. Overall, global competitive moves depend on an MNC's strategy and sector of activity. Some MNCs use both global and local moves, while some of them use mainly global, and some focus on local strategies.

In consumer goods, Poland is often included in global competitive moves. Poland is the battleground for MNCs in computers, sweets, cigarettes, cars, telecommunication and white goods. Poland is the largest market for these products in the region and is a testing ground for expansion to other countries. For example by 1996, Polish consumers had purchased US$1bn of audio and video equipment.[27] The introduction of global products and global advertising campaigns are very common among MNCs. There are groups of products with high globalization drivers, such as expensive investment goods, which must be managed globally and coordinated. But some MNCs, like Unilever, have distinct strategies for countries in the region. A successful strategy of Unilever has been to acquire local brands similar in content but promote them differently throughout the region.

To control sales and unauthorized imports, prices in Poland tend to be similar or lower than in neighboring countries and the EU. Pricing politics include dealing with local duty and tax barriers. MNCs often analyze price to end-consumer ratios to make the appropriate price moves. However there are products and services that are not transferable and prices may differ. For example, Allianz started to offer car insurance at a higher price than in Germany. GSM cellular phone operators have higher activation fees and charges per minute than in many Western countries. However, as foreign investment in network coverage grows, and the market becomes more demanding, prices will go down.

Some global companies have had to make exceptions to their global strategies in Poland. PepsiCo acquired Wedel, the Polish leader in chocolate products. Wedel was a favorite of the old regime, founded before World War II. The company already had very good recognition in Poland and with significant exports and good branding. A few years after PepsiCo took over the company, Wedel lost its leading market position to Cadbury. PepsiCo is preparing to sell Wedel and terminate its activity in chocolate products, because of the lack of its connection to the rest of PepsiCo's activities.

Poland has become the testing region for digital television services. Poles watch more television than any other European nation. After the failure of digital television in Germany, Poland has become a major battleground for companies in digital television. There are two digital platforms, Wizja TV and Cyfra+, competing in digital television broadcasting from 1998. Both televisions will have multi-channel digital broadcasting. Polish National Television TVP will be transmitted in Canal+ package. Wizja TV's main investor, At Entertainment, is the owner of a large cable television network, PTK, and will use that to increase sales. Access to the archives of Polish National Television is a competitive advantage possessed by Cyfra+. The market is quite new and the next few years will prove if there is room for two digital platforms in Poland.

Global Organization and Management

There are a number of different ways to begin business activities in Poland. A foreign company can establish a representative office that can take the form of a Supervisory Office, a Technical Information Office or a Branch Office. A Supervisory Office can be opened on the basis of a contract concluded between a Polish and a foreign company. The sole function of such an office is to supervise the performance of a concluded contract. A Supervisory Office may not conduct commercial or information activities. A Supervisory Office is only set up for the duration of the contract and are used in connection with turnkey or shipbuilding contracts.

The role of a Technical Information Office is limited to disseminating product information and promoting the exchange of know-how. It is prohibited from engaging in any marketing or sales activities, including price or terms of delivery information. This type of representation office accounts for only 10 per cent of the total number of representation offices in Poland.

A Branch Office is authorized to conduct foreign trade activities such as supplying information, advertising, searching for customers, negotiating and signing contracts on behalf of a foreign company. It is also authorized to perform warranty and post-warranty technical service for machines and equipment delivered to Poland by the company and to own a stock of spare parts of this machinery and equipment. A permit to open a Branch or Technical Information Office is issued by the Ministry of Foreign Economic Relations in the area of foreign trade and the State Sport and Tourism Administration in case of tourism, Ministry of Transportation for international transportation and Ministry of Culture regarding cultural services. Foreign companies that have a Branch Office pay taxes on profits made in Poland. Foreign personnel residing in Poland must pay personal income tax. The company's taxable income is determined on the basis of the company's books, which show all transactions concluded in Poland.

The two most popular legal forms of activity are limited liability companies and joint stock companies. There are no restrictions on a foreigner's share of equity in such companies. Setting up a limited liability company does not require more than one investor. As of 1999, the initial capital of the limited liability companies should be at least 4000 New PZL (about US$1400), and 100 000 New PZL (about US$35 000) for a joint stock company. Local partners are not legally required but may be very useful. There are many global and local consulting companies that can help in opening activities in Poland. Furthermore, using local partners is an easy way to test products and make the first adaptations to products, depending on local needs and requirements such as technical standards.

Introducing global management processes into Polish companies is not a problem. For example, McDonald's, General Electric, IBM, Daewoo, McKinsey, P&G, and Sandvik, all take a global approach to their management. From the start, most Polish subsidiaries can be treated as a part of the global network. Participating in global processes from an early stage will help install a global culture, and may prevent some cultural conflicts in future. Generally, there are no problems in installing global culture in administration although some companies have had problems implementing it with line workers. Difficulties arise especially in formerly government-owned companies. Problems with instilling global culture can also occur in rural areas far from large cities.

The approach that an MNC takes in Poland often depends on its company culture and country of origin, and a number of trends can be recognized. Some MNCs, in particular Korean and Japanese, tightly control the activities of their local subsidiaries. They give little autonomy and are closely connected by a global strategy. Korean middle managers are not prepared to work in a multicultural situation and often try to install a Korean culture, whereupon conflicts arise with local personnel. Other companies let Poles take up committee positions were they have more control over certain local activities. For example, Toyota is developing its marketing strategy in Poland, with the help of local personnel. A number of Japanese companies operate via local distributors. For example, CASIO products are distributed in Poland by ZIBI Company, Citizen watches are distributed by Apollo Electronics.

French companies have their own approach in Poland to the delegation of power and autonomy. They all too often rely on expatriates, whether or not they are competent to work as supervisors. Strict control and forcing solutions made in France can have disastrous effects. However, even companies which start with a period of rigid control can become more flexible with time. Other European MNCs, including English, Swedish, Italian and Dutch give much more autonomy to their Polish counterparts. Management by local personnel is a popular solution. Companies from these countries usually give greater autonomy and at times Poland is not even rigidly connected to their global strategy. Because of their geographical closeness, German MNCs have yet another pattern of activity. Polish operations are controlled from Germany, but some autonomy is given to Polish managers who are often promoted into top management positions after the initial stages of development. For example, the manager at Siemens ZWUT was formerly a director of that company. American MNCs give much more autonomy to local branches and include them in their global strategies. Polish managers are included in management, although some MNCs rely on expatriates or foreign managers.

Conclusions

Poland is very attractive to foreign investors in Central and Eastern Europe because of its large market, central location, economic and political stability and rapid progress. Poland initially received half of the investment in the whole region; however, because FDI *per capita* compared to other countries in the region is still low, investors will probably continue to search for opportunities in Poland. The Polish market place is a battleground for MNCs testing their strategies for Central and Eastern Europe. Other factors, like an inexpensive and well-educated workforce and lower cost prices than in Western Europe (common to many countries in the region), are also important considerations for foreign investors. Poland has a lot of potential as a key market, a regional hub, and an outpost to the rest of Central and Eastern Europe. Transformation of the economy is rapidly progressing and many difficulties of the transition period have been solved positively. Large-scale privatization is progressing and will be largely completed by the year 2001. Recent years have shown that Poland is firmly placed on the road from Solidarity to solid economy.

Bibliography

Business Central Europe Magazine http://www.bcemag.com/.

CEAR™ Central European Automotive Report http://www.cear.com.

Chalmers, J., Douglas H., Richardson, B., McArdle, J., and Lavers, J., (1998), Why Tortoises Won't Win: A Survey of Telecoms, *Business Central Europe*, 6 September (54), p. 37–48.

Davies N. (1991), *God's Playground. A History of Poland*. Polish ed. (Krakow: SIW Znak).

Dzierżńska-Nowak (1998), Jasna przysłość czarnej porzeczki? *Gazeta Wyborcza*, 27 June, 148 (5008), p. 25.

Glapiak, E., Wrabiec, P. (1998), W skali hiper. *Gazeta Wyborcza*, 12–13 September, 214 (2817), p. 22.

Kapoor M. (1998), Fruitless delay. *Business Central Europe*, 6 September (54), p. 9.

Klodecka, M. (1999), Relaks z koszykiem, *Gazeta Wyborcza Supermarket*, 5 March 1999, 54 (3052), p. 1.

Koźmiński, A. (1998), *Zarzadzanie miedzynarodowe,* in Andrzej K. Koźmiński and Wlodzimierz Piotrowski (eds), Zarzadzanie. Teoria i praktyka (Warsaw: PWN), p. 537–88.

Margas, D. (1998), Prawie 1,4 mnl użytkowników w tym roku, *Rzeczpospolita*, 6 August, 183 (5043), p. 21.

Morka, A. (1997), W raju RTV, *Gazeta Wyborcza*, 24 September, 224 (2827).

Na poczatek elementy kadluba (1998), *Rzeczpospolita*, 22 July, 169 (5029), p.20.

Nowak, A. and Sopoćko, A. (1988), *Polskie banki a integracja z Unia Europejska*, in Bankowość (Warsaw: Leon Koźmiński Academy of Entrepreneurship and Management), pp. 81–105.

Oktara, L. (1998), Inwestycje wieksze od zobowiazań, *Rzeczpospolita,* 14 August, 189 (5049), p. 23.

Oktara L. (1998), Umacnianie pozycji na rynkach eksportowych. *Rzeczpospolita*, 1–2 August, 178 (5038), p. 25.

PAIZ Polish Agency for Foreign Investment http://www.paiz.gov.pl/.

Pronobis W. (1991), *Polska i świat w XX wieku*. (Warsaw: Editions Spotkania).

Rozwijasie zachodnie i polskie sieci sklepów (1998), *Rzeczpospolita*, 13 August, p. 8.

Rożyński P. (1999), Idealne straty, *Gazeta Wyborcza*, 5 March, 54 (3052), p.22.

Raport o super i hipermarketach (1998,) *Rzeczpospolita* 13 October, 240(5100), pp. 21–23.

Raport o przemyśle meblarskim (1998), *Rzeczpospolita* 4 May, 103 (4963), pp. 12–13.

World Bank Report, December 1997, Washington DC.

Roszkowski W. (1992), *Historia Polski 1914–1990* (Warsaw: PWN).

Stanuch, S. M. and Bartoszwewicz, D. (1998), www.cyberraszyn, Gazeta Wyborcza, Dodatek Lokalny, 15 September, p. 1.

Statistical Yearbook of Republic of Poland 1998 (Warsaw: Central Statistical Office).

Tomaszkiewicz, B (1999), Karty sa, terminali brakuje Eletroniczny pieniadz w natarciu, *Gazeta Finansowa*, 8 March, 10 (27), pp.1, 17.

Tymowicz, M., Kieniewicz, J.and Holzer J. (1990), *Historia Polski* (Warsaw: Editions Spotkania).

US Department of Commerce, Bureau of Economic Analysis.

W cieniu Shella (1998), *Gazeta Wyborcza*, 31 March, 76 (2769), p. 21.

Wyżnikiewicz, B. (1997), Zarabiaja coraz lepiej. *Gazeta Bankowa* 8 May, p. 14.

Zwierzchowski, Z., 1999, Lamanie regul, *Rzczpospolita* 5 March, 54 (5219), p.25.

Notes

1 Norman Davies, The Heart of Europe; A Short History of Poland (Oxford University Press, 1990).

2 Poland: Private Funds in Pension Dash. *Financial Times*, 2 March 1999.

3 Kapoor, M. (1998), Fruitless Delay. *Business Central Europe*, September 6 (54), p. 9.

4 Oktara, L. (1998), Inwestycje większe od zobowiązań. *Rzeczpospolita*, 14 August, 189 (5049), p. 23.

5 *Rzeczpospolita*, 4 May, 1998

6 Central European Automotive Report,

7 Rożyński P. (1999), Idealne Straty, *Gazeta Wyborcza*, 5 March, 54 (3052), p. 22.

8 Tomaszkiewicz, B. (1999), Karty sa, terminali brakuje Eletroniczny pieniądz w natarciu. *Gazeta Finansowa*, 8 March, 10 (27), pp. 1, 17.

9 Wyżnikiewicz B. (1997), Zarabiają coraz lepiej. *Gazeta Bankowa*, 8 May, p. 14.

10 Central European Automotive Report.

11 Margas D. (1998), Prawie 1, 4 mnl uży tkowników w tym roku, *Rzeczpospolita*, 6 August, 183 (5043), p. 2 1.

12 Statistical Yearbook of Republic of Poland, 1998.

13 *Rzeczpospolita*, 22 July, 1998.

14 Stanuch, S. M. and Bartoszwewicz D. (1998), www.cyberraszyn, *Gazeta Wyborcza, Dodatek Lokalny*, 15 September, p.1.

15 Electrim: Polish Company Buys Telecoms Group for $325m. *Financial Times*, 25 March, 1999,

16 Kapoor, M. (1998), Fruitless Delay, *Business Central Europe*, September 6 (54), p. 9.

17 Oktara, L., 1998, Umacnianie pozycji na rynkach eksportowych. *Rzeczpospolita,* 1–2 August, 178 (5038), p. 25.

18 Kapoor, M. (1998), Fruitless delay, *Business Central Europe*, September 6, (54), p. 9.

19 Ref.: Official Act Publication Dz.U. no. 123, It. 600.
20 *Rzeczpospolita*, 13 August, 1998.
21 Glapiak E, Wrabiec P., 1998, W skali hiper. *Gazeta Wyborcza*, 12–13 September, 214 (2817), p. 22.
22 *Gazeta Wyborcza*, 31 March, 1998.
23 CEAR http://www.cear.com
24 World Bank Report, December 1997, Washington DC.
25 Nowak A., Sopoćko A. (1988), *Polskie banki a integracja z Unią Europejską*, in Bankowość (Warsaw: Leon Koźmiński Academy of Entrepreneurship and Management), pp. 81–105.
26 Koźmiński, A. (1998), *Zarządzanie międzynarodowe*, in: Andrzej K. Koźmiński and Wlodzimierz Piotrowski (eds), Zarządzanie. Teoria i praktyka (Warsaw: PWN), pp. 537–88.
27 Morka A. (1997), W raju RTV, *Gazeta Wzborcya*, 24 September, 224 (2827).

4 The Czech Republic – A Window of Opportunity

HELENA HRUZOVÁ and ZDENEK SOUCEK

Overview

THE Czech Republic (CR) is a small country, with a population of only ten million people. But its history as a long time member of the Austro-Hungarian Empire, its geography bridging Western and Eastern Europe, its leading role in economic transition from communism, its cultural prominence, and its well-developed democratic institutions, together make it of priority interest to foreign multinational companies.

Investors within this market can gain both from growing local and regional demands. Up to 1997, the CR was the only country to obtain an A-class investment degree rating by the rating agency Standard & Poor's, making it an important investment center in the region.

According to CzechInvest, a number of interesting features exist in the country that positively distinguish it from others in the region.[1] These included its central geographical position, its extensive infrastructure, a developed domestic supply base which supports manufacturers, a highly educated and cost-competitive work force, a high credit rating, relatively low inflation rate, international business confidence, territorial structure of foreign trade, and a favorable business climate. German and Austrian medium-sized enterprises often favor the country as a place to move their production. Investors from other countries can use their German and Austrian subsidiaries as the intermediaries for investing into the CR.

History

The Czech Republic (CR) is a democratic and industrially developed Central European country. After 75 years of common life with the Slovaks, an amiable division of Czechoslovakia, into the Czech Republic and the Slovak Republic, was officially announced in January 1993. First established in 1918, Czechoslovakia performed promisingly as an independent and economically developed country with a liberal democracy. With the end of German occupation in 1945, Czechoslovakia came under the political sway of the Soviet Union. The political and economic system developed in the next 40 years was based on the centralized economy of a socialist state. Ultimately, the unsustainable economic situation as well as the demands for political change throughout the whole of Central and Eastern Europe, led the Czech Republic to the 'Velvet Revolution' of 1989 and a resignation of the communist leadership.

Political changes and transition to a market economy began in the early 1990s. The state became politically independent from the USSR, a new legal framework was created and parliamentary democracy established. The transition of the Czech Republic into a democratic state was relatively stable compared to other post-communist countries in the region.

The historic development of the CR helps to explain its current economic situation. The nineteenth century saw a period of industrial development when the Czech region became the so-called 'iron heart' of the Austro-Hungarian Empire. In the 1920s and 1930s Czechoslovakia witnessed a successful period of economic development under the framework of a democratic government, and was ranked among the top ten most developed countries in the world. Under Soviet influence, Czechoslovakia underwent a stage of centralized planned economy based on the economic and military doctrines of the Eastern bloc. Together, these stages of development resulted in an unsustainable economic structure in which production was too raw-material and energy consuming as well as environmentally unfriendly.

At the end of 1989, significant political changes were followed by economic reform. Key pieces of legislation included the Bankruptcy Law (1991), and the Law on the Prague Stock Exchange (1992). Other important reforms included the start of a mass privatization program (from 1991), the liberalization of prices, trade, wages (1991) and interest rates (1992), and the implementation of internal currency convertibility (1995). The Czech currency was devalued by 75 per cent in 1991. The exchange rate, stable from 1992 to 1997, was pegged to a basket comprising the Deutschmark and the US dollar, with approximate weights of 65 per cent and 35 per cent, respectively. Restrictive monetary and fiscal policy has included tax reform from 1993 onwards.

Initially, the Czech Republic implemented the most extensive privatization program of all C&EE countries. Several hundred (privatization) investment funds played an active role in this process. These took majority stakes in large, previously state-owned, enterprises. Some funds pushed these enterprises to short-term profitability, even at the expense of long-term stabilizing policies. A number of large, domestic companies also played a major role in the consolidation of ownership. Some foreign buyers also acquired strategic stakes in newly privatized companies.

The market liberalization policy was closely connected to price reform. Prices were deregulated in several stages. The only remaining controls pertain to the price of water, energy, state rented housing and public services including transport. Nevertheless, electricity tariffs are still below OECD levels and residential tariffs are lower than industrial tariffs.

The transition process, the shocks of liberalization, and the collapse of CMEA soft markets, all caused GDP to fall in 1989 and remain low until 1993. It was necessary to re-orient export and import from C&EE countries towards the market economies of the West. In 1994, the economy recovered and dynamic growth culminated in 1995 and 1996.[2] The unexpectedly high growth rate, the fall in inflation and unemployment coupled with an almost balanced state budget, led to speculation at the time, that the Czech Republic was to be the 'tiger economy' of Central and Eastern Europe.

However, by the late 1990s, economic growth had slowed, unemployment rose, and inflation did not decrease. More worryingly, a fiscal deficit appeared, the current account deficit increased to eight per cent of GDP, and external debt rose to $US21.2bn.[3] The main macro-economic indicators for Czech Republic are shown in Table 4.1. The reasons for the economic deterioration included insufficient restructuring and disproportionate economic growth (for example, the growth of labor costs was higher than labor productivity growth). Political and managerial decisions were not made swiftly or efficiently enough, and the pace of reform was too slow. The political situation was not characterized by a united coalition and political opposition grew especially at election times and periods of economic hardship. The Czech economy itself went into recession from 1997, a painful but, some would argue, natural period in the economic cycle.

Table 4.1 Macro-Economic Indicators for the Czech Republic

Indicator	1990	1991	1992	1993	1994	1995	1996	1997	1998
Nominal GDP ($bn)	32.3	25.4	29.9	34.3	37.4	50.8	56.5	56.5	55.0
GDP per capita PPP ($)	9526.0	8721.0	8951.0	9273.0	10760.0	11820.0	12710.0	13120.0	12900.0
GDP (% change)	–1.2	–11.5	–3.3	0.6	3.2	6.4	3.9	1.0	–2.7
Industrial production (% change)	–3.3	–21.2	–7.9	–5.3	2.1	8.7	2.0	4.5	1.6
Budget balance (% of GDP)	n.a.	–1.9	–3.1	0.5	0.9	0.5	–0.1	–1.0	–1.7
Unemployment (%)	0.8	4.1	2.6	3.5	3.2	2.9	3.5	5.2	7.5
Average monthly wage ($)	183.3	129.0	164.0	200.0	239.0	308.0	356.0	337.0	362.0
Inflation (%)	9.7	56.6	11.1	20.8	10.0	9.1	8.8	8.5	10.7
Exports ($bn)	5.9	8.3	8.4	13.0	14.0	21.6	21.9	22.8	26.3
Imports ($bn)	6.5	8.8	10.4	13.3	15.0	25.3	27.7	27.2	28.8
Trade balance ($bn)	–0.7	–0.5	–1.9	–0.3	–1.0	–3.7	–5.8	–4.2	–2.5
Current account balance ($bn)	–1.0	0.3	–0.3	0.1	0.1	–1.4	–4.3	–3.2	–1.0
Foreign direct investment flow ($m)	72.0	523.0	978.0	580.0	1038.0	2732.0	1138.0	1300.0	1400.0
Foreign exchange reserves ($bn)	1.1	0.7	0.8	3.9	6.2	14.0	12.4	9.8	12.6
Foreign debt ($bn)	6.4	6.7	7.1	8.5	10.7	16.5	20.8	21.4	24.6
Discount rate (%)	n.a.	9.5	9.5	8.0	8.5	9.5	10.5	13.0	7.5
Exchange rate (/$)	18.0	29.5	28.3	29.2	28.8	26.6	27.1	31.7	32.3
Population (m)	10.3	10.3	10.3	10.3	10.3	10.3	10.3	10.3	10.3

Source: Statistical Yearbook of the CR 1998; Transition Report 1996, OECD. Business Central Europe Magazine, The Economist Group, http://www.bcemag.com/

The Transition

The transition process, despite its 'ups and downs', has been unique in Czech history and has completely changed the structure of the Czech economy in a number of ways. Particularly visible changes can be seen in ownership structures, in the legal framework, and the contributions of different sectors to the national economy. For example, in 1990, 12 per cent of GDP was created in the primary sector, 48 per cent in the secondary sector, and 40 per cent in the tertiary sector and in 1996 GDP created was 7 per cent, 39 per cent and 54 per cent respectively.[4] This means that the private sector now represents three quarters of GDP. This is the case to 1998, the state had held on to significant shares in a number of large enterprises and 45–67 per cent shares in the four largest banks.[5] If privatization of large state enterprises continues as planned, the private sector will continue to grow.

The Czech economy faces a number of difficult challenges. These include the growing trend of inflation, high interest rates, and a low GDP growth rate. It will be crucial to get the balance of exports and imports right in the next few years, the banking sector needs more foreign investment and privatization, and the capital market needs development. A rise in wages (which is higher than the growth of productivity), slow restructuring of industry, a growing unemployment rate, increasing living expenses, and increasing deficit of state budget, are all hampering the growth of the Czech economy. In November 1998, the European Committee's main criticisms of the CR were: insufficient structural reform, insufficient links between banks and new enterprises, insufficient progress in the protection of intellectual property and personal data, and insufficient regulation of state support.[6]

However, compared to many developing countries world-wide, the Czech Republic is ranked at a relatively advanced stage of transition by the EBRD. The European Union considers the CR (together with Estonia, Hungary, Poland and Slovenia) as a functional market economy that is able to fulfil the economic criteria for EU membership in the short to medium-term prognosis.[7]

External trade

Reintegration into the world economy is an essential part of the transition process. Before 1989, international trade and payments were directed by the government, not by the market. It is important to note that throughout the region, the most advanced countries in industrial restructuring and management, are also those which are most integrated into the world economy.

The Russian crisis

The repercussions of the Russian financial crisis of August 1998 were felt in markets all over the world and inevitably influenced all Central and East

European countries. However, the degree of influence varied throughout the region.[8] The portion of Czech imports that were directly dependant on the Russian market was relatively low ($US772.2m in 1997). Therefore the effect of a weakened rouble was neutral or slightly positive (US$1842m), and temporary losses in exports could be compensated for by imports. The Czech economy is not as closely tied to the Russian economy as it is, for instance, to the German market. Nonetheless, the Czech capital market is much more sensitive to fluctuations of the Russian market than of the German. In the aftermath of the crisis, Czech enterprises could be divided into three groups, according to the degree the Russian crisis affected them. Firstly, there are firms that were not affected (mainly thanks to barter agreements or payments in US$). Secondly, there are firms that were very affected (resulting in book-debts, cessation of exports, increases of stocks of goods, cancellation of orders, delays, losses and limitations of production). And thirdly there are firms that were affected through trade with other, mainly C&EE, countries.

Czech external trade

The territorial structure of Czech external trade (in US$) has changed dramatically during the 1990s.[9] External trade with EU countries reached 60 per cent of total Czech foreign trade in 1998. In 1996, the share of total exports was 58 per cent and the share of imports from EU countries was 62 per cent, which ranked the Czech Republic (CR) second (after Poland) among C&EE countries.[10] Of their total export volume in 1998, EU countries exported over 2 per cent to the CR, and import of Czech goods made up nearly 1.7 per cent of the total EU import volume. It is interesting to observe the way Czech trade (imports and exports) is distributed among different foreign countries. By far the largest proportion of the total value of Czech trade is with advanced market economy countries, who had nearly a 65 per cent share of total CR import value and above 70 per cent share of export value. European transition countries had a share of 30 per cent import and 23 per cent export. Table 4.2 shows the structure of Czech external trade for 1998 and shows the main export and import commodities for the country.[11] Developing countries had a share of 5 per cent (import and export together), and other (planned economy) states had about a 1 per cent share (both import and export).[12]

Devaluation of the Czech currency, lower labor wages, and international co-operation, have all helped to position Czech goods in Western markets. The main export commodities are: fuels, materials and supplies; machinery and transport equipment; manufacturing goods; foods and agricultural goods. A number of large export based companies exist in the CR. For instance, the brewery Budejovicky Budvar[13] exported 70 per cent of its production into 43 countries worldwide.[14] It is the number one beer importer in Austria, number two in Slovakia, and number three in Germany. Budejovicky Budvar has been exporting to the US since 1872 and now reaches tens of per cent of growth year-upon-year in England. The main import commodities are: fuels, chemicals and raw

Table 4.2 Commodity Structure of Czech External Trade, (%)

Commodities	Exports	Imports
Food	3.5	5.3
Beverages and Tobacco	1.1	0.9
Crude Materials	3.5	3.9
Mineral Fuels	3.2	6.5
Chemicals	7.7	11.9
Manufactured Goods	26.4	20.9
Machinery and Vehicles	41.3	39.4
Manufactured Articles	13.2	11.1
Without Specification	0.1	0.1
Total	100.0	100.0

Source: Data by the Czech Statistical Office, Prague 1998..

materials; machinery and transport equipment; manufacturing goods and chemical products; and foods.[15]

From 1995–99, external trade resulted in a deficit but decreasingly so: 160 bn Czech Kroner (CZK) in 1996, CZK140 bn in 1997 and CZK80 bn in 1998. Table 4.3 shows the top ten domestic companies and Table 4.4 shows the ten largest domestic exporting companies in 1997, with export of goods representing 40 per cent of GDP with a slightly increasing trend.[16] Due to its dependency on energy resource imports, the CR tends to be highly sensitive and susceptible to external fluctuations on world markets. To combat this, the country has attempted to diversify its oil resources from 1996. While the foreign exchange reserves saw a positively increasing trend up to 1996, external debt has shown a negative trend.

Table 4.3 Top Ten Largest Czech Companies

Rank	Company	Main Business	Revenues 1997 (in CZK m)	Revenues 1996 (in CZK m)
1	Skoda Auto	Automobiles	90.1	58.9
2	Unipetrol	Chemicals, pharmaceuticals	65.6	68.2
3	CEZ	Electricity	54.8	55.5
4	Ceska rafinérská	Petroleum refinery	39.3	36.6
5	SPT TELECOM, Praha	Telecommunication	34.9	29.4
6	Nová hut	Metallurgy	30.3	23.4
7	OKD	Coal mining	23.9	23.8
8	SKODA	Mechanical engineering	23.5	23.5
9	Cepro	Transport, communication	18.9	18.8
10	Chemopetrol	Petroleum refinery	18.7	6.2

Source: Czech Top 100, 1998

Table 4.4 Ten Largest Exporting Domestic Companies in the CR

Rank	Company	Main Business	Rank	Company	Main Business
1	Skoda Auto	Automobiles	6	Metalimex	Trading
2	OKD	Coal mining	7	SPOLANA	Chemicals
3	Vítkovice	Metallurgy	8	MOTOKOV Int'l	Trading
4	Nová hut	Metallurgy	9	Vítkovice export	Trading
5	Ceské aerolinie	Transport	10	Barum Continental	Tires

Source: Czech Top 100, www.medea.cz/top100, 1997.

The country has succeeded in winning important foreign commissions (such as the manufacture of buses for public service in the US and Canada), and has established major subsidiary companies abroad (such as the subsidiaries of brewery, Pilsner Urquell, in Slovakia, the US and Germany). The CR has also successfully invested capital abroad (Pilsner Urquell has a majority in the brewery, Rabulis in Lithuania and the shoe company C´Svit a.s. Zlín owns 100 per cent of Eurosvit a.s. Moscow), and in manufacturing goods abroad (C´Svit a.s. Zlín operates a manufacturing plant in Southeast Asia).

Foreign investment inflows to the Czech economy

One of the factors limiting rapid progress in transition economies of C&EE is a lack of capital funds and an insufficiently developed market structure. Foreign direct investments (FDI) have a very positive influence on the development of both these factors. FDI can enhance the dynamics of economic development, increase the country's export capability, and induce companies to make modernizing investments. Many investment opportunities exist in a wide range of business sectors.

In 1999, the CR ranked third in FDI inflows among CEFTA countries. By the end of 1997, there were over 47 000 entirely or partly foreign-owned companies registered in the CR, and over 600 foreign-owned manufacturers employing above 50 persons. Large foreign investments were made into a variety of industries, and in particular into telecommunications.[17] Table 4.5 shows some of the largest foreign investments in 1990–97 including US, UK and Asian companies that have chose to invest in a variety of industries, and Table 4.6 shows the scale of these investments and the time period over which they were made. The volume of FDI in the Czech economy reached over US$8.4bn as the cumulative value in the period 1990 to 1997, and amounted to US$1.4bn in 1996 and US$1.3bn in 1997. The orientation of MNCs making such large investments correlates with the traditional industry structure of the country. More than 450 North American companies are represented via their substantial investments into the CR and both international companies (such as Arthur D. Little) and

Table 4.5 The Largest Foreign Investments

Company (in alphabetical order)	Country of Origin	Branch
ABB	Sweden, Switzerland	Electronics, electro-technology
AssiDomän	Sweden	Pulp and paper
AVX/Kyocera	UK	Electronics
Bass Breweries	UK	Beer
Coca-Cola Amatil	Austria	Soft drinks
Daewoo	South Korea	Trucks
Danone, Nestlé	France, Switzerland	Food
First International Computer	Taiwan	Computers
Ford Motor Company	USA	Automotive components
Glaverbel Group	Belgium	Auto glass
Heinz	USA	Consumer foods
Matsushita Electric Industrial Co.	Japan	Consumer electronics
Motorola	USA	Electronics
Pepsi-Cola International	USA	Soft drinks
Philips	Netherlands	Metal parts
Philip Morris	USA	Tobacco
Procter & Gamble	USA	Consumer goods
Showa Aluminium Corporation	Japan	Car components
Siemens AG	Germany	Automotive electronics, telecoms cable, electro-mechanical components
T mobil	Germany and Italy	Telecoms-GSM
TelSource	Netherlands, Switzerland	Telecoms
Toray	Japan	Textiles
Volkswagen	Germany	Cars

Source: Special enclosure by the Czech Investment Agency in business magazine Czech Business and Trade, Praha, 6/1998.

multinational ones (such as Andersen Consulting) are present.[18] Nearly 70 Japanese companies are also doing business in the CR.[19]

Both small and/or medium-sized and large companies have been set up. Both greenfield and reconstruction investments have been made, operating as subsidiaries and joint ventures Table 4.7 lists the top ten largest joint ventures in the CR showing that joint ventures have been made in a range of industries and by a variety of countries.[20] Examples include VW – Skoda Auto, ABB –Vítkovice, ZDB – BEKAERT. A number of joint ventures were eventually bought entirely or partly by their foreign partners. This was the case, for example, for Linde Technoplyn. In retailing, joint ventures are often formed with local partners to operate and co-invest in distribution centers. Examples include fast foods and oil services based on the franchise system such as McDonald's, KFC, Family Frost; Shell, Aral, Agip, and so on. In other cases, distribution centers are set up as private firms supported by MNCs (VW). Some MNCs maintain regional

Table 4.6 Scale of the Largest Foreign Investments in the CR

Rank	Foreign Investor	Time Span	HQ Country	Type of Business	Commitment (CZK bn)	Czech Partner or Subsidiary
1	TelSource (PPT Nederland NV & Swiss Telecom)	1995–2000	Netherlands, Switzerland	Telecoms	37	SPT Telecom
2	Volkswagen	1991→	Germany	Vehicles	25	Skoda Auto
3	Bankers Trust (BT)	1990→	US	Stock market	19	n.a
4	International Oil Company (IOC) (Royal–Dutch/Shell, Conoco (DuPont, Agip)	1995–2000	Netherlands, US, Italy, UK	Petroleum refineries	16	Unipetrol (for Chemopetrol Litvínov & Kaučuk Kralupy refineries)
5	Philip Morris	1992→	US	Tobacco	13	Tabák a.s.
6	Ind. Power, NRG Energy, Nat'l Energy	1996–1999	US	Energy	10	Energetické Centrum Kladno
7	Daewoo–Steyr	1995–2000	Netherlands, Korea, Austria	Vehicles	10	Daewoo Avia a.s
8	Daventree Limited	1995→	Cyprus	Investment	8	Harvard Capital & Consulting
9	Glaverbel (Belgian subsidiary of Asahi Glass, Japan)	1991–1998	Belgium	Auto glass	7	Glavunion a.s.
9	IFC Kaiser	n.a	US	Steel	6	Nová hut
10	Pepsi-Cola International	1994–2000	US	Soft drinks	5	Pepsi's operations in the CR. No Czech partner.
11	T Mobile (Deutsche Telekom Mobile & STEET	n.a	Germany & Italy	Telecoms – GSM	5	Sporitelní kapitalová spolecnost, Telekomunikascní montáže Praha, Podnik výpocetní techniky
12	Bass Breweries	1993–1998	UK	Beer	4	Prazské pivovary, Pivovar Radegast
13	Linde AG	n.a.	Germany	Industrial gas & commercial gases	4	Linde Technoplyn
14	Cokoládovny Partners B.V. Amsterdam	1991→	Netherlands, France, Switzerland	Candy	4	Cokoládovny, Praha
15	AGA Gas, s.r.o.	1991–1998	Sweden	Technical gases	4	No Czech partner

Source: Special enclosure by the CzechInvest Agency in business magazine Czech Business and Trade, Praha, 6/1998.

Note: Data as of March 1996 by Consulate General of CR, Los Angeles, USA, updated as of October 1997 by the Prague Post 1998 Book of Business Lists, Prague.

offices in the Czech Republic (e.g. General Motors Opel division). A range of industrial, research and commercial co-operations have also been established. For example, Boeing and Letecké opravny Kbely, Boeing and engineering company Skoda Plzen, British Aerospace with Saab and Chemapol Machinery, Lockheed Martin Corporation and the Institute for Nuclear Research in Rez, Rubion, Butler and ProntoPlus. Partners have been selected for strategic reasons (such as Boeing and Aero Vodochody). Partnership agreements are based on benefits for all partners. For instance, General Electric Capital Services (GECS) is admired for its rich network of collaborators, good firm reputation and excellent management of its local partner Multiservis. GECS kept the Czech brand and won the trust of local clients. Consequentially, Multiservis could significantly spread its service portfolio and the number of its commercial points.

Table 4.7 Top Ten Largest Joint Ventures in the CR

Rank	Czech Company	Products	Foreign Partner(s)	Partner Country
1	SPT Telecom Praha	Telecommunications	Telsource	Netherlands, Switzerland
2	Skoda Auto M. Boleslav	Cars	Volkswagen	Germany
3	Tabák Kutná Hora	Cigarettes	Philip Morris	USA
4	Elektrárna Opatovice	Energy	National Power	Great Britain
5	Cokoládovny Praha	Chocolates, biscuits	Danone, Nestlé	France, Switzerland
6	RadioMobil, a.s., Praha	Telecommunications	C Mobil	Netherlands
7	Ceská rafinérská, Praha	Petrol	Conoco, Shell, Agip	Netherlands, Italy, USA
8	Linde Technoplyn Praha	Technical gas	Linde AG	Germany
9	Glavunion Teplice	Glass works	Glaverbel	Belgium
10	Cementárna Mokrá	Cement works	CBR Brusel	Belgium

Source: Czech Business and Trade 5/1997, Prague, p. 11, Czech National Bank, H. Pisková

Three investor countries – Germany, the Netherlands and the US,[21] – together represent over half the total volume of FDI made into the CR. Over 96 per cent of FDI comes from OECD countries, and nearly 71 per cent from EU countries. Even though Asian companies have begun to show a growing interest in the CR in the last few years (a good example being the Japanese textile firm Toray Industries which announced a US$150m investment), the investment inflows from Asia are not significant yet.

The highest share of FDI has been made into the communications and the transport sectors (18.2 per cent) mainly by VW and Renault, from 1990 to 1997. Companies such as Philip Morris, Pepsi-Cola, Bass, Danone, Nestlé, and Procter & Gamble have made the majority of the investment into consumer goods, with Matsushita, First Int'l Computer and AVX/Kyocera the major players in the electronic sector. Foreign investment in consumer goods production, trade and service sectors is stimulated by lower labor wages, good access to the local market, and the opportunity to export production.

The automotive components and electronics sectors also attract substantial FDI. Based on CzechInvest research, over one-third of Europe's top 100 car component firms are already assembling or manufacturing in the CR. There were over 100 firms within the country identified which produce automotive vehicles, components or accessories with both domestic and overseas destination. The investment made by VW into the automotive industry has been a spectacular success. Skoda Auto has introduced new models which meet West European standards. Skoda's assembly plant operates in Poland, another is being built in Russia and Belorussia and there are negotiations to open plants in Egypt, India and China. The Daewoo investment into Avia utilized its engineering capabilities to develop a new engine and truck that meet EURO II standards. Daewoo plans to use Avia as the central supplier for Daewoo's global sales network.

Matsushita's investment of US$66m, to build greenfield production plant, represents the first major investment direct from Japan into the country. Matsushita has built its first Central European production factory in the West Bohemian city of Pilsen, which was selected among 27 places in the CR, Poland, Hungary and Slovakia. Matsushita's plant in Pilsen is the second largest in Europe, after its plant in Wales. With this decision, Matsushita is standing along other global electronics competitors such as Philips, Samsung and Grundig which have also decided to locate their plants in Central Europe. In an article published in 1997, the *Financial Times* reported on the experience of the Matsushita factory, highlighting the short time it took to build the factory (only 12 months after the land purchase), the rapid fulfillment of targets (starting daily target was fulfilled in four months), and the reaching of required quality standards (which took only four months, compared to years in other factories). At a time when a range of factories were deep in recession, Matsushita had expanded strongly. Another company with successful investments in production was Kyocera's, British subsidiary of AVX. The company invested in an existing Czech company to produce tantalum capacitors, a component used in mobile telephones. This plant now supplies one-third of the world's market for tantalum capacitors.

In the 1990s, the share of FDI accounted for around 20–40 per cent in capital, for around 20–40 per cent in portfolio investment, and for around 30–40 per cent in other capital investments. A preliminary agreement for more capital inflow has been recently signed with the American investment fund, E.M. Warburg, Pincus & Co LLC, and the company, AIP Group E.M.[22] The American fund plans to buy new equity issued worth US$20m. This ranks the agreement among the top forty foreign investments made into the country. Such investments are evidence that the CR is considered a perspective market offering good stability for long-term investments.

Besides FDI, the Czech Republic is taking advantages of the prestigious financial support given by the IFC, a subsidiary company of the World Bank, which helps to develop the private sector in transition economies. The CR receives among the largest amount of IFC financing in Europe. From 1990 to the end of 1998, seventeen projects, mainly in banking, metallurgy and industry, were financed by IFC amounting to a total of US$305m, and by other banks to a total of US$283m.[23]

Banking is a developed and thriving new sector and banks are bound to play an important role as providers of investment finance. However, the financial sector has undergone several large crises. The investment boom of the early 1990s and interest rate battle among banks resulted in the failure of several medium-sized and large local banks in 1996, which rocked the whole sector. As the state financial institution, the Czech National Bank withdrew licenses, placed certain banks under its administration and in other cases encouraged mergers to consolidate the banking sector. The sector includes several tens of commercial banks; there were eight majority Czech-owned banks, nine partly foreign-owned banks, three entirely foreign-owned ones, 11 subsidiaries of foreign banks and nine branches of foreign banks, in the fourth quarter 1997.[24] The four biggest banks in the CR are: Komercní banka (KB, Commercial Bank), Ceská sporitelna (CS, Czech Savings Bank), Ceskoslovenská obchodní banka (CSOB, Czechoslovak Commercial Bank) and Investicní a Postovní banka (IPB, Investment and Post Bank). Standard & Poor gave a first rating to three Czech Banks (CS, CSOB and KB) in January 1997.[25] Because the foreign presence in banking is already substantial, the CR ended the issuing of new licenses to foreign banks in 1996.[26] Foreign banks are increasingly active on the Czech market with over 22 per cent of the sector assets in 1996, which grew to 37 per cent in 1997. The state's 36 per cent stake in IPB sold to Japan's Nomura in 1997, and a good portion of AGB (Agrobanka) sold to General Electric Capital. From the spring of 1998, the privatization of the other three majority state-owned banks, (CSOB, KB and CS) started. This process is expected to proceed rapidly. Bank privatization is urgent and the sale of the banks will open up the market even further to foreign enterprise.

Foreign MNC participation in the Czech Republic

The markets of C&EE countries are small compared to Asian or Latin American ones, and enterprises operating in the Czech Republic have to cope with the frequent changes and uncertainties accompanying the transition process. A characteristic feature of MNCs entering C&EE markets is that they do not enter each country individually but enter the region as a whole by establishing a regional office in one or more country. Overall, MNCs behave more dynamically in this region than in stable (advanced economy) markets and provide innovative strategies and creative solutions for coping with local conditions.

The social and economic changes introduced in recent years have brought about a new level of competitiveness never before witnessed in the region. And the structure of competitiveness of enterprises has been extremely influenced by MNCs. By the end of 1997, there were nearly 1 630 000 businesses registered in the CR, nearly three per cent of which were foreign-owned or partly foreign-owned. Table 4.8 shows the largest foreign firms operating in the CR.[27] MNCs play an important role in developing the country, by introducing high-tech facilities, restructuring industry, building-up new capacities, and introducing training programs to enhance the skills of local managers.

Table 4.8 Largest Foreign Firms in the CR

Rank	Company	HQ Country	Type of Business	Revenue (in CZKbn)	Staff
1	Skoda Auto	Germany	Auto manufacturing	58.9	16 721
2	Asea Brown Boveri (ABB)	Switzerland	Engineering/ construction	13.2	6400
3	Unilever CR	Netherlands/UK	Household goods	4.4	900
4	Daewoo Avia	S. Korea/Austria	Vehicles	4.2	2270
5	Tesco Stores CR	UK	Retail	4.0	2600
6	Julius Meinl Praha	Austria	Supermarkets	3.7	2700
7	Delvita	Belgium	Supermarkets,	3.7	2200
8	Karosa a.s.	France	Buses	3.2	1737
9	Linde Technoplyn	Germany	Technical gas	2.3	619
10	Robert Bosch	Germany	Automobile equipment	2.0	957

Source: Data by the Prague Post 1998 Book of Business Lists, Prague. Updated October 1977.

A particularly important MNC in the country is VW, which has strategically directed Skoda Auto since 1991. In fact, VW's acquisition of Skoda has been reputed to have not only saved the carmaker and many of its suppliers, but increasingly the Czech economy itself.[28] Skoda Auto is now the largest car producer in Central and Eastern Europe after outstripping Fiat in 1997. Skoda Auto belongs to the top companies in the country ranked by revenues and export volume.

Generally, MNCs beginning operations in the CR have chosen any number of ways to enter the country and have implemented the same or only slightly adapted strategies as in other markets. They regard the CR as a minor or small market, albeit in a developed, compact and stable country. Successful MNCs in the region use a range of strategies and policies, including favorable price/quality ratios, well known brands, advantages taken from a company's global base, fast adaptation for local conditions, professionalism, and so on. If they are less successful, then this can be due to their late entry into the market, a rise in prices, or a low level of investment, amongst other reasons. Mostly, MNCs are the leaders in their core industry and are dominant in any market and any product. They compete in a huge range of industries and sectors, including engineering, car products, PC products and services, cosmetics, cookies, drugs, hotel chains, insurance, consulting and so on. Further opportunities emerge as the government continues to privatize more industries (such as banking). MNCs' actual growth rates range from five per cent to 60 per cent year upon year, and their market shares from four per cent to 20 per cent, depending on the brand and the company.[29]

Overall Globalization Drivers

When deciding whether to enter the CR, among the advantages that MNCs have identified are: the quality of the infrastructure, the industrial tradition, relatively low production costs, a rapidly expanding economy, a competitive environment, and the opportunity to expand into a new, unsaturated market. From a global point of view, the local market is still undoubtedly small, with a population of 10.3 million, with a 52.6 per cent rate of economic activity, which represents 2.84 million households of economically active inhabitants.[30]

Market globalization drivers

Common customer needs

Being in the center of Europe, Czech people have access to information on the latest trends and products. Consumer awareness has increased due to the information boom, growth of external trade, and tourism and the emergence of Western television. Overall, Czech consumption patterns are very similar to those of other industrialized nations. Some MNCs consider the CR similar to Belgium, the Netherlands or Austria. For instance, Toyota's model Carina E, which was designed for the European market, has been very successful on the Czech market. Consumption habits and tastes are similar to those of Central and Western Europe, and especially to German, Austrian, Polish, Slovak and Swiss. Some MNCs have noticed a higher price sensitivity of Czech consumers. New consumer habits can also be seen among the inhabitants of large cities. For example, they choose to make purchases in supermarkets, megamarkets or hypermarkets and there is a rise in the demand of higher quality and extended services (for example, for employed women). The young and middle-aged generation is very influenced by communications and information technology.

Transferable marketing

Transferable marketing, especially of brand names, is generally high. Nevertheless, some marketing adaptations are recommended and amendments are necessary because of language and cultural differences. By law, products sold in the CR need to contain information, in Czech, of ingredients, instructions of use, weights and best-before dates. Foreign brand names are nearly always easily recognized, and due to their reputation, quality, design and marketing, they are often preferred by consumers. However, Czech consumers are sensitive to the quality to price ratio. Consumers prefer quality products that are characterized by measurable technical parameters (such as performance, consumption, etc.) like cars or electronics. When the quality of comparable products is similar or when consumers cannot distinguish the quality of a product due to specific characteristics (such as taste, colour, ingredients, or experience), then consumers make purchasing decisions according to price. This is the case for products such as foods, beverages, medicines, washing powders, cosmetics, and consulting services.

MNCs need to consider if their advertisement campaigns need to be adapted to the CR for increased effectiveness. For example, there are only two examples to date of MNCs using famous Czech sports-people in an advertisement. Mars' product was promoted by J. Zelezný, a world and Olympic champion in the javelin, and Nutella, a Ferrero product was promoted by S. Hilgertova, the Olympic winner in Ski slalom in Atlanta 1996. MNCs should carefully consider if marketing adaptation is needed. There are examples, although not many, of foreign advertisements that were not adapted for the Czech market and were therefore unsuccessful.

Global Channels

MNCs increasingly include the Czech Republic in their regional distribution systems. This is aided by the country's strategic geographic location – as a gateway to the newly emerging neighboring economies which are similar in size, such as Slovakia, Hungary, Lithuania, Latvia, Estonia – and also because of its growing domestic market, expansion challenges and turnover.

Asian companies have shown a growing interest in the CR. Matsushita Television Central Europe built a factory for TV Panasonic. Japan's Toray Textiles opened its production and commercial branch in Prague and is going to build a factory for textile materials manufacturing. The Daewoo-Steyr BV Consortium with Korean Deawoo Heavy Industries Ltd. and Austrian Steyr-Daimler-Puch AG was established in 1994 and became the tender winner during the Czech privatization process. Thus AVIA, JSC became a part of the Korean Daewoo global network. Also, FIC (First International Computer) from Taiwan entered the Czech marketplace through the Czech firm AAC and is planning to make huge investments in the country.

Cost globalization drivers

From the point of view of MNCs, in comparison with developed countries, the CR is cost competitive in labor, utilities and local products. But it is not the cheapest country in Central and Eastern Europe. Wages are lower than in Hungary and Poland, but higher than in Romania, Bulgaria, and Ukraine. Electricity, water, hiring and insurance costs increased several times after prices were liberalized and this trend is likely to continue. However, in general, this has not eroded the cost competitive advantage of MNCs.

It is possible to find cheaper sources of labor or material inputs in C&EE than in the Czech Republic. However, due to the monetary policy, the costs of Czech inputs are still relatively inexpensive or 'favorable'. This comparative advantage has been observed by companies such as VW and Bosch. Wages are several times lower than in Germany while the quality and productivity of University educated technicians are comparable to those in the West. The CR has among the world's highest percentage of graduates in the science and technical fields and the Czech workforce is highly skilled and well educated. There are 23 Universities in the CR having 177 723 students in 1998. Education expenses accounted for 4.7 per cent of GDP in 1997, which was also the average over 1989–97.[31] More than 60 per cent of the Czech population reach secondary education or higher.

Labor

There is no shortage of highly-skilled labor. The results of an international test of 13-year olds in 41 countries (TIMSS), which ranks Czechs second place worldwide at mathematics and science after Singaporeans, is evidence of an excellent education system in the country.[32] However, MNCs often suffer from a shortage of skilled managers and therefore foreign or mixed management is used by some MNCs, such as Daewoo–Avia and VW–Skoda Auto. The level of education and self-awareness of Czech employees has been remarked upon by a number of MNCs (such as the German company Linde Technoplyn).[33] According to a marketing manager at Sybase, 'people are more inquisitive here, demonstrate much more interest in technical novelties and like to test them.' The Czech labor force is quite flexible and adaptable. For instance, nearly 50 per cent of production enterprises with foreign investments operate in three shifts. The country does not have an abundant supply of low-skilled labor. Many people are hired from nearby C&EE countries for lower qualified jobs. The fact that the workforce is insufficient in certain professions and industry sectors, means employers need to recruit foreigners, often from neighboring countries. The number of foreigners granted work and residence permits exceeds seventy thousand, the highest number coming from Ukraine followed by Poland, the United States and Germany. A further seventy thousand Slovaks should be added, but they do not require work permits.

Income differentiation is based on the level of management (top, medium, low), the current position, and the company's income structure. For example, the structure of an average employee's income is roughly: 70 per cent salaries or wages, five per cent social benefits and 25 per cent contributions to social security. Income also varies with industry sector and branch. For example, the highest incomes are found in financial institutions, metallurgy, energy, transport and telecommunications. Income also depends on the status of the enterprise (for example, fast growing or stagnating), on the region or town (the highest incomes are in Prague), the size of the enterprise (higher incomes are in either small or very large enterprises), and the effort of individuals. An average wage rate of labor is rather higher (it might be 100 per cent higher) in MNCs than in comparable local companies.[34]

Because Czech engineers are well-trained, MNCs often cooperate with Czech science/design/engineering enterprises. On the whole, however, research and development is stagnating in the country. R&D in the Czech Republic is faced with a number of serious problems. These include a general shortage of finance, a short-term instead of long-term business policy in local enterprises, and a preference of foreign companies for foreign R&D. These factors are detrimental to Czech researchers and developers. On the other hand, there are some good examples of R&D activities that have returned to the CR or are serving the needs of a new MNC. For instance Skoda Auto or Bosch have moved development activities from Germany to the CR. Ford behaves similarly, and has built a huge R&D center in Autopal Novy Jicin, where mainly Czech technical experts work. Microsoft has also incorporated Czechs into its development teams. According to the CzechInvest agency, nearly one quarter of companies are performing R&D or product development in the CR, while more than half were undertaking either R&D or product development on products sold locally. In the opinion of many MNCs, the Czech government has not done enough to lure home expatriate scientists and engineers, expand R&D, support industrial development, or promote investments in technology projects.

Natural resources

The Czech Republic is not particularly rich in natural resources. The exceptions are limestone, kaolin, glass sand, wood, black and brown coal. Therefore, MNCs entering the Czech market do not consider it a place for production-based activities with intensive consumption of raw materials. Some environmental problems continue in the country due to insufficiently monitored manufacturing, the chemical industry, coal mining areas and coal power stations. The solution of environmental problems is of interest both to the government (environmental standards need to reach European standards, environmental restrictions must be enforced) and to enterprises (environmental penalties exist). A special inspection board (the Work Safety Board) monitors the work environment and its safety. Both foreign companies and MNCs use the Czech Republic's enterprises as partners in heavy industries (such as ABB) and light industries (such as Colgate Palmolive).

Infrastructure

From a strategic point of view, the CR has a comparative advantage in its geographical location, similar to Poland and Hungary. As a landlocked state, it cannot offer the advantages of sea ports. However, it can be reached through German ports (North Sea), Polish ports (Baltic Sea) or ports in the Jadran Sea. The largest river in the CR, the Elbe, is navigable as far as the German port of Hamburg. There are 68 airports located within the CR, including 11 public international ones. The largest is the Prague Airport Ruzyne, which was extended and modernized in 1997 and whose year on year growth rate is 15 per cent. The Prague Airport Ruzyne partners 40 airport companies. Of these, the largest airlines in the CR ranked by passenger counts are: Czech Airlines (CSA), Deutsche Lufthansa, British Airways, Swissair, Alitalia, Austrian Airlines, Air France, British Midland, KLM, Delta Airlines, Hungarian Malév, and LOT Polish Airlines.

There are no bottlenecks in Czech infrastructure or restriction in the supply of electricity or water. Rush-hour traffic is similar to that in highly developed countries. Some traffic lines are under reconstruction or are being extended. There are 9441 km of railways, 55 912 km of roads and 303 km of navigable flows.[35] The railway network is as much as two or three times denser than in West European countries, excellent for distribution purposes.[36] The dependence on foreign-owned transportation is mostly low. The domestic company Cechofracht followed by DHL International, UPS Czech Parcel Service, TNT Express World-wide and Federal Express Inc.-Inspekta are the largest shipping and cargo services. Czech companies started to use the Internet and World Wide Web substantially from 1997, both for commercial and managerial purposes.

Technology role

Further investment is needed in order to increase the country's technological capabilities. Some MNCs transfer technologies to the Czech Republic (for example, Opel, Colgate, and Palmolive). Others are engaged in a production strategy (for example, VW). If they use only one production strategy the level of technology tends to be higher. MNCs implement both high technology strategies (for example technology for production of components), and so called colonial technology (technology for production of cement) and acquisitions, joint ventures and green-field investment.[37] Political stability is the one factor important to investors across the board, a key reason why the CR has been more successful in attracting foreign capital than many other C&EE countries.

Government globalization drivers

Examples of government globalization drivers are: transferring to European regulation and legislation, open friendly foreign affairs politics, applying to join the EU and NATO, actively supporting external trade operations. However, no special incentives or preferential tax/credit treatments are given to MNCs. No discriminating regulations by sector, age, sex, race, political party, ownership, size of enterprise, or economic cooperation, exist.

To provide supportive incentives to foreign corporations, from 1998 the Czech Government has approved the encouraging of interest by large international investors. Incentives, representing a major change in Government policy, include:

- Total documented investment of at least US$10m;
- Sale of suitable land owned by the Land Fund and the National Property Fund of the CR at a symbolic price;
- Implementation of an industrial zone support system;
- Total amount of investment into machinery to equal a minimum of 40 per cent of the total investment;
- Provision of an interest-free loan of up to CZK 100 000 per person for the creation of each job for a Czech citizen;

- Application of investment incentives for extensions of production;
- Implementation of investment incentives for R&D;
- Implementation of investment incentives for environmentally-friendly management systems.

After legislation changes, other incentive provisions will be available for foreign investors, such as tax-free holidays.

Absence of state-owned competitors

The Competition Law of 1991 regulates the competitive environment in the country and is designed to promote the principles of a free market economy, avoid abuse of monopoly power, and provide advice on mergers. According to Czech law, every foreigner has the opportunity to establish an enterprise either with 100 per cent of foreign capital or in Czech partnership, as long as legal requirements are subscribed to. These include the Czech Economic Code, Commercial Code, International Trade Code, Small Business Act, Foreign Capital Participation, Foreign Economic Relations Act, Foreign Exchange Act, Customs Act, and the Taxation Act.

Reliable legal protection

Trademarks, intellectual property rights and international copyright are protected by law and provided by the Industry Property Office, Author Protection Union, Office for Standardization, Metrology and State Examination. All these legal regulations are based on an approximation to the standard in the developed countries and are being harmonized with EU law. Practical protection and raid solution of disputes should be improved rather than Acts themselves. Despite this legislation, a lack of control and enforcement means that some consumer goods, such as clothes, videotapes and cigarettes, are illegally imported and sold by retailers in open-air markets.

From an MNC's point of view, legal protection given in the CR is similar to its local neighbors. MNCs operating in the CR do not think that the Czech Government fears their expansion will crowd out local consumer goods manufacturers. In reality, some MNCs (Coca-Cola, Pepsi-Cola) did worsen conditions for local producers or even pushed them out completely. For instance, the CR's 150 smaller producers of soft drinks have only half the market left to share.

Compatible technical standards

International standards, such as EEC and ISO, and some EU standards are gradually being introduced and implemented. Although slowly, managing product and service quality in the CR is getting better. There have been thousands of certificates of ISO series 9000 issued within the country. A good illustrative example can be given by the final producer Skoda Auto (part of Volkswagen), which is linked to a number of its suppliers and sub-suppliers who are selected on their basis of their certification. Marks of quality are granted to first rate quality product/services, and companies. Awards include 'Czech Made' and the 'Czech Republic Quality Award' (the second based on the National US M. Baldrige Award and European Quality Award). IBM CR is among such awarded companies. Encouragement of competitiveness and an image of quality improvement and control are the main incentives to apply Czech, European and international standards. MNCs have no unusual comments on cooperating with the Office for Standardization, Metrology and State Examination. They consider the Office as being of European standard although believe that it could work faster and be more flexible.

Freedom from government intervention

The government (ministries) take part in business dealings only in special situations, for example when partner(s) for state owned companies are selected. This occurred when Skoda Auto was seeking a foreign partner (VW vs. Renault vs. GM), as with Budejovický Budvar (Anheuser Bush), with SPT Telecom (PPT, the Netherlands and Swiss Telecom), and with IPB (Nomura).

Regulation of MNCs occurs mainly in food, cigarettes, drugs, waste, water protection, consumer protection, monopoly, and advertisement. Production and distribution of food, beverages, tobacco, and drugs have to be directed by law. Environmental promotion and protection refers, for instance, to EIA required by law, S2 required for imports, and destruction of expired drugs. The Anti-Trust Office (the Economic Competition Protection Office) located in the Moravian city of Brno has been established to control monopoly throughout the country. Reduction of customs for drugs or activities, as a signatory state of CEFTA (for example the reduction of customs within CEFTA), are examples of the Czech government's efforts to liberalize trade. Some additional product modification can also be based on government regulations, for example electric voltage (220), drug and food packaging that must be in Czech. Enterprises employing handicapped people have discounted tax duties. Products have to receive a permit for import/export from a state administrative body.

Foreign investment can be made both directly and through indirect investment vehicles. There are no legal obstacles to ownership by foreigners, who may own or lease land. Business undertaking is enabled to foreigners generally under the same conditions and extent as to Czechs.

Regional trade blocs

Trade between Central European countries fell dramatically after the momentous political changes of the early 1990s. The level of mutual C&EE trade, so crucial to the success of all transition economies, led to the establishment of CEFTA in 1992. Thanks to membership in CEFTA, exports and imports between the CR, Hungary, Poland, Slovakia, Slovenia, and Romania rose, facilitated within a free trade zone. The Czech and Slovak Republic also have a Customs Union Agreement. The trade of food and farm products is determined by the sensitiveness and position of agricultural production in economies of CEFTA countries.[38] The share of CEFTA trade for each existing member is still low, although steadily increasing. MNCs operating in the CR and trading in consumer goods, state that CEFTA is very advantageous from the point of duties. The CR is second place in terms of volume traded within CEFTA members, representing about 23 per cent of exports and nearly 15 per cent of imports.[39] The fact that the CR is not yet a member of EU causes some disadvantages, such as duty rates, customs policies, and inflexible registration of certain goods, such as drugs.

Participation in trade blocs

Liberalization of exports and imports began in 1991 with no significant administrative trading barriers. The CR signed a commercial agreement on industrial products with the EU in 1989, the Agreement upon Trade and Cooperation in 1990, the Association Agreement, (called the European Agreement) in 1991 and 1995. These have laid the legal foundations for relations between the EU and Czech Republic. The CR became a member of GATT in 1948, EBRD in 1991, WTO in 1993, OECD in 1996, and NATO in 1999, all bringing it ever closer to entry to the European Union, for which it applied in January 1996. The CR also has a special customs union with Slovakia. It is a signatory country of CEFTA and a signatory member of the IMF. The Czech Export Bank and EGAP are two institutions that help facilitate external trade and control export and import risk. CzechInvest is an institution that can supply information to foreign investors and give them general support including assistance in dealing with possible local partners, local authorities and finding suitable sites to locate green-field investments.

The Czech republic is a compact country with equally developed areas. However, under certain circumstances, a particular area may receive special government attention. For example, areas with high pollution or with naturally occurring disasters such as the worst floods for a hundred years in the summer of 1997. In this case, finance from the state budget was released and flood obligations were issued, houses and roads were rebuilt and state support was given to the area for the next few years. The Government is in contact with the Chambers of Commerce and initiates business events abroad such as Czech Days or business missions world-wide.

Competitive globalization drivers

Despite its size, the Czech Republic is an important market both for local and foreign businesses. For a range of MNCs, the importance of the Czech market rests upon its development on the one hand, and upon the fact the country serves as a gateway for expansion to Eastern markets.

MNCs consider the major business sectors to be: manufacturing, telecommunication, energy, banking and finance. Recognizing the economic and social importance of telecommunications, the Czech Government prioritized its development in the 1990s. A commitment was made by the Dutch/Swiss consortium, Telsource, to modernize the Czech fixed-line telecommunications network. This investment is not only an opportunity for manufacturers of telecommunications technology but will also ensure that the Czech telecommunication system will meet Western standards. Therefore, this strategic sector remains heavily regulated (except for prices). Free competition in voice-transmission services is protected by the Government until the end of 2000. SPT Telecom enjoys a monopoly position for the bearing commitment of high investment in extension and modernization of the sector. Data transmission and other telecommunication services are open to competitors. Communications and transport represents over 18 per cent of total foreign investment.

To survive the growing competition, forward-looking Czech enterprises are making IT investments an important part of their business strategy. Local and multinational IT suppliers are battling for dominance as competition increases. Even in the Czech market, the region's most advanced, most of the top 20 IT distributors enjoyed revenue growth of over 20 per cent in 1995.[40] American Apple, IBM, Hewlett-Packard, Microsoft, German SAP, operate in the country. MNCs consider that information technology plays an important role in the Czech Republic's repositioning in the global market place. The Czech information technology market is worth US$1.2bn, representing Central Europe's biggest market after Russia.[41] Despite the differences among C&EE countries, the IT market is generally split between the HW market of around 70 per cent, SW market of around 20 per cent and services market of around 10 per cent.[42] Motorola concluded a deal with a Czech producer of semiconductors in silicon, Tesla Sezam, in 1997 which makes the country the first one among C&EE countries to attract Motorola investment. Lastly some local Czech companies compete successfully in traditional industries. However they are not globally dominant and the competition they face in Western markets is strong.

Overall Global Strategies for the Czech Republic

Its small size and relatively high costs mean that most MNCs need to treat the CR as a self-sufficient market rather than as a base for regional production (with some exceptions). At the same time, the relatively high level of development allows easy application of most global strategies. Furthermore, MNCs do not see any reason to view the CR as a separate market. Some of them see it in context with Slovakia and other Central European countries. Also, operations in the Czech Republic are becoming increasingly linked with those in other countries.

Market participation strategies

Despite its small size from a global point of view, the CR is large enough to be an attractive market for MNCs from a local and regional aspect. Hundreds of MNCs document this fact by their very presence in the country. Moreover, some of them have chosen to buy majority shares in new distribution firms in the CR, as opposed to other countries in the region. And up to now, they have enjoyed watching this market grow. For example, in 1996 Toyota's sales worldwide total more than 4 000 000 vehicles,[43] with a small proportion of the total sales volume in the CR, 2174 vehicles; Ford sold 10 692 (which makes 14 per cent of market share) and Opel 11 507 cars (12 per cent of market share). The CR also attracts MNCs because its market is a gateway to Central and Eastern Europe, it has unsatisfied demand, relatively low input costs, growth potential, skilled and experienced workers and engineers, good opportunities for foreign managers, acceptable legislation, liberalization policy, and relatively low inflation.

Most MNCs entered the CR between 1990 and 1994 (e.g. Opel 1990, VW 1991, Roche 1992, R.J. Reynolds Tobacco, Amway, Korn/Ferry 1994). However, the Czech market is still open to other foreign companies and MNCs and they are still entering the market (for example Matsushita in 1997). In total, MNCs consider their entrance into the CR to take between one and three years. The collaboration of Czech management is often extensive and important in this process.

To start with, MNCs do not use a unique entry strategy for the Czech market. Some of them (such as Colgate Palmolive) used a general strategy applied in other markets, while others (for example Opel) have applied aggressive strategies. The ways in which MNCs have chosen to enter the CR have varied from company to company. They have included an export strategy (Nina Ricci), through subsidiaries (Coopers & Lybrand), through license production/service (Colgate Palmolive, Conoco), and through joint ventures (VW, Philip Morris). Over time, strategies are often combined.

The most frequently chosen strategies of MNCs have included: export/import contracts, license contracts, setting up a wholly owned subsidiaries, joint ventures with a local partner, or direct investments into production/operations. Particular strategies were chosen for different reasons. These included: the need to supply domestic market demand, to supply neighboring market demand, or for penetrating the market. Many strategies are export driven. A Taiwanese First International Computer considers only five per cent of its computer production in the CR is for the local market, while the rest is for export, mainly to Eastern Europe. A Japanese company, Showa Aluminium Corporation, started its project in Kladno for US$22m and built a new manufacturing factory in 1999. It wants to export the majority of its production of condensers from the CR with Skoda Auto to be supplied by about five per cent of this production. Matsushita also aims to export 95 per cent of what it produces to 30 different countries in both in Central Europe including CIS countries and Western Europe.

Some MNCs consider that local partner-companies and the Government play a vital role in setting up and utilizing distribution channels, and effective sale. As for joint ventures, the government might own a company's shares. This can also happen in regulated industries such telecommunications, nuclear power stations and so on.

Several MNCs were subsequently engaged in direct investment strategies by supporting the establishment of joint ventures in industry of suppliers for local advantages purposes. It is obvious that some MNCs have had to change their market strategies (for example, Roche). The reasons for change were primarily two: due to dynamic change in a development, or due to strategy decision remaking. Both reasons are true for Cokoladovny, which was split into two companies – one selling biscuits and the other chocolates. The split was intended to boost export and reduce reliance of the company on the domestic market. This case is particularly interesting both because of its successful performance as well as its market strategy change. While companies are merging, this profitable company announced splitting in 1998. In 1992, two food competitors on the world market – Danone of France and Nestlé of Switzerland – joined in a former state-owned company Cokoladovny, the biggest manufacturer of chocolate, wafers and biscuits in Central Europe. It held 75 per cent and 55 per cent of the Czech market share in biscuits and chocolate respectively. The company wanted to grow and top management decided the only way to attain growth was through export. However, expanding on the foreign markets was difficult because Danone and Nestlé were competitors in these markets. A very unusual situation was created when this joint venture as a whole competed against the products of each joint venture partner. The split may create new mergers. The first merger may be Danone in the CR with its distribution subsidiaries in Poland and Hungary. The second merger might be Nestlé Cokoladovny and Nestlé Food, importing Nescafe to the CR. Both companies, Danone and Nestlé, not only sought their own strategies but also have a different organizational concept. The former organizes its activities by type of product while the latter organizes territorially.

Local demands have grown and some MNCs have described that growth as extremely rapid. The range of goods on offer is high and one no longer sees customers queuing for goods, as was the case so often in the past. The palette of goods is nearly as colourful as in developed countries and the variety of goods available is both a reflection of customer segmentation as well as an influence on it.

Some MNCs, however, have remarked upon a slow-down in demand in recent years. This is partly due to the growing intensity of competition, stimulated by the level of market saturation and by purchasing power. While for some goods (such as cosmetics and electronics) customer loyalty is low, in other goods (such as cars, consulting, health care, and banking), loyalty is much higher. Purchasing power is still up to five times lower than in Western Europe. A decrease in purchasing power recorded in 1997 and 1998 has influenced the volume of sales significantly. The slowdown is also caused by a rise in people's savings which are among the highest in the region, and higher than in Western Europe.

MNCs in the CR are usually successful. This success can be attributed both to their strategic planning and to the timing of their entrance into the Czech market. Also, MNCs benefit from economies of scale. The majority of MNCs have plans to expand and hire their labor force from within the Czech Republic.

Product and service strategies

Not many products, and even fewer services, need modification for the Czech Republic. Most MNCs try to leverage their strong global products and brands in the Czech Republic In certain cases, some slight tailoring is needed to meet local needs. Some MNCs use only global brands, while others make use of brands developed domestically, for example VW's retention of the Skoda name. Examples for product/service strategy include: 'international reach, local contacts', 'from local player to a world-wide player', 'the most expansion of our global brands'. Mostly, MNCs in the CR want to broaden their market and pursue market share with a limited number of brands (approximately four to nine). For example, VW, and in some cases, Colgate-Palmolive, pursue market share by blanketing the market with a wide range of brands of its products in several main divisions, with subcategories within each brand. No wonder that MNCs focus on advanced technology products backed by their well-established brand names and quality. Some MNCs have designed and developed special products for the Czech Republic, some have marketed only globally standardized products, while the rest have a mix of global and specific. For example, Daewoo's philosophy abroad is to build up an awareness of its trademark and to dispel the stereotype of 'cheap Asian' products.

Successful examples of products/services introduced by MNCs in the Czech Republic have included:

- The launch of new models organized as a one week introductory event. All models took leading positions in their product segments.
- The product has been ahead for the last five years after significant change of product/service quality and distribution channels while keeping reasonable prices.
- A good marketing mix.
- Development of a new market segment and a spread of product further throughout C&EE.
- The right product, right client, right timing, right approach (good contacts to top management and proven local experience).

Unsuccessful product introduction generally occurs when prices are too high and when there is insufficient understanding of consumer behavior, preferences and sensitivities. While many MNCs believe they did not have any unsuccessful product/service, mistakes have been made. For instance, a West European MNC producing food products in the CR missed its success by avoiding a product test. This company concentrated its orientation more on the packaging than the product. MNCs should be very careful when considering their product/service portfolio, particularly in those branches that were underdeveloped until 1990 compared with developed economies, such as banking and insurance. Evidence of a high product/service quality must exist, as people are very cautious.

Activity location strategies

Because of the existence of subsidiaries in neighboring countries in the region, some MNCs' strategies supply only the local markets (such as Ernst and Young). Other MNCs' strategies include both local and export markets (for example Matsushita, and VW). If export is not actually present, MNCs often plan to develop it. This is because the local market is insufficiently large compared to demand in other markets, MNC can increase their capacity, make use of partnership and favorable country costs.

The CR is not widely used by MNCs as a source of natural resources or brainpower. Only exceptional Czech managers have the chance to assert themselves regionally and globally. The Czech market is largely a base for production and distribution centers. For instance, ABB Service Ostrava located in Moravia, a joint-venture between ABB and Elektronavijarny Vitkovice, produces components for electric equipment and provides special services, some of them taken over from the mother ABB, whilst GE Lighting Europe, a producer of lighting engineering, has established its subsidiary in the CR, which is oriented to sales.

Most MNCs plan to increase or expand their production capabilities. The need for close relations and good partnerships with local suppliers means that some MNCs pursue a strategy of vertical integration. Czech enterprises can function both as component producers for MNC's end products (some wrapped up materials or foodstuffs for McDonald's) and as finished goods producers (Matsushita).

Current activities of MNCs are concentrated primarily on supply and distribution, followed by the provision of manufacturing or service operations, and are involved to a lesser degree in design and development (for example, pharmaceuticals distribution, TV sets manufacture, washing machine repair, insurance provision, car components' design). South Korean Samsung, Hyundai and LG Group are represented only by their trade branches, while Daewoo runs factories all over C&EE. Taiwanese FIC began by selling its products on the CR market and decided to start production here. More than 20 per cent of Japanese companies provide production and more than 10 per cent of them cooperate with Czech production partners. Nearly 90 per cent of South Korean FDI is directed toward industrial production. R&D skills are also considered by MNCs in the country, even though to a lesser degree than production or distribution, as in the case of Microsoft or Daewoo. The Avia vehicle has been developed by local design engineers, assisted by English and Italian designers for the body design. Some MNCs have their own distribution channels (cosmetics), some use local affiliates (pharmaceuticals), others distribute through sales and service retail outlets. The reasons for choosing a particular activity location strategy are: market demand, free capacity and skilled labor.

Marketing strategies

While MNCs can broadly apply global marketing strategies, various adaptations are typically needed. The five key marketing success factors in the country are considered to be positioning, promotion, MNCs know-how and local employee training, product/service quality (including both brand reputation and reasonable price), and distribution. For example, a massive media campaign, well-known brand and low price/promotions, are considered as marketing success factors by an MNC selling mobiles in the country. Elements of standardized global marketing mix (product positioning, global brand names, advertising, promotion, distribution, selling methods, sales representatives, service personnel) are applicable from a very high to a medium extent, and packaging is applicable from a very high to very low extent, depending on the MNCs.

Mistakes are made when deciding which elements to include in a particular marketing mix. For example, many MNCs use direct marketing techniques which people are neither used to nor prepared for. An American MNC offering a wide assortment of products via direct marketing did not take into consideration the difference in market size between the US and the Czech market. While the company has 2.5 million distributors world-wide, its unrealistic target was seventy-five thousand distributors for first year operations in the CR, more than it had after five years of operations at home, and more than in Poland, which has roughly four times the population.

Czech customer needs and tastes are very similar to those of their Western counterparts. Furthermore, Czech consumers prefer goods made in Switzerland, the UK, the US, Germany, and the CR.[44] Consumers trust Western-made products more than those made in the East. At the same time, more than 50 per cent consider that the quality of domestic goods is high. In some sectors, such as food, habits have not changed yet and Czechs are very conservative in their tastes (for example they prefer beer to cola).

Compared with the early and late 1990s, the Czech consumer has begun to ask for more sophistication in product and marketing. Simple brand product advertising is not enough any more. Consumers stress quality, want more courtesy by merchandisers and after-sale services.

Since the CR market, including consumer preferences and purchasing power, is still developing, high quality market research is very important for the success of an MNC entering a transition economy.

Advertising strategies play an important role, in particular for MNCs in consumer goods. Even when multimedia is the primary advertising outlet, others are also used (business magazines, newspapers, exhibitions etc.) to ensure a complexity of advertisement. For example, Toyota Motor Czech started with a complex concept of public presentation of its brand based on proven advertisements in Germany. It included television and radio advertisement, billboards, road shows, brand presentation by dealers, journalist days, parties for VIP's, and competitions for the public. Toyota also presented itself as a general sponsor of the Miss CR competition.

The price of MNCs' products is high or even very high relative to local incomes, in spite of MNCs' considerably lower prices compared with those they set in Western Europe, North America and Asia. MNCs that market products such as equipment and vehicles, apply both financial and operational leasing to benefit end-users as leasing is economically effective for both seller and buyer.

Some products are positioned differently from other country markets. The reasons for this include: lower consumer purchasing power and the fact that standard products in the developed countries are considered premium ones in the C&EE region.

In situations when cash-flow is limited, leasing is a very successful way of offering purchasing opportunities and provides an additional service which adds to competitive advantage. Entrepreneurs, like MNCs, use leasing to make better use of operational capital or credit sources. While operational leasing belongs to extended products, it is at a starting point in the CR. Leasing belongs to a very dynamic sector of the Czech economy, and a quarter of all material investments are financed by leasing. Leasing now plays a similar role in the Czech Republic as in Western economies. Its volume has become higher in the CR in 1998 than in Norway, Denmark, Finland, Luxembourg or Greece. However, the volume of real estate leasing is incomparably low (only slightly above one per cent of total leasing operations in the country) despite an extraordinary high demand for entrepreneurs and individuals. More than 200 leasing companies operate in the CR. Leasing of selected commodities and brands is provided only by production companies. Even when MNCs can afford leasing (mainly leasing of technology and transport vehicles), the volume of leasing provided by Czech companies is higher. Private service sector shares makes up more than 45 per cent of leasing volume in the CR. The commodity breakdown is about 60 per cent cars, 20 per cent machines and technology lines, less than 15 per cent trucks and vans, around five per cent office equipment (PCs, copiers and so on). Among the top ten leasing companies in the Czech market are SkoFIN of VW (established in 1992) and Mercedes-Benz Leasing Bohemia, (established in 1995), both with 100 per cent German ownership.[45] According to market research 53 per cent of the Czech population is ready and willing to lease.[46]

MNCs also provide special incentives such as lower priced products, flexibility of payment, fixed service time, product upgrades, preloaded software, free delivery, bonus schemes, after-sale services, model presentation and other promotion activities. Even though it is important, competitive pricing is only one incentive provided by MNCs.

Competitive move strategies

Competition has been increasingly intense for MNCs in the CR and this trend will continue. They face a mixture of foreign and local competition. Local competitors include other MNCs in the same sector and other foreign companies, as well as local large, medium, and small, privately owned firms. Primary competitors are both foreign MNCs (American, European, Japanese, Korean, Taiwanese) and local companies. The automobile market is particularly competitive. Companies need a good strategy to be successful. They predominately use the advantages of low prices, leasing, and supplementary kit as competition instruments. Evidence suggests that competition will continue to be strong among both foreign and domestic companies.

Overall Organization and Management Approaches

The CR's high levels of education and development, and cultural and historical affinity with the West, all minimize the need to adapt MNC organization and management approaches. But management deficiencies accumulated in the Soviet years require significant upgrading.

Despite a relatively flexible labor force and a good knowledge base, MNCs are making significant efforts to improve management processes and skills. Some of them direct attention to experienced managers, others to young people who have not been influenced by any previous practice and who are trained in the most up-to-date way possible, often in the MNCs' home country. Poor internal communication, bad organization and an inability of managers to delegate to subordinates are considered as the major problems in Czech companies. Generally, MNCs always provide training programs for local employees for both knowledge enrichment and carrier development (such as Procter & Gamble, and Coopers & Lybrand). MNCs take care with human resources because labor is an inhibiting factor for company development. But this idea does not mean that human resources are inhibited in applying the strategy of any particular MNC.

If MNCs have established a joint venture with a local partner, it used to be the case that the partner would be engaged in the reorganization. Reorganization applies to factors such as new cultural implementation, new departmentalization, change of HQ, the finance of strategy redesign, staff restructuring and effective training, IT implementation, growth adjustment, productivity increase, incentives modification.

The regional business headquarters office of an MNC incorporates different functions. Marketing and regional sales often operate regionally, but carry out financing only in some cases. Several MNCs set up headquarters in Prague to supervise all Central and Eastern European business. The country managing director for the CR is responsible for overall country operations, including sales and office profitability. MNCs do not pre-stock in large quantities because destinations to West Europe or to another regional manufacturing subsidiary are relatively small. From a management point of view, they centralize all inventory and warehousing in the CR into one or a few centers.

MNCs have globalized, or at least regionalized, their top management teams in the region. Some MNCs implement a policy of managers circulation between developed and less developing countries where they can pick up new ideas or new products. The time intervals are about two to six years. Many foreign executives, mainly those from long-distance countries such as the US, Canada, the Far East and Australia are leaving to be replaced by local talent. The majority of foreign investors entrust management to Czech managers in the country and more than a third of such investors do not employ any foreign labor. As surveyed by foreign investors entering the CR through CzechInvest agency, 68 per cent of such companies were run by a Czech managing director and 44 per cent had no full-time foreign staff in Czech plants. Skoda Auto and Unilever are two examples of those companies that have used an in-house management team. Wienerberger Baustoffindustrie, an Austrian MNC with 160 enterprises in 21 countries, producing bricks and concrete tiles, while having its German market as its Number One, considers its activities in the CR and other Central and Eastern European countries profitable. The company does not have Austrian managers in the CR but assumes the production should be managed by Czechs because the customers are Czech.

The number of local managers occupying top management positions will increase with time. However, Czech management may not be ready yet in such sectors as electronics and telecommunications and in companies with their own business philosophy. A range of MNCs implemented special attempts to improve personnel relations. Teamwork is necessary, constantly and to such a degree that it became a feature of MNCs operating in the CR. There are about several hundred foreign firms in the CR, each with its own personnel policy.

The presence of foreign firms is fruitful not only in an economic sense. They bring new opportunities to the Czech environment – modern products, new jobs, up-to-date know-how, new attitudes to problem-solving, a chance to learn how to understand other cultures and accept them, new working style and morale.

Production/operations efficiency and personnel productivity is ensured by standard measures, by monitoring personnel capabilities, by a range of both tangible and intangible incentives, by regular reporting and assessment. Managers, but mainly employees, are involved in training courses. Some courses are served locally, some internationally.

In many ways native Czech culture is not so different from other European countries. Historically, Czech people have been influenced by Austrian and German managerial practices, which concentrated on discipline and administration. They also like to present their orientation towards the US as a symbol of economic prosperity. The mentality of Czechs can be characterized by a love of improvization, cautious interaction, mistrust and fear of uncertainty, a tendency towards social relations, informal relations and informal groups, an emphasis on working positions and the powers derived from them, low self-confidence, diffusion of work and leisure time and an individual attitude to their work.[47]

Conclusions

The Czech Republic is a relatively small country in Central Europe with a well developed industrial base, advanced agriculture and a relatively well operating infrastructure. In the first years of the reform, 1990–96, the transformation process was one of the most extensive here of all reforming C&EE countries. Key parts of industry have been privatized, the orientation of external trade has been shifted from east to west, trade has been liberalized, the convertibility of the Czech currency has been realized, a banking sector has been established and important steps have been taken towards admission into the EU, including NATO membership. From a political point of view, the CR is a democratic and stable country. One of the most rapidly transforming countries in the early years of reform, economic development slowed from 1996, and a deficit of external trade and of state budget appeared. The value of the national currency, stable for several years, has been slightly devalued. The challenges facing the Czech economy are partly the result of the difficulties of privatization, the bankruptcy of several banks and insufficient restructuring of industry.

MNCs considering the future view the CR becoming a standard European country in the style of Austria, Belgium, the Netherlands or Denmark. They see that it has limited natural resource potential which is, however, not yet utilized to its full capacity, limited human resource potential which is highly economically active and not utilized fully, and limited market potential which is also not yet fully saturated. The Czech Republic hopes to ensure and sustain economic growth and play an important role in the globalization process. Its success will depend on political stability, government policy, enterprise development and the response of individuals.

There are a number of advantages for potential investors seeking to expand in Europe to locate in the CR. Centrally located between Eastern and Western Europe, a company can take advantage of not only the 360 million person EU market but also at least the 90 million CEFTA market and the 110 million former Soviet Union market. Strategies implemented by MNCs in the CR were principally successful and are still being applied. MNCs in a huge range of industries have tried to incorporate technological transfer, human resource development strategies and multicultural philosophies, into their overall corporate/regional strategies.

The CR is an attractive place for MNCs' foreign investments for the following reasons:

- An advantageous position in Central Europe – on the borderline between the former East and West, offering accessibility to nearly 500 million customers in C&EE and EU countries.
- A highly qualified work force in practically all industry sectors and good R&D capabilities.
- Low wages in a comparison with the West European countries, even though higher than in other C&EE countries.
- High adaptability of Czech management to new conditions.
- A national culture which is very similar to West European neighbors.
- A friendly business environment for foreign capital investment.

- A history of good experience of investors from Western countries (for example, VW, Nestlé) and of strong international investments.
- Contacts of Czech managers with the markets of the former CMEA which continue to exist.
- Established infrastructure and supply base.
- Rapid integration into the global business community.

There exist a range of specialized organizations, such as CzechInvest, CzechTrade, Chambers of Commerce, Industrial and/or other Associations, Management Consulting Firms (for example, the Management Focus International Consulting Group) that help foreign investors during their initial entry into the Czech market.

Notes

1 Czech Business and Trade 1/1997, Prague/ 'Economic Environment and Investment Opportunities' by J.A. Havelka. CzechInvest is the Czech national agency for foreign investment. It was established in 1992 to promote the CR to potential investors worldwide, to facilitate the inflow of FDI and to assist foreign investors in establishing and exporting their operations in the CR.
2 See basic economic indicators in the Introductory Chapter.
3 Data of 2nd quarter 1997, economic daily *Hospodadárské noviny* (Praha: 6 October, 1997).
4 1st–3rd quarters, 1996.
5 Data from Fall 1997, dealings with strategic partners started in 1996.
6 P. Zahradník, Prague Securities for *Hospodárské noviny* (Praha: 5 November, 1998), p. 18.
7 Data by the economic daily *Hospodárské noviny* (Praha: 5 November, 1998), source EU, EBRD.
8 Data by the economic daily *Hospodárské noviny* (Praha: 17 December, 1998), source: Salomon Smith Barney.
9 Preliminary 1998 data by the magazine *Czech Business and Trade* (Prague: September 1998), pp 7–8.
10 *Business Central Europe* (Prague: July/August 1997).
11 Data by the *Czech Statistical Office* (Prague: 1998).
12 Data by the economic and business weekly *Profit* (Praha: 21 September, 1998).
13 Economic weekly *Ekonom* (Praha: 15, 1997).
14 Data as of 1996.
15 Data by the *Czech Statistical Office* (Prague), 1998).
16 *Czech Top 100*, www.medea.cz/top100, 1997.
17 Special enclosure by the CzechInvest Agency in business magazine *Czech Business and Trade* (Praha: 6/1998).
18 Consulate General of the Czech Republic, Los Angeles, USA (Embassy of the CR does not guarantee the comprehensiveness of the list).
19 Data as of 1998.

20 *Czech Business and Trade* (Prague: 5/1997), p. 11, source Czech National Bank, H. Pisková.
21 *Czech Business and Trade* (Prague: 5/1998), p. 11, source Czech National Bank, H. Pisková.
22 *Ekonom* (Praha: 44/1997).
23 *Ekonom* (Praha: 45/1998).
24 Rocenka HN 1998 (Yearbook) *Economia* (Praha: 1998), pp. 94–97.
25 Rocenka HN 1998 (Yearbook) *Economia* (Praha: 1998), p. 101, banks, data, and economic daily *Hospodárské noviny* (Praha: 11 November, 1998).
26 Data by The Prague Post 1998 Book of Business Lists, Prague, updated October 1997.
27 Magazine *Moderní rízení*, Praha.
28 *Business Central Europe*, 'Selling to Skoda' (cover story), August, 1999.
29 Data based on the survey run by the authors in 1997.
30 Statistical Yearbook 1996 of the CR, Prague.
31 *Cesky statisticky úrad* (the Czech Statistical Office), Praha.
32 Business Central Europe (Prague: May 1997).
33 *Ekonom* (Praha: 45/1997). The company is bemused that the Czechs are so enthusiastic, they even like to invent what has been already invented.
34 *CEER* (Brussels: February, 1999), p. 19, data source by Egon Zehnder International. The survey did not differentiate between locally hired and expatriate employees.
35 *Vseobecná encyklopedie* (General Encyclopedia), Nakladatelský dum OP Diderot, Praha: 1996, 1997).
36 There are 11.9 km of railway lines per 100 km^2 in the CR, 11.3 km in Germany, around 5 km in Italy, around 6 km in UK, Austria and France, 8.2 in Hungary and 7.8 km in Poland. Extent of the railway net in km lines per one thousand inhabitants is 0.91 km in the CR, 0.50 km in Germany, 0.28 km in UK and Italy, 0.70 km in Hungary and 0.63 km in Poland. Source: daily *Lidové noviny* (Praha: 24 November, 1997).
37 *Ekonom* (Praha: 46/1997), source: CzechInvest.
38 *Czech Business and Trade* (Prague: 3/1996), p. 7.
39 *Business Central Europe* (Prague: May 1997), p. 68.
40 *Business Central Europe* (Prague: March 1997), p. 41.
41 *Business Central Europe* (Prague: March 1997), p. 41.
42 Business Central Europe (Prague: March 1997), p. 42.
43 Toyota Annual Report, 1996.
44 The Prague Tribune (Prague: July–Aug. 1998), p. 13.
45 *Prague Business Journal* (Prague: 3/1997).
46 Economic weekly *Ekonom* (Praha: 45/1997). p. 50.
47 Based on empirical research, by I. Novy in *Czech Business and Trade* (Prague: 3/1998).

5 Slovakia – Seeking Alliances

JÁN MOROVIC

Overview

THIS chapter on Slovakia discusses the historical and economic environment in which the first multinational companies (MNCs) began to operate after 1989. The focus will be on a description of the political, economic and social conditions, and an analysis of individual aspects and motivations for possible activities of MNCs in Slovakia. After an analysis of macro-economic indicators, we introduce the MNCs operating in Slovakia. Their development, results and successes over the last few years may motivate the representatives of global and international organizations to enter the Slovak market.

History

Slovakia is a landlocked country in the heart of Central Europe, bordering Austria (91 km), Czech Republic (215 km), Hungary (515 km), Poland (444 km) and Ukraine (90 km).

Understanding Slovakia's historical background is an important factor when considering globalization drivers and key to understanding the reasons for its current position as a transition economy. For most of its modern history Slovakia was under the domination of other powerful neighbors. The capital of Slovakia, Bratislava, was the capital of Hungary between 1563 and 1830, during which period nineteen crowning ceremonies took place in Bratislava's St. Martin Cathedral. After centuries under foreign rule, mainly by Hungary, in 1918 the Slovaks joined with their neighbors to form the new nation of Czechoslovakia. Following World War II, Czechoslovakia fell under communist sway and became one of the Central European countries under Soviet political and economic leadership. When Soviet influence collapsed in the dramatic events of 1989, Czechoslovakia once again became an independent country with a realigned westward gaze.

The Slovak Republic came into existence on 1 January 1993, as a result of the division of Czechoslovakia. Even though it was carried out in a constitutional and politically civilized fashion, the division has had several negative effects on the Slovak economy. Except for limited independence during World War II, this is the first time that Slovakia has existed as an independent state. Slovakia is undergoing a process of transformation that complicates its integration into larger European and world structures. Recent political and economic developments are characterized by a certain degree of authoritarianism, which is partly a result of decreased living standards and limited social security for certain strata of the population. Furthermore, authoritarian tendencies disrupt the balance among the executive, legislative and judicial powers of the state.

While the Czech Republic was viewed as one of the 'Tiger Economies' and success stories of the emerging region, Slovakia has experienced rather more difficulty in developing a full-fledged market economy. Because its economy is based primarily on heavy industry and a predominance of large, previously state-owned companies, Slovakia has been rather slower in the transition process and is still grappling with many of the core issues of privatization and fundamental structural reform.

Population

Slovakia's population has been decreasing steadily over the course of this century. In the past ten years, the rate of natural increase dropped from 5 per cent in 1989 to 1.6 per cent in 1995, and a similar decrease is foreseen for future years. There are 5.3 million inhabitants in the Slovak Republic of which 86 per cent are Slovaks, 10.5 per cent Hungarians, 1.5 per cent Gypsies, one per cent Czechs, and the remaining one per cent other nationalities. The question of minority rights, essential for the integration of Slovakia into European structures, is also at the crux of the concurrent processes of identification and integration. In terms of religious groups, Slovakia is also diverse, predominantly Roman Catholic (60.3 per cent), with relatively large minority groups of Protestants (8.4 per cent), Orthodox (4.1 per cent), and others (17.5 per cent), as well as atheists (9.7 per cent).

Political organization

The Slovak Parliament – the *National Council of the Slovak Republic* – is a single-chamber body and consists of 150 members elected for a four-year period on the basis of reciprocal representation. The seats are filled by parties that received at least five per cent of the valid votes. Other alternatives are seven per cent in case of a coalition of two or three parties or ten per cent in case of a coalition of three to four parties. The large number of political formations (79 registered parties) testifies to the existence of a truly plural system in Slovakia. Parties that gain more than three per cent of votes can claim state funding. The President of the Slovak Republic can dissolve the National Council when a vote of confidence in the government fails three times within the six months following elections. Except for this eventuality, Parliament can be dissolved when two thirds of its members vote for dissolution. Elections in Slovakia have been free and just and most recently took place in September 1994 bringing into power a coalition government led by *Hnutie za demokraticke Slovensko* (The Movement for a Democratic Slovakia). The Slovak Republic is administratively divided into eight regions and seventy-nine districts. Each region is headed by public administration representatives. A law passed in July 1996 expanded the jurisdiction of these bodies and diminishing that of governments. The distribution of responsibilities between the decentralized levels of state government and local governments is not completely clarified. The Slovak administration does not have a public service law that clearly defines the rights and responsibilities of state employees, which in turn complicates the struggle against corruption.

The Constitution of the Slovak Republic was approved by the National Council in 1992. The Constitution is based both on national and civil principles. This has proven to be decisive for the democratic transformation of Slovakia, as it guarantees the rights of minorities and other ethnic groups in agreement with international standards accepted in Europe. The major principles of the Constitution are the principle of the rule of law, the principle of the sovereignty of citizens, the principle of threefold division of state power, the principle of the priority of legal standards and the principle of equality and the inviolable character of property.

The Transition economy

In terms of economic development the former Czechoslovakia was ranked among the ten most developed countries in the world in the period 1920–30. The socialist era of the *Council for Mutual Economic Help* (COMECON) specialization overloaded Slovakia's heavy metal production and processing industries. This put a serious and destructive strain on the environment. In the economic sphere, the last five years of development brought a gradual separation from the Czech Republic. After 1993, Slovakia had to find a concept of its own economic strategy and to consider the specific conditions of the country. The separation was followed by a period of stabilization that prevented devaluation, inflation and preserved the requirements for economic growth. This was partly brought about by the restrictive policies of the *National Bank of Slovakia* (NBS).[1] Another key point of economic development was the introduction of import and exchange charges. It also increased the competitiveness of Slovak exports that started to be reoriented towards the markets of developed countries.

It is possible to say that, by the beginning of 1997, the Slovak Republic had successfully completed the introductory stabilization period of the transformation process. Recently the focus of the transformation process has shifted from the macro-economic to the micro-economic level. The economy is going through the intensive process of restructuring associated with a decrease in its performance and with the social dimension of the country's democratic transformation. The main macro-economic indicators for Slovakia are illustrated in Table 5.1.

Gross domestic product (GDP)

The loss of traditional export markets caused by the disintegration of the COMECON and the disruption of trade relations with the Czech Republic has led

Table 5.1 Macro-Economic Indicators for Slovakia

	1990	1991	1992	1993	1994	1995	1996	1997	1998
Nominal GDP ($bn)	15.5	10.8	11.7	12.0	13.8	17.3	18.8	19.4	20.4
GDP per capita PPP ($)	n.a.	n.a.	n.a.	n.a.	7260.0	7990.0	8780.0	9540.0	10050.0
GDP (% change)	−2.5	−14.6	−6.5	−3.7	4.9	6.9	6.6	6.5	4.4
Industrial production (% change)	−4.0	−19.4	−9.3	−3.8	4.8	8.3	2.5	2.7	5.0
Budget balance (% of GDP)	n.a.	n.a.	n.a.	−7.0	−5.2	−1.6	−4.4	−5.7	−2.7
Unemployment (%)	0.8	0.0	4.8	12.2	14.8	13.1	12.8	12.5	15.6
Average monthly wage (US$)	178.7	127.7	160.6	175.3	196.4	241.9	266.0	274.4	283.9
Inflation (%)	10.4	61.2	10.0	23.2	13.4	9.9	5.8	6.1	6.7
Exports ($bn)	5.8	3.3	3.7	5.5	6.7	8.6	8.8	8.3	10.7
Imports ($bn)	6.5	8.8	6.9	6.4	6.6	8.8	11.1	10.3	13.0
Trade balance ($bn)	−0.7	−0.5	−0.6	−0.9	0.1	−0.2	−2.3	−2.0	−2.3
Current account balance ($bn)	n.a.	n.a.	n.a.	−0.6	0.7	0.4	−2.1	−1.9	−2.1
Foreign direct investment flow ($m)	18.0	82.0	130.0	200.0	309.0	304.0	295.0	197.0	330.0
Foreign exchange reserves ($bn)	1.1	3.2	0.4	0.5	1.7	3.4	3.5	3.3	2.9
Foreign debt ($bn)	2.0	2.7	3.0	3.4	4.3	5.8	7.8	10.7	11.8
Discount rate (%)	n.a.	9.5	9.5	12.0	12.0	9.8	8.8	8.8	8.8
Exchange rate (/US$)	18.0	29.5	28.3	30.8	32.0	29.7	30.7	33.6	35.2
Population (m)	5.3	5.3	5.3	5.3	5.3	5.4	5.4	5.4	5.4

Source: Business Central Europe Magazine, The Economist Group, December, 1999. http://www.bcemag.com

to a steep economic decline, which caused the GDP to drop by one quarter in the years 1989–93. In 1991, at the beginning of the economic reform the GDP decreased by 14.5 per cent, which had an extensive impact on the living standards of the country's population. In 1993, the GDP decreased again by four per cent. The period of decline in the country's economic activity came to the end after a spectacular turnaround in 1994. This was mainly caused by the increase of export to EU countries and the Czech Republic. An impressive macroeconomic performance was achieved in 1995 when GDP increased by 7.4 per cent. An increase of GDP by 6.9 per cent was achieved in 1996. In 1997, according to the preliminary Slovak Statistics Office's figures, the real GDP has been continuously increasing, with a private sector share of 76.8 per cent. The service and construction sectors were mainly responsible for the increase, with higher domestic consumption also contributing to this growth.

In 1999, the government expects GDP growth to slow from five per cent in 1998 to two per cent, inflation to rise from six to ten per cent, and unemployment to rise from less than 14 per cent to 15 or 16 per cent, but it hopes to bring the budget deficit down to no more than two per cent of GDP and the current account deficit down to between five and six per cent of GDP.

Inflation

The main factors causing the higher inflation rate are higher prices in the health care sector (14 per cent), prices for recreation and culture (9.4 per cent), prices for accommodation, water, electricity, gas and other fuels (9.3 per cent) and also prices for clothing and footwear (8.4 per cent). In 1997, the rate of inflation in the SR has grown for the first time since 1993. Nevertheless, the rate is still one of the lowest among the post-communist countries. The aim of the NBS for 1997 was to keep the rate of inflation below 5.9 per cent.

Legal environment [2]

A special investment tax and other incentives are available in certain regions and industries. All restrictions on entering into import contracts have been abolished.

The following recommendations were made to improve the venture capital industry in Slovakia:

- Increase the personal liability of founders for companies with sufficient starting capital;
- Establish a residential mortgage system to allow entrepreneurs to have at least a little independent starting capital and to take more personal risk and responsibility;
- Define venture capital funds activities including lending money and buying back their own shares;
- Allow convertible or subordinated loans, as well as rules needed to be accepted and enforced by courts for rapidly collecting collateral debt.

Some restructuring rules are needed to permit buying the equity of loss-making companies at a market price that can be below par. Also the rights and fiscal responsibilities of large and small shareholders need more clear definition, especially protection for minority holders.

Fiscal environment[3]

The Slovak government and the Slovak National Bank put in place extremely conservative monetary policies and expect to continue along those lines in the future. This has resulted in favorable macro-economic achievements. The Slovak currency has been internally convertible since 1991. Currently all legal entities must convert hard currency earnings into the local currency, but they may purchase hard currency for payments abroad. The tax rate for corporations is 40 per cent and for individuals it is graduated from 15 to 42 per cent. There is a 15 per cent tax for capital gains and dividends.

The most viable strategy for exit remains the trade sale to already existing partners or to an outside strategic partner. At the present, the stockmarket does not represent a viable alternative, as it is relatively new and inexperienced and the lack of liquidity in the marketplace is reflected in the capital markets. Currently there are two stock exchanges operating in Slovakia: the Bratislava Stock Exchange, which is a traditional stock exchange also trading in credit notes, participation certificates and voucher privatization coupons, and the RM System of Slovakia, which manages a secondary futures market.

Foreign trade

Foreign trade is an important part of the transition process towards a systematic integration with the global economy. This requires the transformation of previously government-led international trade into a free market. Trade liberalization was started in 1991 and an association agreement initiated formal, legal and economic relations between the EU and Czechoslovakia. Slovakia is also a member of GATT, EBRD, WTO, CEFTA and the IMF and there is also a customs union with the Czech Republic.[4]

The 1996 foreign trade deficit reached SK (Slovak crowns) 55 875bn after the collection of the old foreign debts. Exports increased only by 6.1 per cent and a significant revival in 1997 was not expected. On the contrary, imports increased 28.5 per cent. The introduction of import deposits in May 1997 brought a decline in total imports of 8.4 per cent. This represents a decrease of SK2.6bn as compared with May 1996. The import deposits had the most serious impact on the wood and textile industries.[5] Growth of Slovak international trade for 1997 to September reached SK482 744m, an increase of SK45 858m or 10.5 per cent in comparison with the same period of the previous year. The value of exported goods amounted to SK221 754m representing a 12.2 per cent increase in comparison with 1996. Imports to Slovakia between January and September 1997 reached SK260 990m, an increase of SK21 694m or nine per cent over the same period in the previous year. The overall 1997 goods and services exports show an annual increase of 6.1 per cent and imports were reduced by 2.3 per cent. The negative balance of the Foreign Trade amounted to SK27.7m during the period ending with the 3rd quarter of 1997, which is a reduction of 5.5 per cent compared with the previous period of the 1st quarter to the 3rd quarter of 1996.[6]

In 1996 the foreign trade turnover was US20.19bn and the foreign trade deficit was US$2.15bn. Slovakia's foreign trade deficit with Russia and OECD countries has been reduced and there was a surplus with CEFTA countries, mainly with Hungary, Poland, the Czech Republic and Slovenia.

Slovak exports reached a value of US$9.02bn. The main export countries were the Czech Republic, Germany, Austria, Ukraine and Poland. The main export products were consumer goods, machines and machine equipment, industrial products, chemicals, raw materials, natural fuel and foodstuffs.

Slovak imports reached a value of US$11.17bn where the largest importers were the Czech Republic, Russia, Germany, Italy and Austria. The products that made up most of Slovakia's imports were machines and machine equipment, natural fuels, consumer goods, chemicals, industrial products, foodstuffs and raw materials.[7]

The market-oriented economic transition is closely linked to the liberalization of foreign trade and the formation of a competitive environment. According to the Commercial Code adopted in 1991, the foreign trade monopoly of the state was completely abolished. The Commercial Code which came into force in January 1992 permits all entrepreneurs (Slovak natural and legal persons and foreign companies as well) registered in the Trade Register to be engaged in foreign trade activities.[8] The further liberalization of export and import is one of the main principles of the Slovak Republic's foreign trade policy. This principle, however, is optimized through the utilization of market-compatible instruments. It means to promote export as well as the indispensable protection of the domestic market and domestic producers, all in conformity with valid documents of the WTO, GATT, the European Agreement with EU, the Agreement with EFTA (European free trade association) and CEFTA (Central European free trade agreement). The application of market mechanisms ensures the availability of equal enterprise and market conditions to all the entrepreneurial subjects on a domestic and foreign scale.[9]

Domestic companies

A strong tradition of industrial production exists in Slovakia. Over the last 80 years there has been a substantial development in the food processing industry, in particular in the beer, dairy and meat processing industries. In addition to heavy industry, the chemical industry is also among the growing sectors. The production of famous BATA footwear can also be included to the list of top companies, listed in Table 5.2.

Table 5.2 Largest Domestic Companies Active in Slovakia

Company	Yields (SKm)	Employees
VSZ Kosice jsc & subsidiaries	33 420	20 000
Slovak Gas Industry	33 040	5140
Slovak Electricity Ent.	28 210	12 620
Slovnaft Bratislava	28 000	5920
Petrimex	19 790	270
Kerametal	18 840	240
Sipox Ltd & subsidiaries	16 000	7200
Railways of SR	11 710	58 160
Western Slovakia Electricity Ent.	10 820	3820
Central Slovakia Elec. Ent.	10 290	3700
Slovak Insurance Co.	9600	2910
General Credit Bank	7890	6690
Slovak Telecom	7680	15 490
Martimex Martin	7390	200
Benzinol Bratislava	7340	n.a
Eastern Slovakia Elec.Ent.	6840	2160
SNP Enterprise, jsc(Aluminium works)	6820	5700
Matador Puchov	5380	3990
Northern Slovakia Paper Works	4020	3500
Exocel Bratislava	3800	100
Zdroj Kosice	3700	2770
Slovak International Tabak	3600	1500
Hydrostav Bratislava	3410	7390
Duslo Sala	3352	3150
Western Slovakia	3200	2120

Source: TREND Magazine, 1993.

There are two main groups of Slovak companies:

1. A group of small and medium sized enterprises, most of which are start-ups and looking for debt financing. Most of these enterprises cannot build a viable business without an equity injection. The access to funding is limited due to monetary restrictions imposed by the central bank. Successful businesses that were established shortly after the 'velvet revolution' in 1989 are also included into this category. These enterprises are seeking investment to secure further growth and have initiated efforts to attract potential investors. In this group it is more likely that the foreign partner will gain control over the whole joint venture.
2. Larger firms which were privatized or are undergoing privatization. Many of these businesses have attracted foreign investors which support company expansion, modernizing or restructuring. Most of the investments are in the service, banking and insurance sectors.[10]

MNCs in Slovakia

MNCs influence significantly the overall economic development and market structure The introduction of MNCs in Slovakia contributed to the growth of competitiveness and also represented a substantial driver towards globalization and participation in the world economy. That MNCs entered a market simultaneously as a network rather than as an individual company was of particular interest to some of the larger companies in Slovakia. They bring new technology, help with restructuring, show a new perspective on human resource development. MNCs are also able to utilize the conditions of economic transition to achieve fast and efficient growth. Indirectly they also support the development and advancement of other national companies. Appendix A reports on our research on related industries in which MNCs participate.

Foreign direct investment (FDI)

After 41 years of a communist planned economy, the Slovak Republic is developing a market economy and has attracted foreign investment of over US$850m.[11] Although the amount of foreign investment has been slowly growing, this amount is still not satisfactory due to uncertainty about the Slovak government and the process of privatization. The difficulties with obtaining relevant information, the bureaucracy and existing laws have also slowed down foreign investment.[12]

Table 5.3   Development of FDI in Slovakia

Year	Input (SKm)	Growth (SKm)	Growth (%)
1991	1186.00	–	–
1992	6607.10	5421.10	557.0
1993	10755.70	4148.60	162.8
1995	16542.40	5786.70	153.8
1995	21881.70	5339.30	132.3
1996	42300.00	20418.30	193.3
1997	52763.60	10453.60	124.7

Source: Federal Bureau of Statistics, Bulletin of The Bureau of Statistics of the Slovak Republic, Slovak National Bank.

At present, due to the lack of domestic financial support, the investment level does not correspond to the need for restructuring of the economy. Unfortunately, none of the investments have been strategic and sufficiently large. The largest foreign investor is Austria, followed by Germany, the Czech Republic, the United States, the United Kingdom, France, the Netherlands, Sweden and Italy. Table 5.4 shows which countries have regularly been the largest investors into the Slovak Republic.

Table 5.4 Foreign Direct Investment into Slovakia by Country

Country	(% of total)
Austria	22.0
Germany	20.1
Czech Republic	15.5
United States	10.2
United Kingdom	6.2
France	6.0

Source: Statistical Yearbook of the Slovak Republic 1996 – Slovak National Bank, Bulletin of the Statistical Bureau of the Slovak Republic. Business Central Europe Magazine, The Economist Group, http://www.bcemag.com

Table 5.5 shows the largest FDI investments made by companies from different countries and it also lists the sectors into which these investments were made. The largest investments were made in the industrial sector, followed by wholesale and retail businesses and the automobile industry. And Table 5.6 shows the distribution of FDI by sectors of the Economy.

Table 5.5 Main Foreign Companies Investing in Slovakia

Company	FDI (SKm)	Country	Sector
TESCO STORES	1–2	Great Britain	Retail Trade
Volkswagen Bratislava	1–2	Germany	Cars
SLOVALCO	0.5–1	Netherlands	Aluminium
Chemlon	0.5–1	France	Synthetic Fibres
Novácke Chemické Závody	0.5–1	Czech Republic	Plastics
Coca-Cola Amatil Slovakia	0.5–1	Netherlands	Soft Drinks
HYPO-Bank Slovakia	0.5–1	Germany	Banks
OMV Slovensko	0.5–1	Austria	Fuel
Tatransky Permon	0.5–1	Czech Republic	Hotels
Credit Lyonnaise Bank Slovakia	0.5–1	France	Banks
CITIBANK Islovakia	0.5–1	Great Britain	Banks
Bank Austria (SR)	0.5–1	Austria	Banks
KOMERCNI BANKA Bratislava	0.5–1	Czech Republic	Banks
SHELL Slovakia	0.1–0.5	Great Britain	Fuel
FERMAS	0.1–0.5	Germany	Pharmaceuticals
HENKEL-PALMA	0.1–0.5	Austria	Detergents
Molnlycke	0.1–0.5	Sweden	Paper, pulp
Wienderberger-Slovenske tehelne	0.1–0.5	Austria	Bricks
Prvá Stavebná Sporiteľňa	0.1–0.5	Germany	Banks
SAMSUNG-CALEX	0.1–0.5	Korea	Home appliances
PEPSI-COLA SR	0.1–0.5	Netherlands	Soft drinks

Source: Bulletin of the Statistical Bureau of the Slovak Republic, Slovak National Bank.

Table 5.6 Distribution of FDI in Slovakia by Sector

Sector	1994		1995		1996		1997	
	SKm	%FDI	SKm	%FDI	SKm	%FDI	SKm	%FDI
Total:	16 542	100.0	21 888	100.0	42 340	100.0	52 764	100.0
of which								
Industrial production	7810	47.2	9547	43.7	20 100	47.5	23 501	44.6
Construction industry	227	1.4	2378	1.1	1300	3.1	1300	2.5
Business	5156	31.2	7091	32.4	7600	18.0	10 200	19.3
Hotels & catering	534	3.2	536	2.4	600	1.4	400	1.3
Transport & communications	80	0.5	85	0.4	900	2.1	1800	3.4
Finance & insurance	1818	11.0	3430	15.7	10 840	25.6	13 563	25.7
Other business services	820	4.9	745	3.4	800	1.9	1400	2.6
Other public services	98	0.6	211	0.9	200	0.4	300	0.6

Source: Bulletin of the Statistical Bureau of the Slovak Republic, Slovak National Bank.

Turbulent political changes are creating uncertainty for the economic development within individual transition countries. The complex political democratization process is often difficult to separate from economic development. Hence, highly skilled and strategically thinking managers are needed at the forefront of industry restructuring. Multinational companies have been very supportive in helping local companies to develop particular strategies thus benefiting the Slovak economy. There are many examples of how MNCs have been able to utilize local advantages by using the high technical skill of local labor and combining it with regional networks. Slovakia's growing economy, low production costs, low labor costs and stable legal environment, can be advantageous to MNCs. Furthermore, the population is willing to learn and to work hard and there is a cheap and educated workforce. On the other hand the influence of politics on business decisions, the unclear economic policy of the government and the unstable political situation are discouraging factors. The limited market size also has a negative effect on integration of Slovakia into the global market. Table 5.7 on the following page illustrates the key investors in the Slovak economy.

Venture capital

In October 1993 the European Venture Capital Association (EVCA) set up a private equity support program for Central and Eastern Europe. This step was fundamental for the construction of the infrastructure of private equity capital in five countries of Central Europe. Because of the lack of experienced entrepreneurs in Slovakia there has been an urgent need to create and develop a culture which will support entrepreneurial activities. This culture should include such activities as flexible strategic planning, marketing, accounting, discussions between managers and shareholders, and management by objectives. All these

Table 5.7 Largest Investors in Slovakia

Company	Sector
K-Mar	Retail
Volkswagen	Automotive
Rhone Poulens	Chemicals
Novácke Chemické Zálvody	Chemicals
Tatransky Permon	Tools
Degussa	Pharmaceuticals
Henkel-Palma	Detergents
Molnycke	Health/Hygiene
Samsung	Refrigerators
Pepsi-Cola	Beverages

Source: Jan Morovic, Venture Capital Slovakian Studies, Interim Report, September 1996.

activities, which are standard practice in the West, are still in their infancy in Slovakia. Instead the present entrepreneurial culture is based on personal contacts.[13]

The venture capital business in Slovakia is very much in its beginnings. The first venture capital investment fund was the Slovak American Enterprise Fund (SAEF)[14] whose activities began in 1991. In recent years a handful of funds have indicated an interest in investing in Slovakia. In 1994 the majority of equity investments in this country were provided by SAEF. SAEF is the most active investor in small and medium sized, enterprises. The current portfolio consists of more than 30 investments involving a total commitment of approximately ECU20m. The second enterprise fund is the Povazsko Kysucky Enterprise Fund (PKPF). This is a regional investment fund that was established by the Slovak government and the European Union. In 1994, the PKPF was active mainly in Northern Slovakia. Additional members of the venture capital sector are the Seed Capital Company, Tatra Raiffeisen Capital (TRC) and the Tyn Group of Funds. The Slovak Venture Capital Association (SLOVCA) was formed in 1995. Its mission can be summed up in the following points:

- Represent the interests of its members when dealing with the government and other institutions;
- Provide information to those seeking capital for new and existing companies;
- Provide a forum for networking and the communication of views and practical know-how among its members;
- Provide education and training for its members;
- Encourage the highest standards of business practices.

Capital markets

The capital markets were established not as a result of the enormous stock offers from the privatization process. The organized markets in Slovakia are: The

Bratislava Stock Exchange, a market for brokers, The Bratislava Options Exchange, for options and derivatives, and the RM System Slovakia, where all citizens can trade their securities acquired through the voucher privatization. Banking has undergone a very substantial reorganization and is internationally a new sector. In the future, development banks will be playing an important role in the investment finance sector. The Slovak National Bank give licenses to a number of international banks that are already active in Slovakia: Volksbank, Bank Austria, Ing Bank, City Bank, CSOB, etc. A current hot issue is the bank privatization, which will substantially open up the international market for Slovakia. Historically the following domestic banks have been active: Vseobecna Uverova Banka (VUB), Slovenska Sporitelna, PolnoBanka, DevinBanka, IRB, Komunalna Banka.

Country Globalization Drivers

Because of its small size and lateness in transition, Slovakia's globalization drivers are, on average, only moderately favorable to date.

Market globalization drivers

The new high-income class is just being born, creating a demand for special high-level services and luxury products. Although it has a small population of only five million, Slovakia's market globalization drivers have increasingly moved in favor of foreign MNCs. The gradual increase in the income of Slovak consumers, and the concurrent reduction of trade barriers to foreign imports, have created favorable opportunities for foreign businesses in Slovakia. The modernization of Slovak preferences and the change in consumption patterns have also led to a significant increase in Slovak imports of manufactured goods. Due to expanded opportunities for education in foreign countries, overseas travel and multicultural contacts, the younger, more affluent and internationally oriented generation of Slovak consumers are moving in favor of high-quality, Western goods.

Product and service strategies

In terms of product and service strategies, MNCs have discovered that their products and services are eagerly sought after in Slovakia. Since consumers are beginning to increase their consumption of Western goods, MNCs can offer a wide range of products and services. Consequently, MNCs can introduce their global brands with little modification.

Cost globalization drivers

Slovakia has a number of favorable cost globalization drivers.

Favorable logistics and infrastructure

A favorable location in the center of the C&EE region is an important globalization driver of Slovakia. From the transport point of view future fast solutions provide the basis for setting up elements of international transportation networks either by air, rail or by road transport. There are also advantages in terms of a good starting level for IT development because of the practical skills of the local labor force.

As a landlocked country, without the natural advantages of maritime claims and coastline, Slovakia must take extra care to develop its infrastructure. The level of its infrastructure will inevitably have a direct impact upon its effectiveness as a facilitator of international business activity. Failure to cope with the necessary pace of construction of highways has been one of the destabilizing effects from which the Slovak economy has suffered as a whole. Slovakia needs to make better use of its potential based on its geographic position. This is true of both of river, air and road transport, which will also help Slovakia in international relations.

The fact that Slovakia borders five other Central and Eastern European countries, may well work to its advantage as a regional transport hub. However, Slovakia will not become a key player in the region until the development of its infrastructure is exposed to international competition, made transparent, lowered in cost and offers international standards of reliability.

Favorable country costs

Favorable country costs are the key globalization drivers that Slovakia can offer as advantages for MNCs and foreign investors thinking of entering the region.

Labor

After location, the availability of a skilled and low-cost labor force is a strong investment incentive.

Slovakia has a well-educated and skilled labor force of about 3.32 million (1997). Literacy in Slovakia is universal, and most workers are well-educated and technically skilled. Overstaffing is still typical at many Slovak companies, especially larger and previously state owned-enterprises. Many professionals are still relatively poorly paid by US standards. Slovakia has a relative shortage of workers with foreign language and Western-style management skills, and salaries for such workers are correspondingly high. Foreign companies frequently praise the motivation and abilities of younger workers, who often have good language and computer skills and have traveled in the US or Europe. Table 5.8 shows the wage relationship between education and labor in Slovakia. A more detailed examination of the labor structure and legal requirements of employment in Slovakia is given in the Appendix at the end of the chapter.

The low cost and high education level of the labor force are important factors

Table 5.8 Wage–Education Structure for Slovakia

Education	Percentages	Average wage (SK)
Post-graduate	17.77	10 184
Graduate	13.74	9203
Secondary (with A-Levels)	32.27	7751
Secondary (without A-levels)	19.94	7351
Basic	15.36	6706
Other	0.92	8620

Source: HELP Foundation, Project Fernina – Final Report, April 1997.

in economic development as they contribute to the competitiveness of the Slovak market and help to attract foreign investment. However, the population has also had to cope with standards of living that have decreased by more than 30 per cent and extensive property differentiation. These and other social problems including a high unemployment rate in some regions of Slovakia, can lead to dissatisfaction with economic policy and government, which could complicate the democratization process.

Sourcing efficiencies

The energy sector has a special role in the development of the country. There are two nuclear power plants and a number of different water-energy dams. Once all power stations are fully utilized, Slovakia will not only be self-sufficient but will also have a surplus of electrical energy. Further there are some natural resources, e.g. the silicate which can be reactivated for the coal mining industry. Most cost globalization drivers in Slovakia are favorable for foreign MNCs, stemming from very low prices of all means of production and from Slovakia's geographic location in the center of Europe. Slovakia's unfavorable cost globalization drivers are its small population and purchasing power, which result in a small market.

Technical skills

Its level of technical skills provide Slovakia's strongest cost globalization driver. Slovakia shares its intellectual history with Austria and the Czech Republic providing over some centuries a basis for technical education which has resulted in a number of different inventions and patents. The stagnation of research in the period of transition indirectly created a good basis for research scholars to be involved in research and development (R&D) for different MNCs.

Technology role

The Slovakian Government is fully aware that the adoption of modern technologies are key industry sources of higher productivity and quality.

Furthermore, adopting new technology will provide the driving force behind undertaking substantial changes in the organizational structure of individual companies. Efforts are being made to enhance flexibility and strengthen a tailored approach to customers.

Government globalization drivers

All governments since 1989 have expressed openness to foreign investment, and there is no formal screening process. In general, there is an unwritten preference for protecting Slovak domestic entrepreneurs and for restricting sales of privatized state property to foreign buyers. There is no discrimination against existing foreign investments, there are no preferential export or import policies affecting foreign investors.

Slovakia's formal and informal governmental policies have often been quoted as very unhelpful to foreign businesses. The Slovak government wants to be sensitive to domestic key industries in the transition of the industry to a market economy. At the same time it tries to attract businesses. The key to Slovakia's future industrial policy lies in the reform of the financial sector and the deregulation of trade. Slovakia suffers from the perception that the level of property protection is lower than in other transforming countries. In spite of growing interest, some foreign businesses from the United States, Japan and Germany have been reluctant to invest capital or transfer technology to Slovakia. The government plans to enact a series of laws related to intellectual property protection. By doing so, it hopes to attract foreign investment and acquire high technology.

Liberalization – competition law

As required by the Competition Protection Act, the anti-Monopoly Office of the Slovak Republic was created and is based in Bratislava. The office has full supervisory powers and may enter premises and examine legal and commercial materials. The office's functions include the following:

- Approval of cartel agreements and mergers;
- Granting of exemptions from the prohibition on cartel agreements and the withdrawal of such exemptions;
- Prohibition of implementing cartel agreements or mergers if they fall within the restrictions set out in the act and have not been exempt by the office;
- Prohibition of a dominant or monopolistic position in the marke;t
- Issuing of preliminary rulings in proceedings begun by the office and the imposing of fines.

Absence of state-owned competitors

The Commercial Code and the 1991 Economic Competition Act govern competition policy in Slovakia, which generally follows the European Union pattern. The Anti-Monopoly Office is responsible for enforcing measures to prevent uncompetitive situations. Prior to 1995, its decisions were relatively uncontroversial; in early 1995 the Chairman of the Anti-Monopoly Office was replaced involuntarily following a reported decision to block the acquisition by a leading refinery of a gasoline-distribution chain. The acquisition was subsequently permitted, prompting some to question the independence of the Anti-Monopoly Office.

Although the Government is keen to create the right conditions to attract foreign investors, in the past it has been known for government to give priority to protecting the large state-owned monopolies, partly because of their social role as the largest employers. It will be necessary for further harmonization of Slovak competition with EU law into the next millenium. Another measure to improve economic protection is the up-coming law on the regulation of natural monopolies submitted for discussion in late 1999.

Customs duty system [15]

The level of customs duty applied on imported goods depends on the type and country of origin. The average level in the Slovak Republic is 5.4 per cent, but duties on some goods such as cigarettes and alcohol are much higher. Goods are classified in accordance with the European Union's *Harmonized Duty System* but there are a few exceptions. Goods may receive beneficial treatment if they are considered as 'temporary imports', that is, stored in bonded warehouses or in transit to a third country, or they are to be used in the renewal of the country's infrastructure. Slovak customs authorities collect not only customs duties but also the value-added tax (VAT) on imported goods. All imported goods are subject to VAT at either 23 per cent or six per cent depending on the type of product. VAT on permanent imports is payable at the point of entry or at the end of the period of temporary import. Excise duties for imported goods are applied to certain kinds of imported goods such as wines, spirits, beer, tobacco products, hydrocarbon fuels and lubricants. Tax is set as a fixed amount per unit of the product concerned. Charges relate to both import and export.

Tax treaties

Tax treaties to prevent double taxation were transferred from the former Czech and Slovak Federal Republic to their respective independent republics. Tax waivers and allowances implied by these treaties are provided with regard to the principles set out in the treaty and the domestic legislature.

Taxes from transfer of real estate are imposed on the seller, and the buyer warrants the taxes. In the case of a tender or retribution transfer that is a consequence of the execution of a resolution or foreclosure, the receiver of the estate is supposed to pay the taxes. The tax base is the payment for the real estate and has a minimum value determined according to special regulations.

Reliable legal protection – property rights

Matters of protection of property rights fall under the jurisdiction of two agencies. The Industrial Property Office has responsibility for most areas, while the Ministry of Culture is responsible for copyrights (including software). Slovakia is a founding member of the World Trade Organization (WTO) and succeeded to membership in the World Intellectual Property Organization (WIPO) upon the dissolution of the Czech and Slovak Federative Republic. Slovakia adheres to major intellectual property agreements including the Berne Convention for Protection of Literary and Artistic Works, the Paris Convention for Protection of Industrial Property, and numerous other international agreements.

Dispute settlement

There have been no major investment disputes in Slovakia. Slovakia is not a member of the International Center for the Settlement of Investment Disputes (Washington Convention). Slovakia is also a member of the 1958 New York Convention on the Recognition and Enforcement of Foreign Arbitral Awards. Property and contractual rights are enforced within the legal structure, but decisions may take years, thus limiting the attractiveness of the system for dispute resolution.

Slovakia accepts binding international arbitration. The Slovak Chamber of Commerce and Industry has a Court of Arbitration for alternative dispute resolution; nearly all cases involve disputes between Slovak and foreign parties. Slovak domestic companies generally do not make use of arbitration clauses in contracts.

Bankruptcy law

This law – the Bankruptcy and Composition Act (BCA)[16] – was introduced in 1991. It provides three possible exit routes for insolvent debtors. The first two are the debtor-initiated composition processes while the third one covers the traditional bankruptcy termination process. The Revitalization Project of 1993 proposed the fourth exit route consisting of a voluntary composition reorganization procedure initiated by the debtor. In this case a proof of insolvency and approval of the terms of composition are required from the court. The proposal is be dismissed if it fails to meet statutory minimum conditions.

Compatible technical standards

The removal of technical barriers to trade in accordance with the European Agreement includes not only the harmonization of legal regulations and technical standards with EU regulations, but also product conformity within basic safety requirements. A central state administration authority, the Slovak Office for Standardization, is responsible for the issues of technical standardization, metrology and testing, including certification, and is currently preparing a new law on product technical requirements and on conformity assessment to bring technical standards into line with existing EU regulations.

Environment

Air pollution from metallurgical plants currently presents human health risks and acid rain from the predominance of heavy industry is damaging forests. Slovakia cannot afford to ignore the world-wide trend towards environmental protection as dealing with pollution will invariably be one of the prerequisites to acceptance by the EU. The country's industrial base is at odds with many far-reaching environmental reforms and the government will tread a fine line implementing legislation that is both environmentally friendly on the one hand, and not adverse to industrial needs on the other.

Competitive globalization drivers

Slovakia's competitive globalization drivers are comparatively weak. For example, as a relatively small economy and market, Slovakia cannot itself hope to compete in world class R&D in areas that require state of the art technologies. Currently Slovakia hopes to (a) improve the competitiveness and quality of existing products and (b) support research and development that will aid Slovakia's industries to develop the new products demanded on the world market.

The role that MNCs could play in increasing Slovakia international competitiveness is substantial. Slovakia will benefit from the influence of MNCs, by information and technology transfer, restructuring of ailing large companies, retraining workforces and introducing new management techniques. Slovakia is invariably – Seeking Alliances.

Distinguishing factors of the Slovak economy are that it is simultaneously (a) small and (b) a remarkably open economy, in which the contribution of exports and imports to the GDP in 1998 accounted for 137.7 per cent. A small internal market means that companies with a high growth potential, unlike their competitors emanating from larger economies than Slovakia's, have to orientate their strategies in the early stages of their development toward gaining positions in foreign markets where they will be able to place a bigger part of their production.

A high degree of openness of the Slovak economy strengthens its dependence on exports, and on the revenues that it is able to generate through exports. This means that the performance of the Slovak economy and its industries is sensitive to external environment developments, global trends (such as the liberalization of economic relations, technological revolution or demographic trends) in the world economy at large, or to developments in the Slovak Republic's major trading partners economies and in particular in the EU countries' economic space. Economic openness has also had its disadvantages. Demand for Slovakia's key exports was falling in 1999, as Russia and Ukraine slumped and EU growth slowed.

The new government intends to address the economy's ills by giving priority to joining the OECD and EU, cutting government wage and infrastructure spending, boosting some taxes and regulated prices, expanding privatization to companies formerly considered strategic, restructuring the financial sector, encouraging foreign investment, and re-energizing the social partnership with labor and employers. Government officials believe as long as two years may be needed before its structural reforms improve economic performance.

Overall Global Strategies for Slovakia

Multinational companies are beginning to apply some of their global strategies to Slovakia. Some MNC managers have noted that the low cost of land, utilities and labor provide the potential for Slovakia to serve as a production center for components as well as a site for establishing local headquarters. It is also attractive as a gateway to Eastern Europe because of its geographical position. Slovakia has a well-educated, highly-trained workforce well suited for MNCs needs. In spite of this, there is a need for top and medium level managers to extend their skills and to understand European law, market economics on a macro and micro level and general European management practice. Slovakia offers, as its main source of comparative advantage, the possibility for MNCs to adopt and deploy new technologies.

MNC strategies in Slovakia

To research what MNCs are doing in Slovakia, we approached approximately forty MNCs in the country. The selected companies' replies are presented below. We would like to thank them for their cooperation.

Civil engineering industry

Slovasfalt is a cooperation between Ban Holding a.s. from Austria which has a 51 per cent share (SK141.7m) and Doprastav s.p. which holds a 49 per cent share (SK136.1m). This company specializes in road construction, airport surfaces, and so on, and in the past it cooperated in the Gabcikovo Dam project. The average wage in Slovasfalt is SK7000 higher than the country average and the number of employees is changing from 199 to 174.[17]

Communications industry

Alcatel SELTLH is a company that understands the importance of market segments catered for by the state-owned telecommunication operator and private telecommunication operators. Their strategy is to target the market segments internationally in Eastern Europe, mainly in the areas of switching telecommunications equipment and transmission telecommunication equipment. These two products have been important throughout the history of the company. On the other hand, in Slovakia they were unsuccessful in introducing radio telecommunications equipment.

Since Alcatel was established in Slovakia, they have opened a procurement office, creating good conditions for product cooperation, and set up a regional marketing team. In terms of a regional sales force, they were focusing on setting up regional distribution centers. Alcatel have set up their regional headquarters which is responsible for the complete service in Slovakia. Connection with other central and Eastern European countries is through production of switching systems. In the future, Alcatel is planning to create local facilities for research and development.

With respect to *global marketing*, Alcatel is planning to apply their standard product positioning and advertising strategies. The global brand names have high priorities whereas in the global marketing strategy, packaging is at a lower level in their scale of priorities. Promotion is prioritized on a large scale. Since their establishment, distribution and sales within Slovakia have reached a standard level. The company has quite a large sales network with which they try to meet the expectations of the country in terms of the level of service personnel.

The company's *price policy* strategy is similar in Central and Eastern European countries. The company focuses mainly on Central European development.

One of the first *marketing success factors* is in local production. The network of local production companies is very supportive of the image of Alcatel. Within the period of the development of the company from the marketing point of view, the modernization of telecommunications has been very important. This is a process that is also supported by the Slovak government. One of the secrets of the marketing success is the investment in personal customer relations.

With regard to *foreign investment*, Alcatel mention a stable legal environment. The second factor for foreign investment is tax exemptions during the startup period. Also important for them is the high level of quality and the low cost of the labor force. As one of the important discouraging factors, the company is quoting an unstable political situation and refer also to unclear governmental economic policy. In addition the size of the market is also a discouraging factor.

Oil and gas industry

Companies in this industry include OMV, Shell, Mobil, British Petroleum (BP), Agip, MOL, Slovnaft and Nafta Gbely.

OMV is one of the important Austrian economic consortiums. In 1985, the company obtained shares in a number of active oil fields via its subsidiary OMV Libya. In 1990 it already had 350 petrol stations in its ownership. Then it took over the Chemia OIAG holding company and later on, the IPIC (International Petroleum Investment Company) from Abu-Dhabi entered it as a strategic partner. At present OMV has 8500 employees and is active in different regions including the North Sea, Pakistan, Vietnam, Albania, Yemen and Bulgaria.

OMV has seen a continued growth of turnover and profit since 1992. Profit in 1996 grew by 520 per cent compared with 1992 levels. There is also a positive trend on the level of dividends that can also be seen from the 14 per cent growth between 1996 and 1995. In 1990 OMV Benzinol was founded. In 1996 its capital grew from SK174m to SK467m and its turnover in 1996 was more than SK2.5bn. At present there are 25 OMV petrol stations in Slovakia and the company is actively supporting environmentally friendly business practise. In 1995 cumulative wages reached SK13.4m and the sum of total wages and other costs in 1996 reached SK23.2m (this represents a growth of 33 per cent). Average wages in 1996 were SK22 673 which is a growth of 9.8 per cent compared with 1995. According to revenues and profit in 1996, OMV has reached 31st place.[18]

Packaging industry

TetraPak is focusing on the packing of juice, milk and cheese as their important set of commodities for the whole region of Central and Eastern Europe. According to their statement, they are focusing on locating their production operations in Slovakia and also in the Czech Republic.

TetraPak positions their product using a *global marketing policy*. They have standard global brand names and they are speaking about a standard packaging strategy in Slovakia. Advertising is locally adapted to the conditions established by their analysis whereby the whole promotion system is locally adapted. The development of distribution systems is at a standard level within the country. They also have local sales representatives as well as service personnel.

TetraPak did not provide any information about their *price policy* and declared it confidential information.

Three *marketing success factors* were listed by TetraPak according to their priorities. First of all, local marketing is stated as being very important. As the second success factor they list the existence of local front offices, and the third market success factor is said to be market research.

According to the TetraPak experience, they think that the most important factor for encouraging *foreign investment* in Slovakia is the high level of education resulting in well educated staff. They also feel that another advantage is the central location of Slovakia within the region. In Tetra Pak's opinion the most discouraging factors for foreign investment are the size of the country (five million inhabitants), political situation and development, state of privatization, and finally its unclear future direction and development.

Computing and IT industry

IBM Slovakia operates in the whole region of Slovakia and focuses on three vertical market segments – the finance sector, the manufacturing and process industry sector, and the public sector. Out of the vertical segments, they are mainly interested in small and medium enterprises. This MNC has successfully introduced information technology hardware and software components, application software solutions for the above-mentioned market segments, and networking solutions for local and wide area networks. However, they report that they had not managed to introduce retail point of sales solutions and IT outsourcing projects (managed operations).

In the first stage of its operation in Slovakia, IBM has focused on development of software solutions connected with other countries of Central and Eastern Europe, setting up a procurement office related to hardware assembly in the other countries in the region. It has also engaged in recruiting a regional marketing team that works in connection with CAD/CAM solutions in other countries, and a regional sales force in Slovakia. IBM plans to locate new activities, primarily hardware assembly and IT software developments in Slovakia.

IBM considers virtually all elements of its standardized *global marketing* mix, including product positioning, global brand names, packaging, advertising, promotion, distribution, selling methods, and sales representatives applicable in Slovakia. The element of service personnel is only partly applicable.

The *price policy* of the IBM hardware/software products are the same all over the world, only the price of services varies according to the cost of labor. Therefore, in Slovakia the service prices are lower than in Western European countries and higher than in Asia.

Marketing success factors are public relations, targeted events (seminars), and direct marketing (including telecoverage and telesales), being the most significant factors in Slovakia quoted by the company.

Detergent industry

In addition to Henkel Slovensko, there are a number of other companies operating in this sector including Johnson & Johnson's and Procter & Gamble.

For Henkel, the most important market segments are detergents and cleaners, cosmetics and adhesives. After taking over the existing state industry, they have successfully relaunched and modernized some existing brands (for example the detergent Palmex), while introducing international high quality brands (for example the detergent Persil). As far as product introduction is concerned, they have not succeeded in creating new local brands of medium quality.

Henkel located a full range of activities in Slovakia, including research and regional headquarters and plans to locate new activities in the country. In all activities besides research, Henkel cooperates with its branch in the Czech Republic. Production is connected within the company in Hungary, the Czech Republic, Poland and Romania, and regional marketing activities are in Hungary, the Czech Republic, Poland and Slovenia.

Henkel also finds virtually all elements of the standard *global marketing* mix applicable in Slovakia, except for sales representatives and sales personnel.

Its *price policy* in Western Europe is comparable to international brands, while in Central and Eastern Europe it is comparable to local brands.

As *marketing success factors* Henkel quotes unique positioning, competitive benefits and clear advertising as the key marketing success factors in the country.

The company is very positive about *foreign investment* in Slovakia, and lists only encouraging factors, which are mainly low production costs, stable political situation, clear legal framework, and market situation.

Automobile industry

In 1960 Bratislava's BAZ (Bratislavske automobilove zcivody – Bratislava Automobile Company) was set up and then waited until 1990 for an economically strong foreign partner. This partner was found in Volkswagen (VW) whose first factory was founded in Wolfsburg in 1938.

VW will be one of the biggest multinational corporations in 1997–2001. According to an international survey VW is planned to invest DM83.4bn between 1997 and 2001. There was a clear positive growth of investment between 1994 and 1996 from DM5.7bn to DM8.7bn.

VW Bratislava is ranked first according to its revenues in 1996 and it belongs to the most highly valued companies with the highest pretax profits, being ranked 12th (SK500.5m). Furthermore it was fourth among the biggest exporters in Slovakia in 1996 as its exports reached a total of SK17.5bn.[19]

Conclusions

The overall factors identified by the analyzed companies as determining marketing success, are related to local conditions and to a presence in the local market in terms of production, front offices, personal customer relations and public relations.

As far as foreign investment is concerned, it appears dampened by the unstable political situation, an unclear governmental economic policy and a limited market size. Nonetheless, investing in Slovakia does offer the advantages of low production costs, the high quality of the local labor force, the stable legal environment, tax exemptions during the startup period and the country's central location in Europe.

Here, we would like to sum up the trends during the transformation of the political system and democratization in Slovakia. Looking at internal social processes, it seems that they can be described in two parallel dimensions. The first dimension applies to all transforming countries in transition from a centrally planned system under socialist conditions to a market-oriented system. This process is dependent on the conditions of a given country in 1989. In that year Slovakia was a developed industrial state with an industrial output five times exceeding the level which would be appropriate for its size – this was due to its specialization within COMECON. Agriculture was managed in a way which made Slovakia 92 per cent self-sufficient in commodities. The second dimension is related to the establishment of a new state. It brings many possibilities for reviewing and redefining some of the layers of the pre-existing system and it also allows for the possible re-establishment of new pillars of state and social structure. Naturally, the process also carries some risks which include reduced market capacity, increased vulnerability in terms of criminality, and so on.

Political and economic processes strongly influence the economic development of countries which is in essence part of the transformation of the status of ownership. The choice of privatization methods used in the management of these processes is the important factor. The process of voucher privatization was started already in the former SFR, but did not continue in Slovakia after its becoming independent. Privatization in Slovakia continued mainly in the form of privatization by management. In the last few years there has been a strong link between political and economic processes which makes the distinction between the corresponding transformation processes difficult.

As has already been mentioned earlier, the development of Slovak economy has been very good in macro-economic terms. However, the process of transition at the company level met with many difficulties. These included:

- Outdated technology.
- Economic orientation being mainly and often solely directed towards machine industry.
- Slow development of companies due to the behavior of the financial and banking systems.
- Primary and secondary debt of companies.
- The level of knowledge and skills of management in individual organizations.

Just as in other transition countries, there is a relationship between the state-forming nation and other nationalities in Slovakia that requires a well thought out management strategy on behalf of the government as well as an appropriate development of the legislative system. The historic richness of development within the state-forming nation, as well as development in cooperation with other nationalities, provide a positive basis for multicultural development in the whole of Central Europe. Conflicts and problems come about mainly due to a lack of tolerance and magnanimity in solving usually not very extensive problems.

When analyzing internal differences between individual regions of Slovakia, it is possible to see different levels of historic development in the political, economic and cultural spheres as well as in the whole dimension of social change. There are regions which are mainly focused on agricultural development and hence the transformation proceeds at a slower pace. This is due to the lack of finance for the development of rural areas. Other parts of Slovakia bear the marks of heavy industrialization and the problem there is that the inhabitants have no alternatives for employment. Here one would expect a dramatic increase in small to medium size enterprises that would require a strong proactive policy for the overall development of this entrepreneurial sector.[20]

In terms of the global economy the MNCs, which are active in Slovakia, play a role in the global market. Slovakia's future role in the global economy will depend on its ability to attract new strategic investors and make the necessary arrangements to attract foreign direct investment.

Appendix: Labor Laws and Regulations

Social benefits[21]

There are five major security schemes to which both employees and the employers contribute: social insurance, health insurance, employment fund, pension scheme, and social fund. Participation in the insurance plans is mandatory for employees of Slovak legal entities, or registered branches of foreign entities. Expatriates employed by a foreign company which has no presence in the Slovak Republic are exempt from the plan.

Employees decide whether a trade union organization should be established. Virtually all entities with established trade unions have collective contracts. The Collective Bargaining Act regulates collective agreements, strikes and lock-outs. Collective agreements can be negotiated either between a company and the relevant trade union (a current agreement) or between associations of employers and the top trade union representation (a joint agreement). Joint agreements take precedence over current agreements. Issues discussed usually with trade union organizations include the schedule of working hours, working conditions and the environment and paid holidays. Disputes between an employer and employees must be taken to a mediator. If the mediator's intervention is unsuccessful, the union may then decide to call a strike, but it may take the dispute to an arbitrator before doing so. Union members must vote on a strike before it is called. More than 50 per cent of the members must support strike action, and a list of those who will be on strike must be given to the employer on the day before the strike begins. Nevertheless, practically all disputes so far have been settled through negotiation and settled by agreement between trade unions and employers and strikes are still a very unusual phenomenon in the Slovak Republic.

Wage policy[22]

According to official statistics from 1996, the gross average monthly wage in the Slovak Republic amounted to SK7 520 and in private businesses salaries depend on individual policies.

Human resources

At present, there are many changes taking place in Slovak human resources management. One of the prime difficulties is recruiting capable local managers, however this is counterbalanced by high technical and academic skills.

Labor legislation

The key law governing employment is the Labor Code under which all employers in the Slovak Republic are obliged to conclude written employment contracts with their employees. The employment contract usually covers matters such as the gross monthly wage and the length of the trial period that must not exceed three months. Employment contracts can be concluded for limited or unlimited periods whereby unlimited employment contracts may be terminated as follows:

- By agreement;
- Immediate cancellation;
- Cancellation in the course of the trial period.

Cancellation by agreement must be documented in writing and include the reason for cancellation. Cancellation of a contract can be done with written notice either by the company or by the employee. The employee is permitted to give a minimum of two months' notice without giving a reason. Notice can be given by the company when

1. The company (or a part of the company) is disbanded or transferred by government order.
2. The company ceases its activity or a part of the company is transferred to another company that is not able to provide a job for the employee.
3. An employee is not able to work for long-term health reasons.
4. An employee does not meet the preconditions of the job.
5. Other serious reasons, for example substantial restructuring of the company.

The notice period is three months for reasons 1–3 and two months for reasons 4–5. Furthermore the company can cancel the employment contract immediately if the employee is convicted of a crime or the employee continually disrupts normal working discipline. The maximum working time in one week is 43 hours. However, the usual weekly working time is 42.5 hours which includes half an hour for lunch and rest each day. Employees may not work more than eight hours overtime per week with a maximum of 150 hours per annum. Remuneration for overtime that may consist in granting additional time off or the payment of additional wages (for example 100 per cent more for Sundays) may be agreed between the employee and the employer. Some companies permit flexibility in working hours for certain days of the week. On these days the employee works *essential working hours* of not less than five hours per day. The minimum annual holiday is four weeks. Any employee who has worked for 15 years or more is entitled to five weeks' annual holiday. Many employers, including the state administration, are now extending this period by an extra week. The Labor Code also specifies the length of maternity leave that is 28 weeks (37 weeks if the mother is single or gave birth to more than one child). If the mother requests it, the employer is obliged to grant her additional maternity leave until her child reaches the age of three. During maternity leave and additional maternity leave the mother does not have the right to wages but receives sickness benefits. Similarly, employees who are sick are given unpaid time off by the employer and their material security is covered by sickness benefits. Men are entitled to a retirement pension when they reach 60 years of age, while women can retire at between 53 and 57 years of age depending on the number of children.

Unemployment and wages

The average number of unemployed was 328 667 in 1996 of whom 34 000 received unemployment benefits (average SK1 935 per month). The unemployment rate increased to 12.8 per cent by the end of August 1997, in comparison with a rate of 12.75 per cent in July 1997. The total number unemployed was 332 828, of which 21.96 per cent received unemployment contributions from the State. According to a recent report prepared by the Slovak Ministry of Labor, Social Affairs and Family, the productivity of labor rose by 5.7 per cent and the growth of real wages reached eight per cent in the first half of 1997. The average nominal wage amounted to SK8622 during the period of the first half of 1997 – a relative growth of 14.6 per cent. For comparison, the average rate of unemployment during the first half of 1997 was 13.8 per cent.[24] The average nominal wage amounted to SK8963 during the first nine months of 1997. This represents a growth of 13.2 per cent compared to the same period of the previous year. The highest average wage of SK11 268 is paid to employees in the Bratislava region and the lowest (amounting on average to SK7472) to those in the Presov region. Traditionally, the lowest rate of unemployment has been in the Bratislava region. In September 1997 it reached 5.1 per cent and the highest rate of 18.1 per cent was in the Kosice region.[25]

The Slovak Republic supports the creation of jobs through the National Unemployment Office. Financial contributions are offered mainly in regions with a high unemployment rate, for example in Cadca, Dolny Kubin, Rimavskci Sobota, Velky Krtis, Humenne, Lucenec). In 1996 SK2.34bn (US$78m) was planned to be spent on jobs creation.[26] The National Labor Office is an independent institution whose every decision must be agreed on by three parties – the Government (represented by the Ministry of Labor and Social Affairs), the Federation of Employers' Unions and the Associations of the Slovak Republic and Trade Unions. Employment support is given to entrepreneurs in accordance with Law no. 387 of 1996. It can be provided in two ways, as returnable contributions negotiated with the Regional Labor Office and as non-returnable contributions negotiated with the Local Labor Office. Returnable contributions may be granted to entrepreneurs in support of a business project creating at least 100 new jobs and can cover wages or salaries, health care and social insurance, retirement scheme, contributions to the Employment Fund, accommodation, boarding and travel expenses or loan interest. Non-returnable contributions may also be granted and can be used to cover wages or salaries, healthcare and social insurance, retirement scheme, contributions to the Employment Fund, accommodation, boarding and travel expenses. The contributions are made to employers when they create new jobs or when they employ registered unemployed citizens, fresh graduates, minors or citizens above 50 years of age. A decision about the level of support is the responsibility of the board of directors of the Local or Regional Labor Office which consists of representatives of unions, state administration and employment associations.

Bibliography

Agenda 2000: Posudok komisie k Êiadosti Slovenska o çlenstvo v Európskej Unii.

Agenda 2000: The Committee Judgement Concerning the Application of Slovakia for EU Membership, 1997.

Arthur Andersen, Reports of 1996, 1997, 1998.

Bankruptcy and Composition Act, 1991.

Bulletin of the Statistical Bureau of the Slovak Republic.

Business Central Europe Magazine, The Economist Group, http://www.bcemag.com

Chovanec, J., *Path to the Sovereignty of the Slovak Republic* (Bratislava: SIA, 1996).

European Economy, Supplement C (10/11, 1996).

EVCA (1994), Slovakia in 1994. *Central and Eastern Europe Directory of Private Equity Capital Companies*, p. 19.

HELP Foundation, Project Fernina – Final Report, April, 1997.

Hoøková, A. and Vágnerová, S. (1998), Priame zahraniçné investÎcie na Slovensku so zameraním na bankovú sféru, Národná Banka Slovenska.

ING Bank, Slovakia Capital and Monetary Market Report (4/1997).

Karsai, J., Wright, M., Dudzinski, Z. and Moroviç, J. (1999), Venture Capital in Transition Economies: The Cases of Hungary, Poland and Slovakia, in *Management Buy–outs and Venture Capital Into the Next Millennium*, Mike Wright and Ken Robbie (eds.) (Cheltenham, UK: Edward Elgar Publishing), pp. 81–114.

Klevana, L. (1996) Slovakia in 1996. *Central and Eastern Europe Directory of Private Equity Capital Companies*, p. 17.

Monitor of the Economy of the Slovak Republic (3/1997).

Moroviç, J., *Venture Capital Slovakian Studies*, Interim Report: September, 1996.

Morovic, J., *The Political and Economical Factors in the Democratic Transformation of Slovakia*, (Tokyo, Japan: The United Nation University, 26–27 September, 1997), pp. 1–34.

Moroviç, J. (1999), Political and economic factors in the Democratic Transformation of Slovakia: Achievements and Problems, in *The Democratic Process and the Market Challenges of the Transition*, Mihály Simai (ed.) Tokyo, New York, Paris: United Nations University Press, pp. 127–141.

SAEF – Slovak American Enterprise Fund.

Slávik, V. (1998), Zahraniçné investície v ekonomike SR a vybranych ponikateiskych subjektoch diplomová práca, Ekonomická Univerzita v Bratislave.

SNAZIR – Slovak National Agency for Foreign Investment, Business Guide (The Slovak Republic, 1997).

Statistical Yearbook of the Slovak Republic 1996, Slovak National Bank.

TREND Magazine, Top Ten Companies in 1993.

UNU Tokyo Report Conclusion, March 1998 planning to locate new activities in Slovakia in view of the current volume of production.

US Department of Commerce, *Country Commerce Guide* (Slovak Republic: 10/1995).

Viewing Figures – Mediaprojekt, 1996.

Notes

1. Hoøková, A. and Vágnerová, S. (1998) Priame zahraniçné investÌcie na Slovensku so zameraním na bankovú sféru, Národná Banka Slovenska.
2. Ján Moroviç, Venture Capital Slovakian Studies, Interim Report, September, 1996.
3. Judit Karsai, Mike Wright, Zbigniew Dudzinski and Ján Moroviç (1999) Venture capital in transition economies: the cases of Hungary, Poland and Slovakia, *Management Buy–outs and Venture Capital Into the Next Millennium*, Mike Wright and Ken Robbie (eds.), Edward Elgar Publishing, Cheltenham, UK, pp. 81–114.
4. SNAZIR – Slovak National Agency for Foreign Investment, Business Guide – The Slovak Republic 1997.
5. Arthur Andersen Report from 1996.
6. Arthur Andersen Report from 1996, and from 1997.
7. Arthur Andersen Report from 1996, SNAZIR – Slovak National Agency for Foreign Investment, Business Guide – The Slovak Republic 1997.
8. SNAZIR – Slovak National Agency for Foreign Investment, Business Guide – The Slovak Republic 1997.
9. SNAZIR – Slovak National Agency for Foreign Investment, Business Guide – The Slovak Republic 1997.
10. Ján Morovic, *Venture Capital Slovakian Studies*, Interim Report, September 1996.
11. Ján Moroviç (1999) Political and economic factors in the democretic transformation of Slovakia: Achievements and problems, *The Democratic Process and the Market Challenges of the Transition*, Mihály Simai (ed.), United Nations University Press, Tokyo, New York, Paris, 127–141
12. US Department of Commerce, *Country Commerce Guide*, Slovak Republic, 10/1995.
13. Ján Moroviç, *Venture Capital Slovakian Studies*, Interim Report, September 1996.
14. SAEF – Slovak American Enterprise Found
15. SNAZIR – Slovak National Agency for Foreign Investment, Business Guide – The Slovak Republic 1997.
16. Bankruptcy and Composition Act, 1991.
17. Slávik, V. (1998), Zahraniçné investície v ekonomike SR a vybranych ponikateiskych subjektoch diplomová práca, Ekonomická Univerzita v Bratislave.
18. Slávik, V. (1998), Zahraniçné investície v ekonomike SR a vybranych ponikateiskych subjektoch diplomová práca, Ekonomická Univerzita v Bratislave.
19. Slávik, V. (1998), Zahraniçné investície v ekonomike SR a vybranych ponikateiskych subjektoch diplomová práca, Ekonomická Univerzita v Bratislave.
20. Moroviç, J., *The Political and Economical Factors in the Democratic Transformation of Slovakia*, The United Nation University, Tokyo, Japan, 26–27 September, 1997, pp. 1–34.
21. SNAZIR – Slovak National Agency for Foreign Investment, Business Guide – The Slovak Republic 1997.

22 SNAZIR – Slovak National Agency for Foreign Investment, Business Guide – The Slovak Republic 1997.
23 SNAZIR – Slovak National Agency for Foreign Investment, Business Guide – The Slovak Republic 1997.
24 Arthur Andersen Report from 1997.
25 Arthur Andersen Report from 1997.
26 Arthur Andersen Report from 1996.

6 Slovenia – Small is Successful

NENAD FILIPOVIC

Overview

SLOVENIA is a small country in the heart of central Europe bordering Austria, Italy, Hungary and Croatia. Although it has a population of just over two million and therefore a small domestic market and limited potential in terms of global strategic importance, Slovenia is presently one of the most Westernized and prosperous countries of the region. As a member of a number of important international financial and trade agreements, Slovenia has a long history of multinational company involvement.

Slovenia was by far the most prosperous of the former Yugoslav republics, with a *per capita* income more than twice the national average. It also benefited from strong ties to Western Europe and suffered comparatively little physical damage in the dismemberment process. In all regional economic analyses Slovenia keeps a clear lead over the other countries of Central and Eastern Europe. It was indicated as the most competitive among countries in transition by a number of leading sources including publications from the World Bank, European Bank for Reconstruction and Development, and the Global Economic Forum, and consistently receives top ratings from agencies such as Standard and Poors.[1]

An entrepreneurial culture combined with similar consumer behavior and consumption patterns to those in Western Europe, brings Slovenia ever closer to its Western neighbors. An easy entry point for foreign companies to the economies of the former Yugoslavia, and the Balkans in general, many Western multinationals companies have already set up operations here. The future looks bright for Slovenia to fulfil its potential role as the 'Singapore of the Balkans', proving that in the case of this country – Small is Successful.

History

Slovenia is one of the six republics of the former Yugoslavia. It proclaimed its independence in June 1991, was recognized by the European Community in January 1992, adopted as a full member of the United Nations in May 1992, and of the Council of Europe in May 1993. It is also a member of all major international financial institutions and some forty other international organizations. This is the first time in history that Slovenians have had their own independent national state. After settling in this part of Europe as a part of massive migration of Slavic tribes from the seventh to the ninth century, the Slovenian people had their own dukes crowned in Karinzian Field (now part of Austria). After falling under the dominance of the Carolings, Slovenia was largely treated as a part of Austria, with its western provinces claimed by Italy during its expansion between the First and Second World Wars. Since the fall of the Austro-Hungarian Empire in 1919, Slovenia has been part of the Kingdom of Serbs, Croats and Slovenians, which changed its name into Yugoslavia in 1929.

The Constitution of the Republic of Slovenia was adopted on 23 December 1991, exactly one year after the plebiscite for an independent state. In accordance with the new Slovenian Constitution, the highest legislative authority is the National Assembly, with 90 deputies elected for a four-year term of office. The National Council with 40 deputies performs an advisory role. The President of the Republic represents Slovenia and is the supreme commander of the defense forces. The government is the highest executive body, independent within the framework of its competence and responsible to the National Assembly. The Bank of Slovenia became the central bank responsible for monetary policy, exchange rate management and regulation of commercial banking. After a turbulent start the multi-party parliamentary democracy has stabilized and in terms of political risk Slovenia has, in the period from 1996 to 1998, been consistently rated as the least risky country in Central and Eastern Europe.

Slovenia has a population of two million, made up of 19 per cent under 14 years of age, 69 per cent between 15–64 years and 12 per cent above 65 years of age. The rate of population growth is only 0.24 per cent per year, while the birth rate is 11.85 per thousand. Projected growth for the next five years is zero or negative, with effective aging of the population due to increased life expectancy. This fact creates several problems for the economy. The most obvious one is the urgent need to redesign the pension insurance system, since currently it is designed for a significantly smaller ratio of retired to active population. Another problem is the relatively slow inflow of new workers into the local economy, as well as its high possibility of brain drain. On the other hand, the low population growth alleviates to some degree the long-term unemployment problems, since the pressure from new entrants into the labor market is not as high.[2]

The transition

Slovenia has undoubtedly benefited from historically strong ties to Western Europe as well as its close relationship with the former Yugoslav republics and countries of the former Soviet Union. Fortunately Slovenia has suffered relatively little physical damage in the dismemberment process of Yugoslavia and has managed to cushion the economic damage of the recent financial crisis in Russia.

The many political problems, and the war which began in former Yugoslavia in 1991, significantly decreased Slovenia's ability to serve markets in the other Yugoslav republics. The fall of COMECOM also slowed trade with countries of Central and Eastern Europe. As a consequence, in the early 1990s, the Slovenian economy fell into severe recession. Real GDP fell 15 per cent in 1991–92, while inflation jumped to 247 per cent in 1991 and unemployment topped eight per cent – nearly three times the 1989 level. The turning point came in 1993 when real GDP grew 2.8 per cent, unemployment leveled off at about 14 per cent, and inflation slowed dramatically to 23 per cent. From 1994 to 1997 the main macro-economic indicators, illustrated in Table 6.1, remained relatively stable, reflecting government and central bank policy that tried to keep relatively high levels of growth while slowly decreasing inflation. However, due to political compromises made by the ruling coalition, the period of 1996–97 is generally viewed less favorably in economic terms. The GDP *per capita* levels in Slovenia have consistently been the highest in Central and Eastern Europe. Prospects for the subsequent years appear good, with economic growth expected to rise slightly while unemployment and inflation decline slowly.

Table 6.1 Macro-Economic Indicators of Slovenia

	1991	1992	1993	1994	1995	1996	1997
GDP growth rate	–8.1	–5.4	2.8	5.3	4.1	3.1	3.0
Inflation (year end)	247	93.9	22.9	12.3	8.6	8.8	9.4
Balance of payments ($m)	129	926	192	540	-36	39	70
Unemployment rate	8.2	11.6	14.4	14.4	13.9	13.9	14.4
Budget deficit/GDP %	0.0	0.0	0.3	-0.2	0.0	0.3	1.1
Public spending/GDP %		45.6	46.7	46.1	45.7	44.9	46.2

Source: Gospodarska gibanja (Economic Trends), Ljubljana, 292/1998, p.25.

Liberalization of prices and trade, extensive restructuring of industry and banking reform have taken place alongside successful macro-economic stabilization. Even in the former Yugoslavia extensive small-scale private activities existed and almost all small-scale trade and service activities were in the hands of the private sector. Under the new reform program a comprehensive mass privatization of 'socially owned' enterprises was launched. Banks, insurance companies and some of public utilities, like Slovenian Telecom and the rolling stock of the Slovenian Railways, are expected to be fully privatized by the year 2002.

By mid 1998, less than two per cent of companies were not formally privatized and this was mostly because of unresolved claims for denationalization of part of their assets or because they had entered bankruptcy procedures. Laws passed in 1992 governed privatization of 'socially owned' enterprises. Slovenia, like the Czech Republic, opted for 'voucher' privatization. Each Slovenian citizen received up to DM20 000 worth of vouchers called 'ownership certificates'. During the privatization process they were then free to purchase the stock of companies of their choice. Alternatively they could put the ownership certificates into investment funds specially designed to manage them. According to the privatization law, a maximum of 40 per cent of any company's shares had to be made available to current or retired employees in exchange for their certificates. A further 40 per cent of the shares had to be transferred to three funds: 20 per cent to the Development Fund, 10 per cent to the State Pension Fund and 10 per cent to the Restitution Fund. The shares of the Restitution Fund were given as compensation to the individuals who could not get their nationalized property back in kind.

Market and trade commodity price liberalization was largely completed by mid-1994. Prices of all main energy sources are below EU levels but are slowly going up. Under the trade liberalization policy, by the end of 1994, 98 per cent of imports were free from qualitative restrictions. The government is committed to further liberalization and the elimination of all non-tariff barriers. In 1996 Slovenia's export of goods and services was US$10 314m while import of goods and services was US$10 553m. Most of the export was to the EU (67 per cent) which was followed by CEFTA countries (five per cent), EFTA countries (one per cent) and 27 per cent to all other countries (about half of this to Croatia).

Slovenia became a full member of GATT in September 1994. In June 1995 a 'Europe Agreement' for Slovenia was initiated, setting out the terms of associate membership of the EU. In April 1998 Slovenia was officially recognized by the EU as the forerunner among the Central and East European candidates for the next wave of EU membership. However, most analysts predict that the expansion of the EU will not happen before the year 2005. In the meantime, Slovenia has also become a member of the Central European Free Trade Agreement which now plays an important role in regional trade.

The currency convertibility and exchange rate regime, introduced in October 1991, made the 'Slovenian Tolar' the national currency and fully convertible for current account transactions. The exchange rate is floating although exchange rate policy kept the rate *vis-à-vis* the German Mark within a narrow band. In the period 1995–98 the central bank consistently followed the policy of slow real appreciation of the Tolar, reflecting a positive balance of payments. Bank interest rates are liberalized so that the commercial banks are free to set their deposit and lending rates. However, the agreement among the members of the Bank Association effectively created a cartel situation, with banks avoiding competition on interest rates. Due to ongoing problems in the banking sector (such as inherited portfolios of bad debt and the ongoing low efficiency of banks), the spread is high and the lending rates are typically six points above LIBOR. Every foreign currency loan to a Slovenian beneficiary of a duration less than five years requires an interest-free deposit with the central bank at a rate of 40 per cent of the loan amount. This effectively reduces the level of direct loans taken by companies.

Financial institutions and banking reform were introduced in 1991. To reduce the domination of the largest bank, Ljubljanska Banka, the government carved out separate banks from LB's subsidiaries and more recently established private banks emerged. At the beginning of 1998, there were 28 banks in total and the two largest were still state owned. Since September 1994, all banks had to satisfy the minimum capital adequacy requirement according to EU requirements. Bank regulation and supervision is well developed. Non-bank financial institutions are also operating in the country. For example about 20 licensed management companies have set up investment funds that will invest ownership certificates on behalf of citizens. These funds can participate in auctions organized by the Slovenian Development Fund where the shares of privatized companies are sold for ownership certificates or cash.

In September 1994, parliament passed a law regulating insurance companies, placing more stringent criteria on the level of capital at risk. The law required as much as 90 per cent of the capital of any insurance company to be invested in low risk instruments (bank deposits or government bonds) and only up to 10 per cent in other securities in Slovenia. A change in the law was announced by the year 2000, relaxing the 10 per cent limit and effectively boosting Slovenian capital markets. At the same time, the law that requires 50 per cent of insurance company equity to be of Slovenian origin will be removed, allowing for more participation of foreign insurers in the local market. At the beginning of 1998, there were only two foreign insurance companies with joint ventures in Slovenia.

Securities markets are regulated by the Law on Securities (1989 and 1990), the Law on Money Markets and Capital Markets, and the New Securities Market Act of 1994. The Ljubljana stock exchange was founded in December 1989, where stocks, and government, municipal and enterprise bonds, are traded. Although the number of issues is constantly growing, trading volume and stock liquidity remained relatively low except for a few 'Slovenian blue chip' companies. The trading of foreign investors was restrictively regulated to limit the risk of crashes due to speculative trading. Although these restrictions received extremely negative press coverage and were seen as very detrimental for market development (both by international and local investors), they proved to be justified during the 'Russian crisis' in 1998, when the flight of capital out of the region hit other countries much harder than Slovenia.

Slovenia also undertook a comprehensive reform of its tax legislation. Corporate profit taxes are among the lowest in Europe with a rate of 25 per cent, and personal income is taxed from 17 per cent to a maximum rate of 50 per cent. Dividends are taxed at a 25 per cent rate for residents and 15 per cent for non-residents. On 1 July 1999, a two-tier VAT (for the vast majority of products and services expected to be 18 per cent) will replace the traditional sales tax of 25 per cent for goods and 6.5 per cent for services.

The Slovenian pension system was partially reformed in 1992. Social security contributions are made by employers and employees, at a rate of 22.9 and 22.4 per cent respectively, of gross wages. In 1994, the share of expenditure on pensions as a proportion of the GDP amounted to 14 per cent and this is growing yearly. A new reform, placing most of the burden on a personal pension saving scheme, is expected by the year 2000.

In May 1995, trade unions and the Chamber of Economy (representing employers) signed a new Wages Agreement, according to which wages will be indexed imperfectly to inflation. The agreement also specifies both a maximum and minimum wage and has been renewed with modifications every year.

Average cost per hour in Slovenia in 1998 was DM9.70 (US$5.40), still significantly lower than in the majority of EU countries but higher than in other Central and East European countries.

Domestic companies

After the breakup of Yugoslavia and the collapse of COMECON, Slovenian enterprises temporarily lost the market of former Yugoslavia and much of Central and Eastern Europe. This forced them to look for new markets elsewhere even for products previously mainly sold in Yugoslavia. Such a new beginning was difficult and the majority of companies had to make significantly changes – cut costs, reduce production volume, as well as change their product and service mix. Eventually most firms found new markets in Europe, especially Germany, Austria and Italy, while keeping some sales in the breakaway republics of former Yugoslavia, especially in Croatia and Macedonia.

Most companies producing consumer goods achieved export levels in the EU only by moving to lower level market segments, while a small number were able to establish their own brands. The opposite is true locally, in the markets of former Yugoslavia and some Central and Eastern European countries, especially in the pharmaceutical, food and fashion industries. Here Slovenian companies are often positioned in top market segments and still compete strongly with global leaders. Their experiences from 1991 onwards have taught companies to establish a balance between risk exposure and the benefits achieved in these 'soft' markets. This has proved to be a wise move as the 1998 Russian economic crisis has shown, when direct exposure of Slovenian companies, with a few exceptions, was not significant. The crisis affected the Slovenian economy only indirectly, through the slow-down of the European economy as a whole, with an estimated drop in GDP growth of less than 0.5 per cent.

Restructuring of companies was on average faster, less painful and produced better results for those with industry type customers, since they were already more used to global competition and tough market conditions. The global trend towards an increased share of services and a diminishing role of manufacturing is obvious in Slovenia as well. The largest Slovenian companies are illustrated in Table 6.2 and the largest exporting companies in Table 6.3. Slovenian service providers, on average, still remain quite local and very few cross national borders.

Along with the restructuring of existing companies, many start-ups developed successfully in the period after 1991. Some of them achieved high growth and gained international recognition. For example, the software subcontractor Hermes Softlab from the start enjoyed an annual growth rate of 50 to 100 per cent and now has subsidiaries in four countries and was named among the best entrepreneurial companies in Europe and the best young company in Central Europe. It is interesting to note that entrepreneurial activity in Slovenia was not hampered by the highly unfavorable conditions in financial and capital markets, which were characterized by high interest rates and a lack of venture capital.

Table 6.2 Largest Slovenian Companies

Rank	Company	Industry	Revenues in 1997 (in US$)[1]
1	Revoz	Automobile	976 630
2	Petrol	Oil derivatives	800 912
3	Elektro-Slovenija	Power transmission	544 287
4	Mercator	Retail	452 637
5	Gorenje GA	House appliances	384 775
6	Telekom Slovenije	Telecommunications	344 637
7	Krka	Pharmaceuticals	329 619
8	Lek	Pharmaceuticals	300 481
9	Slovenske Zeleznice	Railways	294 225
10	Merkur	Trade	292 531
11	SCT	Construction	288 531
12	Sava	Tire and rubber	242 069

Note: [1] Recalculated based on US$1 = SIT160 (Average exchange rate in 1997).

Source: Poslovni almanah Gospodarskega Vestnika (Business Almanac of Economic Weekly), Ljubljana, 1999, p.41.

Table 6.3 Largest Slovenian Exporters

Rank	Company	Industry	Export in 1997 (in US$)[1]
1	Revoz	Automobiles	813 494
2	Gorenje GA	House appliances	332 100
3	Krka	Pharmaceuticals	237 144
4	Petrol	Oil derivatives	207 119
5	Lek	Pharmaceuticals	205 206
6	Sava	Tire and rubber	186 850
7	TUS-Prevent	Protective clothes	114 175
8	Talum	Aluminium	113 262
9	Impol	Aluminium	98 706
10	Acroni	Steel	91 262
11	Mura	Fashion	88 287
12	Iskraemeco	Electrical meters	82 756
13	Iskra Avtoelektrika	Car components	79 781
14	Cinkarna Celje	Zinc products	79 037
15	Iskratel	Phone exchanges	72 031
16	Industrija usnja Vrhnika	Leather products	70 000
17	Julon	Textile	68 687
18	ICEC Videm	Paper pulp	68 075
19	Eta Cerkno	Electrical	66 006
20	Danfos Compressors	Compressors	59 350

Note: [1] Recalculated based on 1US$ = 160 SIT (Average exchange rate in 1997).

Source: Poslovni almanah Gospodarskega Vestnika (Business Almanac of Economic Weekly), Ljubljana, 1999, p.76.

The combination of Slovenian self-reliance (deeply rooted in the culture), and a lack of educational and promotional activities in the field of venture capital, has probably resulted in start-ups being one of the least exploited opportunities. It is likely that foreign investors are discouraged by the many problems plaguing venture capital projects elsewhere in the region.

Foreign investment in Slovenia

Foreign companies were present in the Slovenian economic sphere throughout its recent history. In the period from 1945 to 1965 they were effectively restricted to selling small amounts of goods through local representatives. Legislation gradually changed to allow for foreign direct investment (FDI), although only after 1991 was it fully liberalized. Only a few sectors are still restricted (like banking, insurance or education), while most industries and service sectors MNCs have the freedom of selecting the optimal model for doing business in and out of Slovenia.

Slovenia has the reputation of a country that has not been particularly agile in the competition for FDI. This is reflected in the relatively unfavorable public opinion towards FDI and consequently the cautious moves made by the government to promote Slovenia as an attractive investment destination.[3] The situation has started to change with Slovenia coming closer to the EU, but will most probably never reach the levels of enthusiasm displayed in the Baltic Countries or in Hungary during the early days of transition.

Despite more than thirty years of MNCs direct presence in Slovenia, the cumulative FDI by the end of 1997 amounted to only US$2120m, less than the larger Central European countries. However, in terms of FDI *per capita*, Slovenia comes second, only after Hungary. By the end of 1997, 4937 companies were fully or partially owned by foreigners, of which 605 were subsidiaries of foreign companies operating in Slovenia, 1959 companies fully owned by foreigners, and 2373 companies partially owned by foreigners. About one third of these companies had owners from other republics of former Yugoslavia. Many of the rest were small companies, typically importing and selling products of equally small owners – companies from Italy and Austria, or producing components for them. Foreign investment has entered Slovenia from a variety of different countries, as illustrated by Table 6.4, with the largest amount from neighboring Austria.

Most of the MNCs operating in Slovenia decided to start by opening a sales office, rather than viewing Slovenia as a potential production site. The majority of those involved in production came into Slovenia before it became independent (thus at that time serving the bigger local market of former Yugoslavia). Also, it is interesting to notice that, despite an excellent infrastructure offered by Slovenia, very few MNCs located their regional headquarters here. Instead, the majority of the 'first wave' decided in favor of Vienna or Budapest, with Warsaw, Prague and Moscow gaining importance in the mid-nineties.

Table 6.4 Top Countries of Origin of FDI in Slovenia

Rank	Country of origin	% of FDI
1	Austria	31.4
2	Croatia	17.0
3	Germany	12.9
4	France	7.1
5	Italy	7.0
6	USA	6.9
7	Great Britain	4.4
8	Other	13.3

Note: On 31 December, 1997.

Source: Ministry of Economic Relations and Development, Trade and Investment Promotion Office, Internal data, Ljubljana, 1999.

However, there are several industries where MNCs are making a major impact. The most obvious one is the car industry, with Renault operating a large production and assembly facility in a joint venture company – Revoz and several others, like PSA or Johnson Controls, having component manufacturing. MNCs own almost all the paper mills in Slovenia, as well as by far the largest company in the tire and rubber business – Sava (majority ownership by Goodyear). OMV, the Austrian oil company, through its joint venture with Istrabenz, the small Slovenian gasoline retailer, managed to seize more than 25 per cent of the market share from Petrol, the local gasoline retail monopolist. The largest foreign direct investments made into Slovenia are illustrated in Table 6.5.

Although two local pharmaceutical companies, Krka and Lek, have an extremely strong positioning in Central Europe (in 1996 and 1997 they had larger cumulative sales in Poland than all the US companies put together), all the major pharmaceutical companies have their subsidiaries in Slovenia and compete for the local market. They were not able to gain much of the market share in the first years of their presence, but they did change the nature of the competition (and drove up the salary level of sales representatives whom they tried to hire away from Krka and Lek).

Another example is retail: a growing number of foreign retailers is evident, such as Kastner & Ohler, Bieme Italy, Big Bang, Baumax, Interspar, and most recently one of the leading French retailers, Leclerc. A competitive environment and better supply structure has emerged in the past few years as foreign retailers have opened their shopping centers throughout the country. As a result domestic retailers, like Mercator, Emona, Merkur and Kovinotehna, have become more active and increased their level of services. It is interesting to notice that the restructuring of retail went on regardless of the level of demand, even in the periods of significant economic slow-downs. In fact, an increase of products on offer contributed to spurring new demand. In general the majority of Slovenian business people predict more direct involvement of MNCs, not only in selling their products but also in taking over local manufacturers and service providers as Slovenia joins the EU.

Table 6.5 Largest Foreign Investors in Slovenia

Rank	Company	Foreign Investor	Country	Industry Sector	Investment (US$m)
1	Sava Tires	Goodyear	USA	Tires	100
2	Sarrio Slovenija	Sarrio	Italy	Paper, cardboard	75
3	Revoz Novo Mesto	Renault	France	Cars	54
4	Papirnica Vevče	Brigl & Bergmeister	Austria	Paper, cardboard	45
5	OMV-Istrabenz Koper	OMV	Austria	Oil, derivates, trading	32
6	ICEC Videm Krško	ICEC	Czech Republic	Paper, cardboard	30
7	TovarnaSladkorja Ormož	Cooperative Cosun, SFIAR	Netherlands, Italy	Sugar	26
8	Krka Novoterm	Pfleiderer	Germany	Isolative materials	23
9	Henkel-Zlatorog Maribor	Henkel	Austria	Washing powder, cosmetics	23
10	Tobačna Ljubljana	Reemtsma, Seita	Germany, France	Cigarettes	21
11	Messer Slovenija	Messer Griesheim	Germany	Industrial gases	18
12	Kolektor Idrija	Kirkwood Industries	USA	Collectors	17
13	Iskratel Kranj	Siemens	Germany	Telecommunication equipment	16
14	Treibacher Schleifmittel, Ruše	Treibacher Schleifmittel	Austria	Metal processing	n.a
15	Droga Portorož	E.D.&F.Man. Coffee	UK	Food	13
16	Sava Kranj	Goodyear	USA	Tires	13
17	MGA Nazarje	Bosch-Siemens	Germany	Small appliances	10
18	Alpos Šentjur pri Celju	Nova Hut	Czech Republic	Metal products	9
19	Tomos Koper	Alusuisse Lonza	Switzerland	n.a.	9
20	Carrera Optyl Ormož	Carrera Optyl	Austria	Spectacles	8

Source: Ministry of Economic Relations and Development, Trade and Investment Promotion Office, Internal data, Ljubljana, 1999

Country Globalization Drivers

The gradual liberalization of Slovenia's markets have opened up the potential of this small but stable country to foreign investors. Regulation and protection in many areas of business law, labor and technology are nearing international standards as Slovenia plans to join the European community, offering investors a business environment that is very similar to its Western European neighbors. Culturally Slovenia is very close to Austria which makes product adaptation simpler and marketing techniques easily transferable.

Market globalization drivers

Although throughout the last three decades Slovenia has consistently been one of the most developed parts of Central and Eastern Europe, since gaining independence from Yugoslavia it has suffered from an extremely small local market of only two million inhabitants. A survey of potential foreign investors showed that this was the second most important obstacle to investment (the first being unfriendly government policy[4]). However, the size of Slovenia's domestic market is to some extent compensated by a relatively high GDP *per capita* (65 per cent of the EU average) and comparatively high spending *per capita*, especially in some industry and service sectors like pharmaceuticals, cars, insurance and tourism. The economic decline and the gradual recovery in the last decade created a somewhat more dispersed social structure of the population, but with a relatively large middle class category of consumers used to high quality goods and a growing number of those who can afford higher price levels.

Common customer needs

Comparing consumption patterns of 1985 with those of 1995 show changes in the consumption behavior of the general population, illustrated in Table 6.6. In terms of percentage of total household expenses, less is spent on food, clothes, footwear, drinks, cigarettes and housing, and more on transportation (the single biggest change), personal care, education, culture and entertainment. It is interesting to note that the level of personal savings has gone down sharply, partly because of economic hardship and even more because of a change in attitudes. Leading economists and sociologists unanimously predict further change in consumption structures towards patterns typical for mature post-industrial economies like Germany or Austria. This suggests that MNCs should enter the Slovenian market offering goods and services similar to these two countries.

The new proliferation of paid satellite TV channels or Internet service providers is a typical MNC reaction to opportunities having arisen from changes in consumer behavior. Slovenia is among the leading countries in Central and Eastern Europe regarding the number of Internet connections *per capita*. More than 40 per cent of the population (particularly the urban population) have access

Table 6.6 Patterns of Household Consumption

	1985 (% of total)	1995 (% of total)
Food	25.0	23.1
Clothes, Footwear	9.2	7.3
Transportation	8.9	16.2
Housing	15.6	15.2
Education, Entertainment	4.1	6.5
Household equipment	3.8	4.8
Beverages, Tobacco	5.2	4.2
Personal care, Cosmetics	3.5	6.7
Other expenses	3.9	5.6
Loan repayment	1.7	2.6
Savings	19.1	7.8

Source: Statistical Yearbook of Republic of Slovenia, Ljubljana, 1986 (p. 474, 475), 1996 (pp. 234, 235).

to satellite TV, either by owning a receiver, or being hooked on to cable TV carrying satellite programs. In terms of consumer market communication, this obviously adds to fast product information transfer and globalization of the offers. For example, internet shopping is gaining ground at a similar rate to that in more developed European countries.[5]

Global customers

Naturally, some marketing adaptations are required for the region. These are due mostly to language and to a surprisingly high level of regional cultural difference for such a small country. For example, western Slovenia is strongly influenced by Italian culture, while the northern region is influenced by Austrian. This is particularly true in the case of weaker or newly established regional brands. Although there are differences in consumption patterns in Slovenia in terms of the percentage of total household expenditures, the level of consumer awareness, buying habits and even purchasing power closely parallel those in the neighboring regions of Italy and Austria.

Transferable marketing

Marketing methods are highly transferable because many MNCs have been present in Slovenia since the mid-sixties. At the same time, Slovenians go shopping in Austria and Italy several times a year as it is less than one hour's drive from most parts of the country. Many Slovenians also regularly watch Austrian and Italian television. Therefore, MNCs face a market that can, from a consumer behavior point of view, be qualified as developed and fairly close to West European markets.

Given the small size of the Slovenian market, many MNCs avoid large investment into fully localized brands and use promotion and packaging prepared for several Central European countries at once. However, Slovenian customers usually view such marketing approaches as inferior, since they see themselves as more advanced than many countries of Central and Eastern Europe. The same applies to (rare) attempts made to use old Western commercials. Realistically, in the era of global communications, the largest market introduction delays affordable are at most a couple of weeks (for example, for a new model of car).

Cost globalization drivers

In terms of cost globalization drivers, Slovenia offers multinational companies the opportunity to operate in a business environment that is more developed and less risky than many other countries of Central and Eastern Europe. Although Slovenia is not the least expensive of the C&EE countries in terms of labor and business infrastructure, production costs are still lower here than in much of Western Europe. Such strategies can be cost-effective, especially for export orientated companies.

Economies of scale

Slovenia is a small market, which by itself does not provide for economies of scale. For this reason companies have to focus on export and so the manufacturing plants set up by MNC are also export oriented, covering mainly the markets of neighboring European countries. Some MNCs use the geographical position of Slovenia, its regional image as a market leader, and its well-developed business infrastructure to include Slovenia in their regional business systems as a logistical hub. With the decrease of political tensions in the Balkans this can be foreseen as a growing trend contributing significantly to regional economies of scale.

Infrastructure

Slovenia has better developed infrastructure and logistic facilities than most of Central and Eastern Europe. The road and railway network is being rapidly upgraded, while connections by air (there are three international airports), sea (the port of Koper), telecommunication and Internet network are on an international level. Road transportation is extensively used in Slovenia and traffic demand exceeds the capacities of the existing road network. Increased traffic flows are expected due to transit traffic. For these reasons Slovenia has given a clear priority to the highway construction through the adoption of a national program of highway building. Within the framework of this program, the construction of East-West highways (connecting Italy to Hungary) is well underway to be completed by the year 2000. The completion of North-South highways (connecting Italy and Austria to Croatia) is expected to take another two to three years. The railway network will also be fully modernized by this time.

Country costs

Slovenia is definitely not a source of cheap labor or raw material inputs. The country has very limited natural resources (some hydro power, coal and wood), agriculture which is catching up with Western Europe and suffering from similar problems often aggravated by high state subsidies and (due to small population size) a fairly limited labor pool. Therefore, it strives for international competitiveness on the basis of its high skill employees and knowledge workers. The other main source of competitive advantage is its business infrastructure, which has a solid cost/performance ratio and offers virtually any service required by MNCs.

Labor

Although in the period from 1992 to 1997 the level of unemployment has remained between 8 and 15 per cent, this hides deeper structural problems in the labor market. Slovenia traditionally depended on imports of unskilled labor from other parts of ex-Yugoslavia (especially from Bosnia and Kosovo). As a consequence, most Slovenians prefer staying unemployed to taking menial jobs. On the other hand, there is a constant lack of certain skilled labor profiles (like welders and specialized construction workers) that also used to come from the south (typically from Croatia). All this affects low-end salaries, which are much higher than those in other countries of Central and Eastern Europe. Many MNCs considering whether to locate production in Slovenia as a low cost country see this as an obstacle, although the average salary corrected to purchasing parity is still estimated to be at 60 per cent of the Austrian level.

For some professional profiles (marketing, finance, line managers, general managers) demand constantly exceeds supply. This has contributed to the development of headhunting services and is pushing the entry salaries and other benefits higher. However, job switching is not as frequent as in the USA or some parts of Western Europe, since people tend to stay loyal to employers offering them possibilities for personal development and professional challenges. Surveys consistently show that salary is the only fourth or fifth most important reason for changing the company.

Education

The situation of the high school and university educated labor market is somewhat better, particularly in providing technical profiles. High quality people are available, although the best of these command salaries above those in less expensive EU countries. However, the level of skills offered is very high and matches any global standard and is still less expensive than in many Western countries. For example, the gross salary of a top class software engineer is roughly the same as a medium-to-high class engineer in the US and about 20 per cent less than the medium-to-high class in Germany, with the added advantage that Slovenians have somewhat higher productivity and better individual potential.

Slovenia has two universities, which offer courses in all subjects up to PhD level. There are also some independent educational institutions, like the International Executive Development Center, a leading business school in Central Europe, which offer PhDs. There are various research institutions which have a long tradition of cooperation with corresponding institutions in more developed countries and with the industry. Foreign MNCs can effectively use the capabilities and expertise of the country's R&D institutions as well as existing research facilities available within industry as a prospective base for extending their own research and development activities. For a small country like Slovenia, brain drain is a high potential threat. It has until now not been a major problem primarily because of the solid local opportunities and closely-knit society. At the moment Slovenia imports experts, mostly from the developed countries, not only for the needs of MNCs but also for local companies, at a slightly higher rate than it is losing its own experts.

Business infrastructure

The cost of business infrastructure is relatively high compared to some other, less developed countries and is slowly approaching the average cost in the EU, with estimates that it will reach the EU average costs by the year 2002. In 1997 some costs were still below average, especially electricity and gasoline, while others were above average (telecom services, some insurance and banking services). Deregulation of state monopolies, banking and insurance foreseen for the period 2000–02 is expected to assure a better cost position in most elements of business infrastructure and provide better services.

Technology role

Regarding the existing level of technology in companies, the biggest differences can be observed between different industries (rather than within a certain industry). On average, levels are higher in the pharmaceutical industry (matching world standards), machine building and electrical industries, while they lag somewhat in food, textile petrochemical and electronics. The experience of some MNCs who acquired production facilities in Slovenia show that major improvements come from a better organization of operations, rather than additional investment in technology. One general area, which seems to require investment across the industries, is a more aggressive use of information technology for various business purposes within companies. Evidence at the company level shows that the most successful Slovenian-based companies do not try to exploit cost advantages but rather develop technological innovations or new product design. Specific examples can be found in pharmaceutical products, laser machines and prefabricated houses.

Government globalization drivers

The Slovenian government is, like much of C&EE, keen to attract foreign direct investment and is an active member of a number of regional and international trade organizations. Privatization and liberalization of the economy, however, have proceeded with an eye to protecting national interests as well, and creating domestic companies that can compete abroad.

Regional trade blocs

Slovenia is a founding member of the WTO and a full member of CEFTA. Slovenia has a number of free trade agreements, on the basis of which 85 per cent of Slovenian trade is realized. The Association Agreement between Slovenia and the EU enables duty-free and quota free export of all Slovenia's industrial products to the market of EU member states within two years, and introduces reciprocity.

Common marketing regulations

Marketing practices are to a large extent unregulated, although the influence of consumer and professional regulating associations are growing. This has resulted in some recent restrictions, as well as public denouncement of unacceptable practices. Tobacco products cannot be advertised, directly or indirectly, except for 30 days in the period of launch of a new product and only with proper health hazard information. Less stringent restrictions apply to the advertising of spirits. Deceiving advertising practices, like false claims on product properties or customer benefit or multilevel (pyramid) marketing schemes, are strictly forbidden. In cases where the Advertising Association finds an advertisement to be professionally, ethically or esthetically inappropriate, it publicly asks the advertiser to remove the advertisement from the media.

Favorable trade policies

Slovenia has legislation for direct foreign investment, which is fully compliant with EU norms. Foreign legal entities or physical persons may establish companies in Slovenia and enjoy full national treatment. Indirect investment, such as the issuing of bonds or the activities of private investment funds, is also permitted and regulated. Usually, the senior management of local companies may not consist entirely of foreign nationals. At present foreign individuals may not own land but this is a constitutional restriction that will be largely lifted when Slovenia joins the EU. Profits, whether in the form of dividends or otherwise, can be expatriated freely. The transfer abroad of fully convertible currencies is not subject to central bank approval. Most foreign investment proposals do not require government approval. Where private property is expropriated, compensation must be paid at the time the expropriation is carried out.

Full texts of laws affecting investment, which are drafted by legally trained personnel, are usually available and published within one month of being passed. Important court decisions are published or accessible to practitioners. Independent and comprehensive legal assistance is available, particularly in Ljubljana. The cost of sophisticated legal assistance on investment matters is generally affordable to local investors. Overall, investment law is administratively well supported. Legally prescribed registers for interests in respect of land or security exist. Records are usually current within three months. Regulation of limited-liability companies and joint-stock companies may take up to six months. The investment law is also well supported judicially and private parties generally believe that courts will recognize and enforce their legal rights, including those against state parties. It is reported that court officials have at times refused to authorize registration of the real property interest of foreign-owned Slovenian companies although they are legally required to do so. The waiting period for a commercial case to be heard on its merit by the court of first instance is rather long, usually one year. Although practitioners rate Slovenian courts highly, they are viewed as lacking the financial resources necessary to handle complex investment disputes. Courts automatically recognize foreign arbitration awards, without re-examination of their merits.

Reliable legal protection

Legal protection of intellectual property is effective and follows EU standards. The size and quality of the police force in Slovenia is considered sufficient for the role attributed to it by the law. In a number of areas, especially in capital offences, Slovenia has among the lowest crime rates in Europe. However Slovenia has an extremely bad road safety record, with about 500 people killed in traffic accidents each year.

Compatible technical standards

Since becoming involved in the process of candidacy for EU membership, Slovenia has actively streamlined its overall legislation with EU requirements. Therefore the few concerns there have been about the lack of transparency of some laws related to business, were expected to have ceased by the year 2000. The same is true in the area of standardization, where the situation already matches that in the majority of EU countries. Slovenia is rapidly consolidating technical standards used in the country. In the past, most of the standards were derived from DIN (German standards), which makes the current transition easier. ISO 9000 is widely diffused and more and more companies are engaged in self-assessment activities of the type proposed by the European Quality Award model.

Freedom from government intervention

As noted above, the government is reducing the number of state-owned companies, and creating more room for foreign competition in almost all business sectors. However, the liberalization of the legal framework for FDI has not been accompanied by additional economic policy measures to make Slovenia more attractive to foreign investors. Instead, emphasis has been put on the protection of national interests rather than on an active role for foreign investors. As for most governments in the world, the Slovenian government has a record of getting excessively involved in certain business affairs. An example of this was the government intervention in the sale of a local turbine manufacturer to ABB, against previously reached agreements. Another example was the prolonging of the bidding period for the second concession for a mobile phone network to create a better starting position for local interests. The government is also seriously concerned with regulations and practices related to the protection of the environment. These are rigorous, since Slovenia tries to protect its reputation as one of the unspoiled natural beauties of Europe. Foreign investors are put under strong pressure to go beyond the obligatory levels of environmental safety and become involved only in entirely environment-friendly projects.

Competitive globalization drivers

With a population of around two million, Slovenia is not able to create a global strategic importance for MNCs in any industry in terms of becoming a large source of revenues and profits, a home market for global customers or competitors. Slovenia, as already mentioned, is an export-oriented economy and it is the regional expertise and local connections provided by Slovenian companies that many MNCs value when attempting to reach regional customers. Furthermore, Slovenia has been able to sustain its image of regional economic leader, which adds to its importance.

The presence of foreign competitors has had a number of positive effects on the performance of domestic companies, although very few have become major international players themselves. Those that have, were typically competing in a niche market, like Elan, the second largest ski manufacturer. However, since the Slovenian market is not viewed as strategically important, the level of competitive pressure MNCs face is not as high as it could be elsewhere. A good example is the recent change of the market situation in the mayonnaise business, where Nestlé lost more than 40 per cent of market share in favor of Hellmann's (CPC) because the change of local partner was poorly handled.

Overall Global Strategies

MNCs from Europe, United States and the Far East have pursued a number of different strategies in entering Slovenia. Firstly, some viewed Slovenia as a small although interesting and profitable market without linking their strategy here to any other country in the region. Secondly, others saw Slovenia as a stepping-stone in their advancement through Central Europe and especially into countries of the former Yugoslavia, many of which were committed to doing significant business in Slovenia prior to its independence. Finally, a group of MNCs recognized Slovenia as a country offering an attractive platform for their broader (typically pan-European) operations. The choice of strategy had nothing to do with the region the MNC originated from, but was to some degree influenced by the type of industry and especially with the overall global strategy of the respective MNC.

Good examples of the MNCs treating business in Slovenia as a self-sustainable operation coordinated from regional headquarters are various service providers ranging from McDonald's to Bank Austria. The same approach is widespread among MNCs, which in the short run do not plan to extend their operations in Slovenia beyond sales, like Electrolux, Procter & Gamble or Toyota. All of them achieve respectable business results in terms of market shares, brand position and country operations profitability, but limit their interest to that.

A very different approach was taken by Johnson & Johnson and Goodyear, two MNCs interested in using Slovenia as a regional platform. Johnson & Johnson created regional headquarters in Slovenia, from which they provide full logistical and some sales and marketing support to national organizations. Goodyear bought a majority share in the largest local manufacturer of tires and technical rubber (Sava), through which they effectively gained much higher quality access to (Central) European markets as well as technically excellent, low-cost production platforms (many products are shipped back to the USA).

Finally, illustrations of pan-European thinking are Renault and Daewoo. Renault has for many years cooperated with a small, local car assembly company. In the late eighties it had formed a joint venture with Slovenian owners of that company, primarily intending to serve the markets of former Yugoslavia. After Slovenia gained independence, Renault did not pull out although sales in the region were drastically reduced. Since the JV proved to be capable of producing cars at the European level of quality and productivity, Renault made additional investment and currently uses the plant as one of its major production facilities. To delineate the goals and responsibilities more clearly, they have split the JV into two companies: the production unit with pan-European responsibilities and the commercial unit responsible for sales and after-sales support in the region. Daewoo, on the other hand, chose Slovenia as its logistical entry point to Central Europe and established major strategic cooperation with the port of Koper as its logistical base.

Market participation strategies

All the observed MNCs took an active role in the Slovenian market, most often directly through their own subsidiaries or joint venture partners and in some cases in a more removed, arms-length dealing with strategic partners. Since they saw the potential of the Slovenian market as fairly limited, their investment into the commercial side of operations was also quite limited. Typically they were not ready to make any major effort in acquiring substantial market share if they could not obtain it through self-sustainable local operations. In cases where MNCs were in some other ways involved in Slovenia, they often managed to combine their regional interests and skilfully use them as a leverage for their commercial operations, be it simply for marketing purposes or to gain some concessions from local government, banks or other partners.

Consequently, the absolute strength of the market position of the observed MNCs depended on several factors. Most notably it was related to the structure of the competition in the existing Slovenian market. In the cases where a strong local competitor already existed, the results were without exception significantly less impressive. Examples of this are Electrolux (with Gorenje retaining more than 60 per cent of the local market) and Bank Austria (who did not come into the top ten banks in Slovenia). When the MNC entered a very fragmented market or there were no significant local players (or the MNC entered the market through alliance with the local player) the results were much more significant. Examples of this are Goodyear (acquisition of Sava, highly concentrated market), Renault (Revoz joint venture, an early entry to the market which only later became highly fragmented) and Johnson & Johnson (market leader in the majority of the niches it competes in, within a highly fragmented market).

Global products and services

Few of the products offered by MNCs are in any way specially adapted to the Slovenian market. The only adaptations common for most products are related to language, be it in terms of packaging, instructions for use, required certificates/ guarantees or, in some cases a localized brand name. Again, the rules are mostly dictated by the nature of the industry and the global practices of the respective MNC. For example, cars remain global products while retail bank services have to be largely localized. Nestlé uses only global and regional brands (since it sees Slovenia as too small and too close to Austrian consumer behavior to justify the development of local brands), Goodyear kept the acquired Sava's brands and Henkel developed local brands for Slovenia and exported it to other countries of former Yugoslavia.

Despite the small size of the market, the range of the products and services on offer usually includes the majority of the portfolio available in other continental European countries. In the case of some consumer goods companies like P&G or Electrolux, the selection of products at least covers all the relevant categories and price brackets.

Activity location strategies

MNCs find Slovenia to be an interesting location for a number of different activities in the value-added chain. Apparently, the least attractive part of the chain is sourcing of raw materials, since Slovenia has no distinct advantages in this area.

Direct subcontracting of R&D is rare and only a few of the observed MNCs had major R&D activities in Slovenia. However, a good example to the contrary is Siemens, which through its joint venture with Iskratel uses more than 200 highly qualified software and telecom experts in the R&D activities for new phone exchanges. Quite a few MNCs use Slovenian partners as component providers and leave the whole subsystem design to them. Mura, the largest fashion producer and big exporter, is not only producing for the Hugo Boss brand, but also designing the garments and producing cloth. Hermes Softlab is a large subcontractor of Hewlett Packard, responsible for design and implementation of major parts of HP's system software. Elan is designing and manufacturing skis for several other major ski manufacturers and the same is true for Gorenje house appliances.

All the observed MNCs which have production units in Slovenia are moving towards heavy local purchasing. The majority of MNCs were not confident about local suppliers upon starting their operations in Slovenia. However after working closely with the suppliers and clearly communicating their design, cost and quality expectations, they invariably found them to be ready to improve and reliable enough for developing long-term partnerships. Although, in general, Slovenia does not qualify as a low cost platform, most MNCs were for the time being capable of finding satisfactory cost/benefit conditions. Renault's plant in Slovenia was for several years in a row the best quality plant in the Renault system and very close to the top in terms of productivity. Obviously, the cost structure in Slovenia is easier justified for high added-value products, so such products are typical for those MNCs which produce in Slovenia.

Another activity often located in Slovenia is regional sales and marketing, especially for the Balkans, but sometimes also for other countries of Central Europe. Johnson & Johnson has its regional headquarters with all the supporting activities in Slovenia. Pfleiderer, one of the biggest European producers of insulation materials for the construction industry, has a large plant in Slovenia. The production is sold to various countries of Europe (C&EE countries and also to Germany, Austria, Italy and Greece) and the management of the Slovenian subsidiary is responsibility for sales and marketing. MNCs often recognize the excellent connections which Slovenian companies and individuals have developed in the past with other Central and East European countries and try to use these in their regional sales activities.

Global marketing strategies

Products and services are in almost all cases marketed in the same way as they would be in any other country in Europe. Even though the content is sometimes locally flavored, the approach to the marketing process is the same. Marketing was often among the first areas where MNCs hired local talent, either as employees or subcontractors, since they felt the need to be in tune with local feelings. In general, the majority of global marketing material is applied without problems.

Market research and advertising are two areas where MNCs did have a major impact on Slovenian marketing practices. The entry of MNCs contributed to the proliferation of services in these two areas, a higher quality of local providers as well as their integration in global networks. Nowadays every large advertising agency in Slovenia is affiliated with a major foreign group. The vast majority of advertising money is spent on TV, while printed media, billboards and radio have less of a share.

MNCs have had less of an impact in other areas related to marketing. For example, distribution channels were already relatively well developed, product assortment was rich in almost all product categories and prices were close to European levels. MNCs did not make major adjustments in their own product portfolios or pricing strategies. It was interesting to notice that even when MNCs were in the position to win market shares through aggressive pricing, almost without an exception, they avoided starting price wars and were satisfied with smaller market shares. Companies offering expensive products (like Renault or Daewoo) were usually ready to engage in trade financing, so that they, their dealers or their local financial partners could offer leases or credit lines.

An interesting paradox appeared in relation to pricing and the distribution channels: several MNCs selling consumer products (like Electrolux) sold their products through many parallel channels, including some foreign retail chains. Since the foreign chains were typically buying through their headquarters, they were often getting better conditions compared to local chains, even when selling significantly lower quantities in Slovenia. This led to several bitter disputes and created problems for the MNCs relationships with local distributors.

Competitive move strategies

Perhaps because the attractiveness of the local market is limited, competition in almost all business sectors is quite intense, although not cut-throat. In almost all sectors there is a mix of local and foreign competition (except in those like the car industry) and local competitors often dominate the market. For example, one of the largest European players in the insurance industry, Alianz, has a ten times smaller market share than the local insurance company Triglav, the market leader.

Given the limited size of the Slovenian market, one would expect first mover advantage to be significant. However, it is difficult to trace MNCs trying to achieve such an advantage. On the contrary, some local companies were able to build effective defense against MNC entry. For example Petrol, the market leader in fuel distribution, secured three lines of defense. Firstly, they lobbied the government to get a state requirement that the fuel retailers have to have major storage capacities in Slovenia (any new one is subject to the consent of local government, virtually impossible for the MNC to get). Secondly, they have built a dense network of gasoline stations not leaving many profitable locations for MNCs. Thirdly and most notably, they increased the level of services on their own stations to a level where MNCs have no real advantage.

Sound advice to MNCs thinking of entering the Slovenian market would be to carefully consider the level of activities of the main local players. Those industries where Slovenian companies have achieved a significant level of international activity (like pharmaceuticals or home appliances) tend to offer fewer opportunities. The same is true for market segments where a Slovenian company, which has already undergone significant restructuring, has a predominant market share: for example Droga in the coffee business or Zavarovalnica Triglav in insurance, both of which have over 60 per cent of the market. The investments typically needed to bring down such market monopolies are often hardly justifiable, since within a limited market, battling for market share can result in an overall drop in profit margins.

Somewhat more opportunities exist for MNCs in fast moving consumer goods and other sectors with low brand loyalty, since the promotion and distribution systems of many Slovenian competitors are still not perfected. This can be especially interesting when considered in combination with the further strengthening of foreign retail chains. Some services, like tourism, are also potentially profitable, especially if the current level of service is still mediocre. Finally, some market opportunities exist in industrial goods, where both the Slovenian companies and the MNCs which have production platforms in Slovenia constantly seek for supplier improvements, and where Slovenian component and subsystem manufacturers, especially in high-tech areas, have difficulties in keeping up with global competition.

Global Organization and Management

MNCs in Slovenia use a number of different organizational models and it would be difficult to establish which one is dominant. Historically, joint ventures were the most popular since in the period from the mid-sixties to the end of eighties they enjoyed some legal advantages, primarily related to foreign trade and hard currency availability. After the liberalization of the economy at the beginning of the nineties, other possibilities became at least as attractive. Most of the newly established companies are subsidiaries 100 per cent owned by the mother MNC, especially if they only take care of the sales and marketing side of the operations (like Electrolux or Procter & Gamble). It should be noted, however, that given the small local market, many MNCs decided not to open a subsidiary in Slovenia but instead cover it from another country (typically Austria, like Nestlé). Those opting for production or other more extensive activities in Slovenia often decide to do it through a joint venture with a local partner (like Renault for production or IBM for a regional customer support center). Recently, some of them (like IBM) bought out their partners and converted the company into a fully owned subsidiary.

Legally, the majority of executive board members of a company registered in Slovenia must be Slovenian nationals. The majority of smaller Slovenian companies have only one executive board member, who in that case must be a Slovenian. Many MNCs who want to keep a foreigner as the executive director work around this restriction by nominating Slovenians to 'secondary level' managerial positions while formally giving them the title of executive board member. Although the number of Slovenian managers with top-level responsibility is growing, the dominating practice, especially in larger operations, is to have an ex-patriot managing the local MNC company. Another two areas often directly managed by a foreigner are finance and controlling. When Goodyear bought a majority share in Sava's tire production unit, they moved the previous general manager to the post of marketing manager and brought in their own general manager (previously financial manager, Goodyear Europe) and financial manager.

Most of the observed MNCs justified their use of ex-patriots by two factors: lack of skills of local managers in the area of general management and need for excellent understanding of the company's 'way of doing things' from day one. Both arguments sound relatively weak. Quite a few local managers could be available whose quality does not fall short of international standards and who have significant international business experience. On the other hand several of the observed MNCs, who have been in Slovenia for more than five years (like Renault), still keep a foreigner in the post of general manager. There are other examples, however, like Papirnica Vevče (paper mill fully owned by Briegl & Bergmeister), where after three years an Austrian manager was replaced at the top of the company by a Slovenian. Several MNCs used only local managers for their company in Slovenia. There was no obvious connection between local performance of an MNC and the nationality of the general manager. Finally, quite a few Slovenian managers pursued international careers successfully although most of those who decided to move out of Slovenia found their job themselves, rather than by being promoted out of Slovenia. A 'glass ceiling' still seems to be present. In the long term, appreciation of local talent will probably develop in parallel with an understanding of the nature of the competitive advantages offered by the Slovenian business environment.[6]

Almost without exception, those MNCs that gained control of existing companies in Slovenia went through major restructuring, which often included significant downsizing of the work force. This was often difficult to implement due to the relatively high protection of unionized workers and the strength of unions in general. However, it was never a total obstacle for those MNCs who decided on it, like Renault or Briegl & Bergmeister. Some companies made dramatic shifts in the structure of employees. For example Iskratel had more than 80 per cent semi-skilled and skilled employees when it started talks with Siemens about a joint venture. Now more than 50 per cent of the employees have a university degree. In all the positive cases both sides put a lot of energy in achieving a change for better. Companies where labor relations were entirely dominated by downsizing suffered problems with company culture and employee motivation. Most European MNCs had little or no problems in extending their culture to the Slovenian subsidiary or partner. As expected, more problems were experienced by companies with relatively arrogant, self-centered cultures.[7] Somewhat surprisingly, most US companies had very few problems in transferring their culture, as opposed to their Far Eastern counterparts.

Conclusion

Slovenia appears to be making a solid economic recovery, fulfilling the promise it showed at the time of Yugoslavia's break-up. It was by far the most prosperous of the former Yugoslav republics, with a *per capita* income (and many other indicators of economic and business development) of more than twice the Yugoslav average and significantly above other Central and East European countries. It has benefited from strong ties to Western Europe and suffered comparatively little physical damage in the dismemberment process.

Despite being present in Slovenia for more than thirty years, with the fall of the Berlin wall and in the context of changes in Central and Eastern Europe, quite a few MNCs have decided to reconsider their position and strategy. They have not always acted upon the positive past experience with Slovenia and in several cases have retreated, decreasing the intensity of their presence. There are a number of reasons for this. The overall confusion in the region forced MNCs to adopt centralized, 'controlled-risk' approaches that did not allow for local experience to be fully recognized. The practice of changing regional managers (and even the location of regional headquarters) quite frequently added to poor transfer of experience. Finally, the political and economic changes in the region meant that Slovenia lost much of its appeal as an effective bridge to the rest of Central and Eastern Europe. Opportunities in other countries were often seen as more promising and Slovenia was treated as a less significant, albeit relatively mature, market.

The internal market of Slovenia is very small and is therefore not the major reason for foreign investment. In spite of this limitation Slovenia has been able to attract some foreign investment, though not as much as expected or needed. Major reasons given by foreign investors for their activities in Slovenia include a past record of successful cooperation, the high quality of people, level of business infrastructure, level of products and services, as well as the potential for effective sales and marketing in other countries of (especially Central and Eastern) Europe. Even though the level of FDI in Slovenia is not high in terms of absolute numbers, the presence of MNCs is already strong and will certainly grow. The combination of an open market, its similarity to neighboring Austria, north-east Italy or Croatia, a highly developed business infrastructure, an excellent position in the region as well as its drive towards full EU membership, make investment into Slovenian operations, even considering the associated risks, increasingly attractive for the MNCs.

Bibliography

Bank of Slovenia Annual Report (Ljubljana: 1995).

Bilten Banka Slovenije; Ocena ZMAR, May 1996.

Business Central Europe, May 1998.

Dobovišek, Amada. Zakonodaja ovira tuje vlaganje, revija Manager, no. 3 (Ljubljana: 1997).

EBRD Transition Report, 1995.

EBRD Transition Report, 1996.

Furland, Franc. Tuj lastnik hoče na čelu domorodca, Manager, no. 11 (Ljubljana: 1997).

Gospodarska gibanja, Ekonomski Institut Pravne Fakultete, no. 292 (Ljubljana: March 1998).

Gospodarski Vestnik, number 29/XLVI, Ljubljana, 24 July, 1997.

Mrak, Mojmir. *Infrastructure Investment Need in Slovenia*, ib revja, {t.12, letnik XXXI, December 1997.

Natek, Karel. Discover Slovenia, Cankarjeva Založba (Ljubljana: 1993).

Poslovni almanah Gospodarskega Vestnika (Ljubljana: Business Almanac of Economic Weekly, 1999).

Potočnik, Janez Marjan Senjur, Franjo Stiblar: Strategija gospodarskega razvoja Slovenije, ZMAR, 1995.

Slovenia Country Book: USA Immigration Software, http://www.theodora.com/

Slovenia and Economic Outlook, home page of SKB banka d.d., http://www.skb.si/

Sočan, Lojze. Tuji kapital – kako ga pripeljati v Slovenijo; Gospodarski vestnik, 18 (Ljubljana: 1997).

Statistical Yearbook of Republic of Slovenia for 1986 and 1996, (Ljubljana: Statistical Office of Republic of Slovenia, 1986, 1996).

Notes

1 A good example of detailed benchmarking can be found in 'Economies in Transition Competitiveness Report', IMD Lausanne, 1993. Although not recent, it still presents a relatively accurate overview of the situation in the region.

2 It should be noted, however, that unemployment in Slovenia is predominantly structural in origin.

3 The negative attitudes towards foreign investors is not generally xenophobic in origin. Most foreigners living in Slovenia claim they do not feel excluded from the social life or are being ill-treated in any way. The reasons for such attitudes can most probably be found in the fear that placing full control over business development into foreign hands would decrease the level of responsibility towards the local community and destroy the delicate social balance in the areas such as employment policy or environmental protection. Heavy restructuring carried out by some of the first foreign entrants, although necessary from a business point of view, in general lacked the trust

of local management displayed by their foreign counterparts, as well as some highly publicized examples of fraud that have reinforced negative opinion.

4   The term 'unfriendly government' denotes the lack of enthusiasm displayed by the government and the parliament in proposing and passing legislation and accompanying regulations that would actively favor FDIs. However, an MNC can establish itself in Slovenia as a fully owned legal entity in a fairly straightforward manner, without excessive red tape. The time needed depends on the circumstances, but would typically be less than 30 days. Roughly the same time is needed to obtain a working permit or get clearance for buying a Slovene company.

5   By the end of 1998 all major Slovenian banks were offering their services over the Internet or were preparing to introduce this offer in the first half of 1999.

6   An example is the developing appreciation of Slovenian management education: the premier Slovenian business school, the International Executive Development Center, was invited to join the first wave of EQUIS accreditation together with 17 other top European institutions.

7   Some French and German companies have experienced such problems. For example Renault's subsidiary in Slovenia is having constant problems in attracting and keeping top local talent. The company culture is generally considered as hostile towards Slovenian managers and overly favoring French ex-patriots. Although this was only somewhat detrimental in running the large production plant, it created more significant problems in Renault's commercial operation in Slovenia.

7 The Baltic States – Trading Hub

R. JUCEVIČIUS

Overview

THE three Baltic States – Latvia, Lithuania and Estonia – together have a population of only seven million, but their importance lies in their geographic position on the Baltic rim between Scandinavia and Russia. The Baltic States also possess an existing developed transport and communications infrastructure that is able to cope with a heavy load of transit traffic. So the Baltic States have been heralded as an investment highway, the gateway through which funds from investors would flood into the emerging markets of Central and Eastern Europe. The Baltic states have been described at various times as 'the turnstile' of Eastern Europe, and 'the trade window to Russia.'

Since their impressive growth in the early 1990s, culminating in 1997, demand from developed economies has slowed while the shocks caused by Russia's economic crisis has dented the record growth figures. Estonia, Latvia and Lithuania are all showing signs of losing growth momentum. The reason for the slowdown is predominately the contraction of the Russian economy which has adversely affected all Baltic exports, and particularly agriculture.

The future challenge for the Baltic countries is to become even more competitive, embrace further political reforms and to tackle practical barriers to investment. The Baltic governments could take small but significant steps towards modernizing by streamlining customs controls, strengthening financial regulations and overhauling corporate governance rules for their companies. Such moves would not only greatly reduce their dependence on Russia but would leave them in a much better position to attract further foreign investment.

History

Lithuania, Latvia and Estonia have much in common, both historically and culturally, with Germany, Poland, and the Nordic countries. On the other hand, the Baltic States were under Russian occupation for nearly 200 years, from 1790–1991, and under Soviet influence, providing the bridge to Russia. Since restoring their independence in 1991, these three countries are attempting to re-establish links with Western European countries. The next few years of the transition process, the degree and extent of its success will be decisive for the region's future role.

Describing the Baltic Basin, which spans the wealth of Scandinavia and the emerging markets of Estonia, Latvia, Lithuania, (and Poland), Percy Barnevik stated that 'from the Mediterranean to the Arctic Sea, no region in Europe stands to gain more from the dramatic changes that have taken place in Europe.'

All three Baltic states, Lithuania, Latvia and Estonia, are situated on the Eastern coastline of the Baltic Sea and many similarities exist among them. However, Lithuania, Latvia and Estonia have developed independently and their history has only taken a parallel turn from the start of the 20th century. Lithuania is the first state to be mentioned in historical documents (in 1009), and the country became prominent in the thirteenth century. Having arisen as a distinct state, it expanded through wars. In the fourteenth century it became one of the largest states in Europe. At the end of that century, when the Great Duke of Lithuania became the King of Poland, the Commonwealth between these countries, under the domination of Poland, was established for almost 400 years.

This period was not as favorable to the other Baltic States. Latvia and Estonia were occupied by Germany and Sweden until the eighteenth century, and later by Russia. Only in 1918, after the Soviet Revolution in Russia, a number of new and old states in Europe began to restore their national identity and Lithuania declared itself a Republic (16 February 1918), Latvia (on 18 November 1918) and Estonia (24 February 1918). In the twentieth century the historical paths of these countries have become much more similar. In 1920–40, the Baltic States became independent entities, in 1940–41 they came under Soviet occupation, in 1941-44 they were under German occupation, in 1944–91 under a second Soviet occupation, and finally from 1990–91 they proclaimed independence and foundation of democratic institutions. In 1993–94, the Baltic States witnessed the withdrawal of Russian military forces, and in 1995 signed an association agreement with the EU.

Despite the political turmoil, the social unrest of the whole C&EE region, and finally the complete collapse of communism, the political situation is and has been relatively stable in all three Baltic countries. Even the occasional crises of government do not greatly affect the main thrust of government policy, with all the influential political parties pursuing similar goals moving towards a developed market economy, accession to the EU and NATO. But Latvia and Estonia face problems with Russia because of their refusal to recognize the large Russian minority groups who have been resident in those countries from Soviet times.

Geographically, the Baltic States hug the eastern coast of the Baltic Basin. All three countries also share borders with parts of Russia, Lithuania and Lativia border Belarus, and Lithuania borders Poland. The populations of the three countries are relatively small. The 1998 estimates were only about 3.6 million in Lithuania, 2.3 million in Latvia, and 1.5 million in Estonia, just over seven million in total. Since 1989, the annual population growth of all three countries has also been negative, with -0.45 per cent in Lithuania, -1.41 per cent in Latvia, and -0.99 per cent in Estonia, and population growth is set to continue its decline. This decline is due primarily to mortality rate exceeding birth rate, and the fact that many native Russians are leaving, especially from Latvia and Estonia.

Of the Baltic's total population, 70 per cent live in urban areas. The population of Lithuania is almost homogeneous with 82 per cent Lithuanians and only eight per cent Russians and Poles. The situation is different in Estonia, and especially in Latvia, where the native populations are 65 per cent and 55 per cent respectively and the rest are Russians or Russian-speaking minorities.

A difference between the Baltic countries is the level of economic concentration in the capital cities. Of all enterprises operating in Latvia, 55 per cent are concentrated in the capital city Riga and of the enterprises in Estonia 52 per cent are in Tallinn. In comparison, less than 20 per cent of all enterprises in Lithuania are in Vilnius, suggesting that over all, the different regions in Lithuania are more equally developed than in the other two countries.

Table 7.1 GDP levels of the Baltic states

Lithuania	1990	1991	1992	1993	1994	1995	1996	1997	1998
Nominal GDP ($bn)	n.a.	n.a.	n.a.	2.7	4.2	5.9	7.9	9.6	10.7
GDP per capita PPP ($)	n.a.	n.a.	n.a.	3681	3409	3612	3853	4164	4425
GDP (% change)	−6.9	−5.7	−21.3	−16.2	−9.8	3.3	4.7	6.1	4.5
Latvia	1990	1991	1992	1993	1994	1995	1996	1997	1998
Nominal GDP ($bn)	n.a.	1.0	1.5	2.2	3.6	4.5	5.1	5.6	6.4
GDP per capita PPP ($)	5472	5118	3463	3070	3213	3312	3515	3920	4136
GDP (% change)	−3.5	−10.4	−34.9	−14.9	0.6	−0.8	3.3	8.6	3.6
Estonia	1990	1991	1992	1993	1994	1995	1996	1997	1998
Nominal GDP ($bn)	n.a.	0.6	1.0	1.7	2.3	3.6	4.4	4.7	5.3
GDP per capita PPP ($)	4778	4433	3992	3803	3834	4171	4449	5082	5456
GDP (% change)	−8.1	−13.6	−14.2	−9.0	−1.8	4.2	4.0	11.4	4.2

Source: Business Central Europe Magazine

Economic prospects

Because they were closely tied to the USSR, the economies of the Baltic countries experienced dramatic declines after the fall of the Soviet Union. The traditional markets of the USSR were lost, while entering and competing effectively in Western markets proved beyond many local companies. On average, production dropped by more than 50 per cent, and in some industries, such as electronics or machinery, by almost 85 per cent and many large factories were closed down.

Privatization of state property began in 1991, with the approach to privatization slightly different in each country. Latvia and Lithuania opted for voucher privatization and introduced certain limitations on foreign investors, especially in strategic industries like telecommunication and energy. Estonia set almost no limitations on the extent of its privatization and was willing to sell almost any industry to local and/or foreign investors. This prevented many

Estonian companies from collapse, and the recession of the national economy was considerably lower here than in the other two Baltic countries. Privatization of the largest Lithuanian companies was started in 1996. By the end of 1998 almost all of them were sold to foreign investors, with the exception of Lithuanian Airlines which is still state-owned.

There are no significant differences in the industrial structure between the three countries. The biggest companies are concentrated in the power industries, telecommunications, chemical industries, transport, and oil (petroleum products and gas). The largest companies in all three countries, the industry sector and turnover (in US$) are listed in Tables 7.2, 7.3 and 7.4 below.

The fastest growing industries are telecommunications, banking, transport, intellectual services, hotels, and restaurants. The telecommunications industry is developing faster than any other due to foreign investments, which are on a similar level in all three countries. Banking is best developed in Latvia and Estonia and weaker in Lithuania, but in all three countries there are at least two strong banks with considerable foreign capital. After the Russian financial crash, all these countries experienced a banking crisis, with the most painful in Lithuania. The strict policies of the National Bank of Latvia helped to avoid a deep crisis in the country.

There are a number of sectors that are not developed yet and are far from their potential. Much investment could be made in the entertainment industry, for example, where there are major differences between the capital cities and rural regions. The situation is best in Estonia, which has a stronger tradition in the service sectors in general due to the country's proximity to Finland. Entertainment, restaurants and hotels are prospective sectors for investments where, although the number of businesses is growing, there are still many opportunities. The market for intellectual services and consulting is another underdeveloped sector where there is a lot of potential. But customers need to be educated about the advantages of such services. Local management consulting companies are competing successfully in some areas with global consulting firms.

The main foreign trading partners of all Baltic States are Russia, Germany, Poland, and the Scandinavian countries. The economies of these countries depend on raw materials, gas and partially, on oil from Russia. Textiles, ready-made food, wood, mineral fertilizers, and electricity are the main export goods, while machines, equipment, raw materials, and most of the sophisticated consumer products are imported from the West. For example, among the main export items of Lithuania in 1997 were mineral products, especially fertilizers ($682.2m or 17.8 per cent of total exports), textiles (16.2 per cent), machinery and equipment (12.2 per cent), and chemical products (9.2 per cent). The main imports were mineral products, especially oil (18,3 per cent), machinery and equipment (17.1 per cent), transport equipment (11.4 per cent), and chemical products (9.4 per cent).

Foreign investment in the Baltic states

According to a survey conducted by the Lithuanian Development Agency at the start of 1998, the factors most discouraging foreign investment are insufficient competence of local officials on the municipal level, an unstable system of taxation and business regulations, complicated customs procedures and a small market size. The factors encouraging investors to choose this region are related mainly to the highly qualified and cheap labor force, favorable geographic position, and many free areas (market segments) for development.

Total foreign direct investment (FDI) was about $1 billion in each country, the highest in Estonia, followed by Lithuania, and then Latvia. The structure of FDI varies slightly in all three. In 1997, the biggest investors in Lithuania were the USA (27 per cent), Sweden (12 per cent), Germany (11.3 per cent), United Kingdom (7.4 per cent), Luxembourg, Denmark and Ireland (six per cent) with 30 per cent of investments made by other countries. In 1996, the biggest investors in Estonia were Finland (33 per cent), Sweden (22 per cent), USA (eight per cent), and Russia (five per cent). However, in 1997 the most active investors were companies from Norway (40 per cent), Finland (20.4 per cent), and Switzerland (17 per cent). Most of the foreign investments are concentrated in the capital cities, in Vilnius (40 per cent), Riga (60 per cent) and Tallinn (70 per cent). 'Foreign investments are significant for these countries, but they are not very large by international standards, and not very large compared to the potential' says Mr Gronbjerg, at the Council for the Baltic States. This potential of the Baltic States is yet to be realized.

Table 7.2 Largest Lithuanian companies

Rank	Company	Industry	Revenues 1997 (US$m)
1	Mazeikiu Nafta	Oil refinery	607
2	Lietuvos Energija	Energetic	513
3	Lithuanian Railways	Transport	178
4	Lietuvos Dujos	Energy	169
5	Lietuvos Telekomas	Telecommunications	166
6	Ignalina Nuclear Power Station	Energy	157
7	Stella Vitae	Sale of gas	134
8	Lukoil-Baltija	Petroleum products	123
9	Achema	Fertilizers	112
10	Lietuvos kuras	Petroleum products	78
11	Ekranas	Electronics	78
12	Lifosa	Fertilizers	77
13	Kraft Jacobs Suchard Lithuania	Confectionery	62
14	Dirbtinis Pluostas	Chemistry	60
15	Baltijos Automobiliu Technika	Electronics	51
16	Lietuvos Aero Linijos	Airlines	51
17	Alytaus Tekstile	Textiles	48
18	Klaipedosjuru kroviniu kompanija	Sea transport	46
19	Klaipedos Maistas	Food	44
20	Snaige	Mechanical eng.	43

Source: Based on author's research.

Table 7.3 Largest Latvian companies

Rank	Company	Industry Sector	Revenues 1997 (US$m)
1	LUKoil Baltija R	Petroleum products	279
2	Latvijas kugnieciba	Transport	274
3	Latvenergo PVAS	Energetics	262
4	Turiba CS	Trade, catering	190
5	Latvijas dzelzcels VU	Transport	161
6	Ventspils nafta PVAS	Transit	142
7	Lattelekom SIA	Telecommunications	132
8	Latvijas gaze PVAS	Energetics	126
9	Skonto (grupa)	Banking	125
10	Dauteks PVAS	Chemical products	63
11	Rigas siltums AS	Energetics	62
12	Latvijas finieris AS	Wood processing	57
13	Kurzemes degviela AS	Petroleum products	54
14	Latvijas privatizacijas agentura VAS	Privatization	52
15	Liepajas metalurgs	Metallurgy	47
16	Latvijas balzams	Alcoholic beverages	47
17	Pareks banka	Finance	40
18	Lauma	Light industry	36
19	Baltic Sugar Company	Trade	36
20	LatRosTrans	Transit	35

Source: Based on author's research.

Table 7.4 Largest Estonian companies

Rank	Company	Industry Sector	Revenues 1997 (US$m)
1	Eesti Energia	Electricity	193
2	Eesti Polevkivi	Oil industry	116
3	Eesti Merelaevandus	Marine trade transport	79
4	Eesti Telefon	Telecommunications	67
5	Hansatee	Marine transport	67
6	Kreenholmi Grupp	Textile	66
7	Neste Oy Eesti	Petroleum products	58
8	Kiviter	Chemical products	57
9	ETK Hulgi	Trade	54
10	Tallinna Sadam	Sea port	54
11	Tallinna Soojus	Energetics	52
12	Eesti Raudtee	Railway transport	51
13	Eesti Gaas	Gas supply	43
14	Rakvere Lihakombinaat	Meat products	37
15	Eesti Mobiiltelefon	Mobile communications	37
16	Polva Piim	Dairy products	35
17	Liviko	Alcoholic beverages	35
18	EK	Petroleum products	34
19	Tallinna Kaubamaja	Trade	34
20	Nitrofert	Chemical products	32

Source: Based on author's research.

Estonia is a very aggressive investor in the neighboring countries. In 1996, its FDI abroad was $26.7 (*per capita*), while at the same time it was $7.8 in the Czech Republic, $0.4 in Latvia, and $0.3 in Lithuania. The main investors abroad are MNCs which have established their headquarters in Estonia planning to expand into other countries of the region. Estonia has been chosen for its location and because at the start of the 1990s it created the best conditions for investors. The Estonian bank Hoiupank is one of the most active investors in the financial markets of Russia, Latvia and Lithuania.

Scandinavian companies have made a number of particularly successful investments into the beer brewing industry. Baltic Beverages Holdings (a Prips-Hartwall Company) from Sweden and Finland, and Carlsberg from Denmark bought the biggest beer brewing companies in Latvia and Lithuania. After re-engineering the local companies, production has increased dramatically and these companies are now considered as the base for expansion to neighboring markets, including Western ones. Local brands of beer are dominating in the region again.

One of the biggest investors in Estonia is the Singapore-based textile and paper producer, Tolaram Group. By January 1998, its investment in this country totaled $40.8m. The Group's major investment was into production of textiles (Baltex 2000 in Tallinn, Baltex Non-Wowers in Viljandi and Lotus Colors in Sindi) and paper, in the Horizon paper and cellulose plant in Kehra. The Group is one of the Estonia's biggest exporters, selling 94 per cent of production internationally. The Estonian Investment Agency recognized it as the country's best investor of 1997. The Tolaram Group has also acquired some Latvian textile companies, the biggest of which is Tolaram Fibers. In 1998, Tolaram group bought the biggest Lithuanian textile company in Alytus. In the second half of 1998, the Tolaram Group faced serious problems due to the crisis in Russia. It cut production by 20–25 per cent, but its exports to Western markets did not decrease. Excited by the prospect of producing high quality fabrics at a very low price, the German company Calwer Decken und Tuchfabriken AG bought the textile plant Liteksas. Of total output achieved, 80 per cent of products were sold in Western Europe. But in 1998 the company faced serious problems. The holding company from Germany, Calwer Decken, went bankrupt and Liteksas was taken over by German AEG and DEF.

Activities of Baltic companies

The Baltic countries, and especially Lithuania, were leaders in the electronics industry in the former Soviet Union. After the collapse of the USSR, most of the factories in this industry lost market share because the main customers were from the defense industry. Companies have found it difficult to find new markets in the West. Very few of them succeeded, and the rest went bankrupt. One of the largest electronics companies in the region, Vilniaus Vingis, has established co-operative relations with Philips, and especially with Samsung, producing deflection yokes for televisions. Business partners were pleasantly surprised with the quality and cost of the components and tripled their orders in 1997. Samsung decided to build a plant for the production of computer displays, investing $300m and choosing Lithuania out of all the C&EE countries, despite the fact that other countries proposed more favorable investment conditions. Quality was the main criterion. However, the economic problems in Asia in 1998 have affected these plans, and the project has been suspended.

Country Globalization Drivers

The main advantages of the Baltic country's globalization drivers are related to their geopolitical situation. This includes their proximity to the Russian market, good knowledge of the Russian language, and understanding of Russian traditions and mentality. This enables them to be a kind of bridge between Eastern and Western markets. Their weakest points are a very small population, low level of the economy, and a lack of cooperation between the three countries. They try to achieve their goals on their own.

Market globalization drivers

Market globalization drivers are mixed for the Baltic States. MNCs recognize that a major advantage in setting up operations in the Baltic States is the minimal need for product adaptation. The populations are largely multilingual and acquainted with Western practices as well as Eastern/Russian ones. Although customers tend to be much more price sensitive, common customer needs are becoming increasingly similar to those in developed countries.

Because of their geographic location and historical links with other European and Nordic Countries via the traditional trade routes of the Baltic Sea, the people of the Baltic countries have a more Western mentality than many other countries of the former Soviet Union. There are essentially no differences in consumer behavior in these countries when compared to Western Europe. The main difference, and therefore disadvantage, from the point of consumer spending power is a very low GDP rate *per capita*. In 1997, it was $3220 in Estonia, $2645 in Lithuania, and $2300 in Latvia. It only takes a brief glance at wage comparisons to gauge the gulf between the aspirations to become modern Western economies and the reality.

An advantage of the Baltic States over other countries, is the highly qualified and multilingual population. Almost everybody speaks Russian in addition to their native language. In Estonia most people can also communicate in Finnish, and in Lithuania in Polish. German is better known by elderly people while English is taught in all schools, and it is the common unofficial language broadly used in business communication and intellectual circles.

Customer needs and tastes have experienced dramatic change over the past decade. Of course, due to the lower economic level and less sophisticated consumer needs, there are differences in buying preferences. For example, price often plays more significant role than quality of product. On the other hand, a shift in buying preferences is becoming apparent. Products made in China or Turkey, for example, are no longer as popular as they were in the early 1990s. Approximately every third inhabitant in the Baltic countries owns a car, most of which are made in Germany. The majority of these are, however, second-hand and seven to ten years old.

From the marketing point of view, the Baltic countries are in a way specific. First of all, essential differences exist between the population living in urban areas, especially in the largest cities, and the population of the rural areas. Secondly, some important differences can be observed between Estonians, Latvians and Lithuanians and the Russian minority. Most Estonians are Protestants and have much in common with people from other Western countries, especially from Finland. Lithuanians are largely Catholics, and their attitudes towards different aspects of work and behavior have much in common with those of Polish people. Even though the Lithuanian and Polish languages are different, a common historical heritage shared by these countries serves as one explanation for this situation.

Corporations acting as global customers are entering the region and often establishing their regional headquarters, one for all three countries or even for a broader region. For example, Kraft Jacobs Suchard established their headquarters in Kaunas, Lithuania for the Baltics and Belarus. However, in most cases MNCs prefer to establish their headquarters in Tallinn, Estonia (for example Coca-Cola and Toyota) and to start expansion to other countries. The reason for such a strategy is the fact that Estonia has created more favorable conditions with minimal restrictions for foreign investors and that its government institutions are more entrepreneurial and flexible in their cooperation with investors.

There is no active presence of global channels of distribution in the Baltic countries. Probably the main reason is the small market. To cover this market, the companies employ the strategy of using local distributors. It would be difficult to find any regional distribution chain because these countries still cooperate minimally with each other. At the same time, the channels of local distributors are developing very quickly in all three countries. For example, the leading distribution chains in Lithuania are Vilniaus Prekyba with 32 supermarkets or big grocery shops, IKI, and Sanitex.

Customers prefer to buy imported products and the services of well-known companies and are cautious towards local brands. The situation is different in the food market, where consumers prefer local products. Textile and furniture products could be another but less obvious example. When buying industrial products, especially electronics, significant attention is paid not only to brand names, but also to the country of manufacturing. The prices for the same product under the same brand name in different countries of origin can differ by 25–50 per cent. For example, automatic washing machines made in Germany cost about 30 per cent more than those produced in Spain. Colgate from England is 50 per cent more expensive than that produced in Poland. Most types of Western style marketing are transferable to the Baltics, but the use of American slang should be avoided. On the whole, European approaches fit better.

In the former Soviet Union, the Baltic countries were leaders in several industries. However, with integration into the Western economy it would be difficult to find even one of these industries. An exception is (aviation) gliders, where Lithuania maintains a leading position. The gliders are built by the local constructors and are in high demand on the world market.

Cost globalization drivers

Cost globalization drivers are mixed for the Baltic States, the Baltic States cannot contribute substantially to the sales volumes of MNCs or global economies of scale. The small size of their markets is the main reason why many companies focus on export. The Baltic countries can provide a gateway not only to the much larger, albeit underdeveloped, markets of Belarus and Russia, but also the large emerging Polish market and the developed markets of Scandinavia. Furthermore, MNCs rate the Baltic State's labor force as of a high standard, and good logistics and a reasonably developed infrastructure help facilitate trade.

The Baltic States have no strategic natural resources, except the limited amount of oil in Lithuania and a possible oil deposit in the Baltic Sea, and so cannot provide them to MNCs coming into the region. However they produce agricultural products, especially milk and meat, which are internationally competitive in quality and price. Kraft Jacobs Suchard, which exports a major portion of its confectionery, uses mainly local materials. Consumers in Russia, especially in Moscow and St. Petersburg, express high demand for food products produced in the Baltic countries. The main problem is the comparatively high price of these products due to trade barriers between the Baltics and Russia.

The Baltic countries provide favorable and well-developed logistics. Baltic seaports connect these countries to the rest of the world. The largest port is Ventspils in Latvia, with 75 per cent of all cargo in Latvia carried through this port. It is extremely important for the Latvian economy, because almost 60 per cent of all income to the national budget comes from operations of this port. Lithuania has established itself as the regional transport hub. The Klaipeda port in Lithuania is the only ice-free port on the Eastern Baltic Coast. Klaipeda was declared the EU's regional priority port and received funding from the EBRD and the European Investment Bank to finance major upgrades and expansion projects.

There are good road and railway connections between these ports and the Russian market. The Trans-European highway 'Via Baltica' from Helsinki to Warsaw is under construction. It will connect the capital cities of five countries – Finland, Estonia, Latvia, Lithuania, and Poland. Due to this, transit is an extremely important economic sector in the Baltic countries. The main destination countries of transit via the Baltic States are: Russia, the Baltic countries, Germany, and Poland for Lithuania and Latvia; Sweden and Finland for Latvia and Estonia; Russia and the Baltic countries for Lithuania and Latvia;

Denmark and Belarus for Estonia. Finland and Sweden are the main countries of destination of the Baltic States transit. The gross weight of the Baltic States transit is 800 748 tonnes. Lithuania is the main country of transit. Local transport companies take on responsibility and risk to carry goods to the markets of CIS. This is a considerable support to Western companies doing business in Russia and related countries.

The EU's Transportation Commission declared the two routes running through Lithuania among the ten priority transport routes connecting Scandinavia with Central Europe, and East-West routes linking the huge Eastern markets with the rest of Europe. Lithuanian trucking companies haul nearly the same tonnage to the West as do Russian companies, Russia being a country 262 times Lithuania's size. Latvian and Estonian trucking companies are not far behind. Lithuania especially plays an important role in the European transport infrastructure connecting Northern Europe (Finland, Northern Russia) with the rest of the continent (the North–South corridor), and Western Europe with Russia and other CIS countries (West–East corridor).

In 1996, the Lithuanian Parliament adopted the law of Free Economic Zones in Kaunas, Klaipeda and Siauliai. A Group AOI NV, consisting of Danish, Belgian, Dutch and Irish companies, won a tender for the development and management of FEZ Kaunas. AOI NV is in strategic alliance with the Port of Antwerp, Belgian Railways, Dutch Arcadis, information technology group Artemis, and so on. FEZ Kaunas and Siauliai have international airports, FEZ Klaipeda has a seaport. Aircraft of all types can land without any restrictions at Siauliai airport. FEZ incentives include a corporate tax holiday for five years and a 50 per cent tax reduction for the following ten years; no customs, VAT, excise, road, or real estate taxes.

In comparison with the rest of the former Soviet Union, there are significant positive differences in the standard of roads. The quality of the main roads in the Baltic States is not lower, and often higher than the rest of C&EE. The best roads are found in Lithuania, and in 1989 almost 40 per cent of all highways of the Soviet Union were found in this country.

Local telecommunication systems are not of particularly high quality. They call for urgent modernization, although even the most remote areas of these countries have phones. In 1996–97 Estonian and Latvian Telecoms were privatized by Telecom Finland and by Swedish Telia. This allowed both countries to begin the modernization of their telecommunication industries earlier than in Lithuania. In 1998 Telia of Sweden and Sonera of Finland acquired 60 per cent of the Lithuanian Telecom for $510 million with obligations to invest an additional $210 million to fully digitalize the Lithuanian telecommunication infrastructure.

There is a reasonably high level of telephone usage. Compared to the European average of 50 per cent, in 1997, Lithuania had 24 per cent usage, Latvia 25 per cent usage, and Estonia 27 per cent usage. Wire telecommunication companies in each country retain monopolistic status. However, mobile telecommunications are one of the fastest developing sectors, equipped with the most advanced technology, and each country also has from five to eight companies providing licensed mobile communications and paging services. There are some important differences between Estonia, Latvia and Lithuania. While Estonia has a higher number of mobile phone users, 10.5 per cent of the population, there is not a complete network coverage. Furthermore, most users are concentrated around the capital city and in other important business centers. The situation is similar in Latvia with 4.8 per cent of the population with mobile phones. In comparison, in Lithuania, were 6.5 per cent of the population use mobile phones, there is practically complete network coverage throughout the country.

After building a $450 million electricity line to Poland, Lithuania is going to become the most economical option in Central and Eastern Europe for import and export of electricity between the East and West. This will be the first direct electricity power link between the former Soviet network and the West. The contract for building this line has been signed with the Power-Bridge, a consortium of the US companies including CalEnergy, Siguler Gulf, Siemens, Duke Power Engineering and the Stanton Group.

The Baltic States can offer cost advantages for MNCs, providing a relatively inexpensive and highly trained labor force. In 1997, the gross average wage and salary in Lithuania was $188, in Latvia $225, and in Estonia $237 per month. There are essential differences in the average salary in different industries. For example, the highest average salary in Lithuania is paid to financial mediators and bankers ($404), government administrators ($391), and chemical product manufacturers ($283). The lowest salary is in agriculture ($109), and hotels and restaurants ($153).

Although the Baltic countries are not developers of new technologies themselves, they use and adopt them extensively. However, this process only began recently from 1994–95. Only a few local (and most of the foreign) companies at that time invested in new technologies. At present, the emphasis on modernizing technology is a priority for most companies. Radical changes are taking place in telecommunication, advertising, wood processing and other industries. Many companies, especially those doing business internationally, are receiving the ISO 9000 quality certificates.

PCs and the Internet are widely used by the local companies. Although before the early 1990s there were almost no computers at all in the three countries, at present most companies are equipped with modern computers. However, the level of computerization, and especially of the network development, is still not sufficient. Estonian companies and other organizations are the most active Internet users in the region.

Government globalization drivers

In the early stages of transition, government globalization drivers were the key factor that distinguished the three Baltic countries from each other. Initially the best conditions for foreign investors were provided by Estonia which reaped the economic rewards of rapid liberalization and mass privatization. Latvia and Lithuania, rather slower to embrace full market reforms than their neighbor, have gradually introduced incentives of their own. There is competition rather than co-operation for FDI between the countries. Since the Baltic States compete largely on offering stable conditions for foreign countries wishing to operate in the region, government globalization drivers will be key in determining which country is the winner for attracting FDI in the future.

Government globalization drivers remain extremely important in creating advantages for foreign investors throughout the C&EE region. Probably the best conditions for foreign investment are found in Estonia, where the mentality is generally more Western and government bodies are more managerial-oriented and competent. Close cultural relations with Finland in the Soviet era helped Estonia to train qualified managers earlier than Latvia or Lithuania. Because of this, professional management consulting started in Estonia 15 years earlier than in the other two countries. This country was the first and the most active in developing and implementing strategies for attracting FDI, being sensitive to investors' needs, and working hard to create a positive image of the country in the world. This was probably the main factor which determined that Estonia was recognized as a leader of economic development in the region and was invited to start negotiations with the EU. The differences in economic development between the three countries are not so significant .

Investment agencies exist in all three countries to facilitate and increase the flow of foreign investment. Unfortunately, there is no cooperation, only competition between them, and investors would not receive useful information about business conditions or opportunities in the other two countries.

Bilateral agreements about investment promotion and protection have already been signed with more than 30 countries, among them all EFTA countries. Agreements on the avoidance of double taxation are in effect with more than ten countries in all three Baltic States, and the negotiations have been completed approximately with the same number of countries. The USA, Canada and Australia have granted these countries trade preferences under their Generalized System of Preferences (GSP). All three countries have ratified the main international Conventions on recognition and enforcement of foreign arbitrage awards, intellectual property, and others. However, even if decreasing, there are still problems about enforcing the protection of trademarks and intellectual property rights. However, there were very few complaints about this from foreign companies.

The Free Trade Agreement between the Baltic countries and the EU came into effect in 1995, and there are various free trade agreements between a single Baltic country and other countries. From 1995 these countries became associate members of the EU with goals to full EU membership. The Baltic States do not participate in any political or economic trade bloc with the former Soviet Republics overall, but each of them has signed many different independent formal agreements. This creates more favorable conditions for international business and trade.

Table 7.5 Largest Foreign investors in Lithuania

Rank	Investing company	Country	Lithuanian Company	Industry	Invested (as of January 1998) (US$m)
1	Motorola	USA	Omnitel	Telecommunications	63
2	Tele Denmark A/S	Denmark	Bitë GSM	Telecommunications	45
3	Millicom East Holding B.V. Philip Morris International Finance Corporation	Luxembourg USA	Philip Morris Lietuva	Tobacco products	38
4	Den Norske Stats Oljeselskap	Norway	Lietuva Statoil	Petroleum products	32
5	Shell	Great Britain/ Netherlands	Shell Lietuva	Petroleum products	29
6	Neste Oy OY	Finland	Neste Oy Lietuva	Petroleum products	25
7	Euro Oil Invest S.A.	Luxembourg	LUKoil Baltija	Petroleum products	24
8	Calwer Decken und Tuchfabriken DEG; International Finance Corporation (IFC)	Germany	Liteksas & Calw	Textiles	24
9	The Coca-Cola Export Corporation	USA	Coca-Cola Bottlers Lietuva	Soft Drinks	22
10	Partec Insulation Finnfund; NEFCO; EBRD	Sweden/Finland Great Britain	Partec-Paroc	Construction materials	20

Table 7.5 Largest Foreign investors in Lithuania (cont'd)

Rank	Investing company	Country	Lithuanian Company	Industry	Invested (as of January 1998) (US$m)
11	Baltic Beverages Holding (a Pripps-Hartwall Co.)	Sweden/Finland	Kalnapilis	Brewing	16
12	Odense Steel Shipyard Ltd.	Denmark	Baltijos laivu statykla	Shipbuilding	16
13	Kraft Food International	USA	Kraft Jacobs Suchard Lietuva	Confectionery	15
14	Tuch Fabrik Wilhelm Becker	Germany	Eurotextil	Textiles	15
15	Siemens AG	Germany	Baltijos Automobili Technika	Electronics	14
16	Svenska Petroleum AB	Sweden	Genciu nafta	Oil Processing	14
17	Iceland Health Company; Icelandic Pharmaceuticals; Iceland Prime Contractor Pharmalude; Swedfund International	Iceland/Sweden	Ilsanta	Pharmaceuticals	12
18	Pemco International	Norway	Pemco Kuras	Petroleum products	11
19	Woodison Trading Ltd; Osman Trading AB; Dilan Trading NV; Lancaster Distral Inc.	Ireland/Sweden/ Netherlands/ USA	Klaipedos nafta	Oil Terminal	11
20	Ochoco Lumber	USA	Ochoco Lumber	Wood Processing	10

Table 7.6 Largest Foreign investors in Estonia

No.	Investing company (country)	Estonian company	Industry	Invested (as of March 1998) (US$ m)
1	Telia AB (SWE) Telecom Finland (FIN)	Eesti Telefon	Telecommunications	81
2	Telia AB (SWE) Telecom Finland (FIN)	EMT	Mobile telecommunications	51
3	Asean Interests Ltd (Singapore)	Tolaram Group	Paper products, textile	42
4	Tschudi & Eitzen (Norway) Stanton Capital Corp. (USA)	Eesti Merelaevandus	Sea transport	39
5	Atlas Nordic Tsement, Nefco, Finfund (FIN)	Kunda Nordic Tsement	Cement production	31
6	Coastal Holding Inc. (USA)	E.O.S.	Oil terminal	30
7	Radiolinja OY, (FIN) Helsingin Pyhelin OY (FIN)	Radiolinja	Mobile telecommunications	26
8	Statoil (NOR)	Eesti Statoil	Sale of fuels	22
9	Millicom Int. (LUX)	Ritabell	Mobile telecommunications	22
10	Neste Oy (FIN)	Neste Oy AS	Sale of foods	21
11	Coca-Cola (USA)	Eesto Coca-Cola Joogid	Soft drinks	19
12	BCI Group Ltd	Balti Cresco Investeerimisgrupi	Investments	17
13	Hebeda Tra AB (SWE) ThomestoSverige (SWE)	Toftan	Sawmills	13
14	Boras Vaferi AB (SWE)	Kreenholmi Group	Textile products	11
15	Svensk Leca (SWE)	Fibo Exclay Eesti AS	Gravel	11
16	Elcoteq Network (FIN)	Elcoteq Tallinn	Electronic components assembly	10
17	Gazprom (RUS)	Nitrofert	Chemicals	9
18	OSF Portfolio Investment(DK)	Loksa Lavaremonditehas	Shipbuilding	9
19	Oy AGA AB (FIN)	Eest AGA AS	Gas products	9
20	Akzo Nobel (NLD)	ES Sadolin	Paint manufacturer	7

Foreign investment is allowed in all industries, except national defense and matters affecting the security of the country, such as the production of narcotic substances; the manufacture of weapons, and the establishment of gambling enterprises (the later is being discussed intensively). Property is protected from expropriation and it can only be appropriated under extraordinary circumstances and only when compensated for promptly at market value in hard currency. Repatriation of profits derived from foreign currency earnings is not restricted.

The US political institute 'Heritage Foundation' has published its survey data declaring the Estonian economy to be the most liberal among all the Baltic States. According to the Heritage Foundation, the economies of Lithuania and Latvia are still protectionist. The 'black market' is most deeply rooted in Lithuania. The most favorable taxation system is in Estonia, while in Lithuania government plays the most significant role in regulating prices and wages.

The number of state-owned companies is constantly decreasing, and they are slightly different in every Baltic country. Even if the general level of privatization is similar in all three countries, in 1997 there was a greater number of large state-owned companies in Lithuania than in Latvia or Estonia. These included Telecom, shipping and insurance companies, power stations, and airlines. Most of them were privatized in 1998 or 1999. For example, Amber Teleholdings (Swedish Telia and Finish Sonera) have privatized the Lithuanian Telekomas. Williams International of the United States purchased 33 per cent of shares of the largest Lithuanian oil conglomerate – consisting of Mazeikiai Oil Refinery, Birzai Oil pipeline and Butinge Oil Terminal – for $150m with agreement to invest an additional $300m in equipment and infrastructure. Norway-based Frydenbo Holding AS, together with Lithuanian consortium Azovlit, purchased 93 per cent share of a ship repair yard VLR for $21m. In 1997, VLR renovated 128 ships from 18 countries.

All three Baltic countries are now using the same strategic approach in privatizing their big companies. They are looking for strategic investors, who are willing to invest in the modernization of these companies and are able to develop a market for their products. The pace of activity has lead Hansa Investment, the Baltic investment bank to predict 'in two years, up to 80 per cent of the leading companies in the Baltics will either be owned by Western corporations or have a Western corporation as a strategic partner'.

Due to their shared Soviet heritage, some technical standards of these countries are different from those of Western Europe. But the situation is rapidly changing because of the intensive implementation of Western technologies. All new technical requirements and regulations are comparable with those in the EU. A number of companies granted a quality certificate ISO 9000 is constantly increasing in almost every industry sector. There are strict requirements for the quality of imported goods, especially food and drugs. Importers of pharmaceutical products are required to register the drugs at the Ministry of Health before obtaining their import license. The same applies to marketing regulations: there are almost no essential differences.

Competitive globalization drivers

The Baltic countries, with a total population of 7.7 million and relatively small consumer purchasing power, are not able to create global strategic importance for MNCs in almost any industry. They are not leading countries in terms of customers, competitors or innovation. The main importance is their geopolitical location – they serve as a bridge between the West, Belarus and Russia. Foreigner investors recognize that it is much more risky to do business in Russia than in the Baltic countries, while local Baltic entrepreneurs and managers are well acquainted with the formal and informal rules of doing business in this region and have enough experience to take such risk.

Overall Global Strategies for the Baltic States

MNCs strategies in the Baltic States are to a degree determined by their individual goals and degree of privatization in their industry. The strategies used are often not dissimilar from those used in other countries. In many cases MNCs have acquired, then privatized and modernized, local companies. The establishment of joint ventures is a common strategy for initial entry.

Market participation strategies

In the long term, many MNCs entering the Baltic countries have their eye on supplying the much larger markets of Russia, Ukraine and Belarus as well as the smaller markets of the Baltic States. However many multinationals have reaped substantial rewards in the smaller Baltic markets since even the more developed sectors, where the basic infrastructure already existed (such as telecommunications), were far from fully saturated.

At the start of the 1990s, the most typical strategy used by MNCs entering the region was the establishment of joint ventures with local partners or participation in privatization. With legal changes and increased privatization, many investors can own a 100 per cent share in almost any industry, in all three countries. This is the case with Kraft Jacobs Suchard, Siemens, and Philip Morris in Lithuania; Shell, Statoil, Neste Oy Oy, ABB and others, in all three countries.

The largest investors in the Baltic countries operate in telecommunications (Motorola, Telia, Ericsson) and petroleum products (Williams International, Statoil, Shell, Neste Oy). These companies were among the first to enter the market. Despite the fact that telecommunication systems in the Baltic countries were much better developed than in any other country of the former Soviet Union (coverage in Estonia was 29 per cent, in Latvia 25 per cent, and in Lithuania 24 per cent), in the early 1990s it was extremely difficult to make a phone call abroad because all connections had to be made via Moscow. Omnitel, a joint venture between Motorola and Lithuanian Telecom, was established in 1991.

Although the first operator in mobile telecommunications was Comliet from Denmark and Luxembourg, Omnitel soon became the industry leader. Its main goal was to create a modern telecommunication system in Lithuania. The first step was to build a satellite station. But no one, not even Omnitel, expected its success. At the end of 1997, it had 100 000 subscribers from a total number of 197 000, while the second biggest competitor, GSM Bitë, also a Danish-Luxembourg company, had only 70 000 subscribers. Along with the development of infrastructure, the annual market growth in 1997 was 4000 per cent. In 1998, Omnitel expected to double its number of clients. This is because even with this dramatic growth, the market in mobile telecommunications is far from saturated in Lithuania and is still behind Estonia. In 1997, nine per cent of the population in Estonia were using mobile phones in comparison with only four per cent of the population in Lithuania, which is more than twice the size. Currently, mobile phone coverage is almost total in Lithuania. Most subscribers in Latvia and Estonia are concentrated in and around the capital cities.

Kraft Jacobs Suchard (KJS) entered the confectionery industry where local companies, especially Latvian, were well positioned and considered among the best in the former Soviet Union. KJS acquired one of the leading Lithuanian companies in the industry, replaced old technology and, with an intelligent marketing strategy, became a leading company in the industry. At the start of 1998, the company was awarded an ISO 9000 quality certificate. The majority of its products are supplied to the Baltic States and Belarus. The competition is tough because while the market potential is only slightly more than 50 000 tons, Estonian, Latvian and Lithuanian companies could produce three times that amount. There are other competitors like Master Foods, Leaf, Cadbury, Fazer (Sweden), Nestlé, and so on, operating in this market.

Since ABB, one of the world's largest power engineering companies, came to the Baltic States, it has identified and pursued major opportunities to develop a profitable business for local and export markets. The original operations were established to sell ABB products in local energy markets. Later ABB entered into joint ventures with the local energy service companies, to service and repair energy equipment. An ABB joint venture in Latvia has three factories in the main cities – Riga, Daugavpils and Liepaja. In other Baltic countries, ABB has not yet bought manufacturing plants and concentrates on distribution and product development, especially in Lithuania. Regional management quickly recognized that the location was excellent for serving Scandinavian and North-West European markets.

Philip Morris was the first foreign company (in 1993) to express interest in participating in privatization of the state property of Lithuania. Before that, only voucher privatization took place. There were nine other companies from different countries willing to buy the biggest tobacco plant in Lithuania, Klaipedos Tabakas. The company has signed an agreement with the government that another small factory will be acquired in the next three years. At present, Philip Morris Lithuania is the main producer in the industry, throughout the Baltic region and Belarus. Not only does it supply to these markets, but it also exports to Western countries.

Coca-Cola is the fastest growing company in the region. It entered the Estonian market in 1992, Latvian in 1993, and Lithuanian in 1994. It is a leading company in the industry in the Baltic States with a total market share of 55 per cent. In Estonia it has almost 75 per cent, and in Lithuania 40 per cent of the market. The company has its own bottling plants in all three Baltic countries but still does not use a franchise system. The main reason for this is probably the lack of trust in the ability of local companies to maintain total quality. Initially, Coca-Cola intended to buy a part of the biggest beer brewing company in Lithuania and to reorganize it. However, negotiations fell through partly because the local management were not willing to risk their own interests.

PepsiCo was active in this market in Soviet times but later lost its position. However, in 1996 Pepsi-Cola General Bottlers (PCGB), related to PepsiCo, entered the market from Poland, expanding with a new strategy. The company's future plans are very ambitious. PepsiCo considers the Baltic region as the spearhead for expansion to the East. However, at present the main task is to capture a part of the market from Coca-Cola and local brands.

Petroleum product companies like Shell, Statoil, Neste Oy, and LUKoil are very active in the region and working very profitably. The main suppliers of oil products are local state owned companies, to be privatized. Statoil and Neste Oy use oil produced in their own plants, while Shell imports oil mainly from Russia.

Product and service strategies

MNCs have had both successful and failed products in the Baltic States markets. In general, MNCs introducing new products or services into the local markets have almost always been successful while entering a sector in which there is already a strong local competitor proves much more risky. Redevelopment of local products by a foreign MNC is another successful approach.

Consumer products and services do not require any more adaptation to the needs of local customers than in other Central European countries. Existing differences are related to the lower level of consumer education about particular aspects of products. For example, in a food industry, consumers often lack knowledge about nutrition. Differences in consumption exist between different segments of the population. For example, Kraft Jacobs Suchard and retailers of coffee have discovered that Estonians prefer ground coffee, Lithuanians prefer coffee beans, and Latvians prefer instant coffee, while the Russian population of these countries generally prefers tea.

All consumer markets can be divided into four major segments. First, the upper income segment usually buys modern and expensive products and services. Price does not play an important role in determining consumer choice. Efforts to adapt products to this segment would probably fail because these people prefer original brands and would not buy a modified product. On the other hand, a changing trend in their behavior can be observed as they have started to pay more attention to the value-to-price ratio. Local brands are not popular in this segment.

The second segment covers educated consumers with moderate incomes. They usually pay more attention to quality but are price sensitive because, in general, their economic wealth is much lower than that in Western countries. Analysis of strategies of MNCs in the Baltic countries shows that these companies try to supply cheaper products or to charge lower prices than in the West. On the other hand, some groups of products are even more expensive than those in the West, for example, clothes or leather products.

The third segment consists predominantly of elderly people with a very limited income. They are conservative users, do not know foreign languages and do not understand product instructions. The government has not passed a regulation yet about the requirement to print product information in the native language, but most foreign companies have already started doing this and print instructions in several languages, usually including Russian, Lithuanian, Latvian, Estonian, and sometimes Polish. By the year 2000, this will become the formal requirement in all three Baltic countries. Consumers from this segment usually buy local products or those made in Eastern countries which are generally much cheaper.

A recent shift in preference towards local products made in partnership with Western manufacturers can be observed in all three groups. For example, 77 per cent of Estonians prefer local food, provided it is of equal quality to imported goods.

The fourth segment covers the younger generation whose expectations and preferences are very similar to those of their Western counterparts. Because of new opportunities to visit the West or to watch Western television, they are well acquainted with current worldwide trends. However, due to the small size of this market segment, companies face some problems assuring broad product choice, and prices for fashionable goods are usually higher than in the West.

The situation is slightly different with industrial products. Sometimes it is difficult for local companies to implement modern technologies or equipment into the existing production system, which is usually old. The economic situation does not allow them to buy all the necessary equipment, and changes in construction or other parameters may be necessary. MNCs have developed a strategy to overcome this problem. They can create a product development group or, what is more common, enter into strategic alliances with local companies. For example, the Dutch company Festo, producing pneumatic elements, has established the alliance 'Four in One' with another Dutch company Hitech, supplying hydraulics, and two Lithuanian companies GTV (which designs and produces engineering systems and measuring equipment), and Elinta, (which designs, replenishes and assembles the production automation and technological processes and solves problems related to electronics and its programming).

The introduction of air-handling equipment and ventilation products group by ABB was considered as the most successful for this company in the region and has the highest rate of growth. It was introduced as a systems product including packaging and aftersales services offering the highest value for the company. In comparison to this, because of insufficient adaptation to local conditions, overly high prices, and lack of negotiations and interest from ministry officials, efforts to introduce power network control systems failed.

One of the most successful product introductions has been made by Kraft Jacobs Suchard (KJS), who introduced chocolate under the Lithuanian brand name 'Karuna'. It was the first time in the Baltic countries that a product developed by an MNC, under a local brand name, has been supported by highly professional marketing know-how. A group of other products were developed later under the same brand name and have captured more than half the total market in the region. This strategy now has been applied in other, former Soviet, countries. For example, the trademark of chocolate 'Korona', has become a dominant brand in Ukraine. In comparison, the chewing gum 'Hollywood' was an unsuccessful introduction by KJS because this product had to compete with other strong brands already present in the market, and KJS eventually decided to get rid of it.

Multinational companies that have entered the region have often enhanced the quality of local production. A good example is the electronics plant Elcoteq Tallinn AS, owned by the Finnish Elcoteq Network. Under the management of a foreign multinational, in only a few years, this local company developed from basic subcontracting work in assembling companies into a high technology manufacturer of end products. The company now produces a new generation of mobile phones for the Swedish company Ericsson, which are distributed worldwide under the label 'made in Estonia'. By 1997, the turnover of Elcoteq Tallinn was $18.6m.

Consulting companies like Arthur Andersen, KPMG and others, stress that their most successful branches are related to financial and auditing services which are the branches of consulting services underdeveloped in the region. Local companies are not able to compete with them on quality but are successful at competing on price. In general however, foreign companies, their local partners, and banks prefer to use the services of an internationally recognized auditing company. All Big Five consulting firms are present in Estonia, Latvia and Lithuania and the demand for this type of service is constantly growing. Some foreign business consulting services are rated substantially higher on the global market place than they are perceived by local management. Therefore the efforts to introduce them have not been successful yet. In training and consulting, local companies are less disadvantaged because they have a better knowledge of local conditions.

Activity location strategies

MNCs in the Baltic States conduct little R&D, some manufacturing, a lot of distribution and significant amounts of local marketing. Many MNCs use the Baltics to site 'offshore' factories to serve Russian and other CIS markets.

For example, ABB chose to locate its activities in the Baltic countries to reap the advantages of positioning themselves in a geographical location, with good transport infrastructure, linking them to the much larger markets of Russia, Belarus and the Ukraine. With all the industrial companies in the region undergoing restructuring, ABB can consolidate a market with good growth potential. The company uses slightly different strategies in Latvia and Lithuania. In Latvia it has acquired three manufacturing plants, while in Lithuania, as well as in Estonia, it has mainly regional marketing and sales offices in different cities.

The location strategy of KJS involved developing the outsourcing system in the region after the company established a large confectionery plant. At present, almost all the main raw materials, with the exception of coffee, are purchased from local suppliers. Combined with the re-engineering of local plants for the production of milk products, and thus increasing their ability to supply quality products, the KJS plant has become a procurement center for other plants operating in Western Europe. The research department is coordinating its activities with headquarters in Munich, Germany. The technology created in KJS Lithuania has also been implemented in other plants. Development of new products is realized at the local plant cooperating with KJS in Switzerland, Germany, and Scandinavia. Because most of the possible activities of the company are already located in this region, it is not planning to locate new ones but is concentrating on expanding production by using local resources. Food, and especially the milk industry, is among the most attractive industries for foreign companies. The biggest problem at the moment is the inefficiency of small farms, which make operating costs higher.

The Baltic countries can be considered a prospective area for locating and expanding engineering, high skill assembly and service activities. Development of products in connection with adoption and design of software is a common strategy. IBM had sales representative offices in the Baltic countries for a long time, but in 1996 the company decided to establish a factory for assembly of high-end Power PC line and other personal computers. IBM worked to set up a local company, Sigmanta, to be the main production center in Central and Eastern Europe, as well as in the CIS. It should become the main source of IBM computers sold in the Baltic States, CIS, and Central Europe.

Branches of the international electronics giant Siemens have established their factories or regional distribution and marketing teams in all three Baltic countries. Probably the most successful decision was to establish a plant for producing automotive electrical systems. The plant produces about 26 000 units (cables) of various products a week. Before 1997, the plant supplied components to Volkswagen, Saab, and Renault, but now it produces exclusively for Renault's Megane model. Other business units of Siemens actively participate in the modernization of telecommunications systems in the region. They have established regional marketing teams for medical equipment.

Global marketing

There are almost no limitations to using global marketing techniques and approaches in the Baltics. However, because consumers are still relatively unfamiliar with differences between a broad range of similar products, it is better for MNCs not to promote too many products at once, but concentrate on one or two and on consumer education.

All companies can use their global brand names, advertising (especially European) and to some extent pricing and sales methods. This has been highlighted by many MNCs as one of the major advantages of this region. Because of the lower level of consumers' economic wealth, price plays a more significant role in these countries than in the West. Some differences between consumer behavior do exist between the three countries. For example, Estonian customers pay more attention to quality and less to price when compared to Lithuanian and Latvian consumers. This is especially true in the case of industrial consumers.

Advertising plays an extremely important role in the development of the marketplace because most products are new to local customers, and they may be unfamiliar with their features and often have not yet developed product loyalty. Television advertising is one of the most efficient methods of reaching consumers. The participation of local 'stars' is often a successful strategy. For example, a well known Lithuanian singer advertising beer under the brand name 'Dvaro' has been so successful, that this new brand has become the most popular in Lithuania. For many people this brand even became a synonym of the entire company. Another example is Coca-Cola's advertising campaign for Sprite, which used a well-known Lithuanian former NBA basketball player, Marciulionis. The brand name has become the most popular in the country, even more popular than Coca-Cola itself. MNCs understand that using American slang or obscure expressions unknown in the region make advertising considerably less attractive. Another key factor making Coca-Cola's marketing strategy successful, is that it does not try to promote a broad range of products, but concentrates on only three. Coca-Cola, Sprite and Fanta, rather than promoting a broad range. In comparison, Pepsi-Cola General Bottlers (PCGB) which is much less successful, is selling six key products.

Most MNCs emphasize the necessity to educate local consumers. For example, KJS faces the problem that consumers are not familiar with nutrition, and do not understand the meaning of special nutritional value symbols on food products. Other MNCs face similar problems. Masterfoods made a mistake when they tried to promote products like 'Whiskas' and 'Chappy' in 1993–94. Many people found the campaign insulting because of their economic situation and the fact they were not used to the idea that pets should receive their own special 'gourmet' meals, foreign pet-food being more expensive than products sold for human consumption. Although Masterfoods was one of their best clients, certain television programs began to avoid their advertising and the company had to change their strategy. There were no similar problems when advertising other products, such as Mars, Snickers or Bounty.

In most cases, although not always, the prices for similar products are lower in the Baltic States than in the West. This is mainly because the production savings made on labor, logistics and other costs, as well as price sensitivity of consumers. The price level is different in all Baltic States. Usually it is a little higher in Latvia and Estonia than in Lithuania, again, to a degree, reflecting the relative economic spending power of consumers in these countries.

Global competitive strategies

Because their markets are small in size the Baltic countries are not usually included in a globally coordinated sequence of competitive moves. Company strategy is mostly conducted on a local basis. Within the local market, competition can be strong. For example Coca-Cola and Pepsi-Cola are facing strong competition from the local brand Selita. Similarly it is not proving easy for the multinational company, Calwer Decken, to fight the local company, Drobe, for the same customers in the textile industry. Only in a few situations can the Baltic States contribute to the global competitive strategies of MNCs, and all of them are more or less related to the plans for further expansion to the Eastern markets, especially to Russia and Belarus. Good examples are ABB, KJS, IBM, DFDS Transport and others. Asian Investments Ltd., owning some companies in Estonia and united under Tolaram Investment, have started to expand their Estonian-based textile, paper and property business into Latvia and Lithuania. The same strategy was used by Toyota and several other MNCs.

Global Organization and Management

Local resources and country infrastructure in all three Baltic States are generally favorable for establishing businesses in this region. All kinds of qualified staff can be found and used on a full scale, especially after brief training. In some sectors, however, such as catering and services in hotels and restaurants, the quality can vary dramatically. MNCs should consider training to ensure high standards

For the most part, though, the labor force is ranked highly by MNCs operating in the Baltic States. In 1995, the Siemens plant BAT was ranked number one in the Siemens automotive division in the Time Optimization process and two-year-old program designed to solicit innovative suggestions from employees to improve plant efficiency. The overall quality of the labor force was also ranked as one of the best. A similar situation is also found in other companies, such as ABB. This is not to say the labor force is generally better in the Baltics, only that low labor costs give companies the opportunity to hire people with much higher levels of education, to do a job which is usually done by less educated people in the West. In a number of companies 100 per cent of the staff have a university degree or higher level of technical education.

For most MNCs, it is not a problem to find qualified local managers. Executives of MNCs stress the level of competence of local managers and especially of technical specialists, considering this one of the main advantages of the Baltic countries. Therefor the majority of MNC business units are fully managed by local management. Initially it takes some time to retrain key specialists in Western factories, after which they can work independently. For example, KJS, which has a regional headquarters, is run entirely by Lithuanians. This is also the case with Calwer Decken, AGA, IBM, Statoil and many others. In other cases, companies have some senior managers from the country of origin, mainly CEOs and financial managers, and perhaps at the start, a marketing manager. This is the case with Siemens, Philip Morris, Kemira-Agro, Neste Oy Oil, and McDonald's, among others.

Metsa–Serla, a Finnish packaging company, provides an excellent illustration of how, and why, multinationals should consider using local management. In 1997, Metsa–Serla bought the Lithuanian company Medienos Plausas for producing packaging material, replaced the local management staff with Finnish managers, and reorganized the company. The result was a considerable decrease in sales. Customers shifted to other local companies who used more flexible and locally sensitive sales policies.

Admittedly, one of the major weaknesses of local managers is their lack of strategic thinking and teamwork practices. Project management capabilities are also not well developed. The biggest gap between Western and former socialist education in business and management was in finance management and marketing. This is still a problem in these countries, although MNCs can get around this by hiring the best specialists.

The Baltic countries could easily become a good source of global managers. Most graduates from local universities speak not only their native language (Latvian, Lithuanian or Estonian), and Russian and English, but also one or two other European languages. Because of the historic links with the Scandinavian market, local people have been exposed to Western culture and business practices. Business and management education is best developed in Estonia. Both Estonia and Latvia have private higher education business schools and even private universities, along with the state-owned ones that offer business courses. In Lithuania establishing a private university is much more complicated and therefore there are only private business schools and state universities.

There are essentially few cultural differences between the Balts and other Europeans. For example, Estonian people have much in common with Finnish, and Lithuanians with the Poles. This makes using similar marketing or management approaches in these countries easy.

For a number of important reasons, in most cases, MNC headquarters for the region are established in Tallinn, Estonia. Firstly, this is because Estonia was the first among the Baltic countries to actively target foreign investors creating favorable conditions for them. These conditions were not necessarily only economic in nature. They often began by removing bureaucratic barriers that could slow down or discourage foreign companies from entry. Estonia was greatly advantaged by its close relations with Finland in the Soviet period, which meant a substantial number of well-trained managers already existed there. Consequentially, Estonia was recognized by investors as the Baltic country where government and municipal officials did not create problems for foreign investors (unlike in the other two countries). Of course, the situation in other two countries is rapidly improving, especially at the government level, though less so on the municipal one.

The structure of ownership varies according to the individual company and sector. For example, all mobile telecommunication companies are fully owned by foreign investors in Lithuania, while in Estonia they are only 49 per cent foreign owned. Petroleum companies like Shell, Statoil, and Neste Oy have full foreign ownership in all three countries. Ownership structure depends very much on the company's initial entry strategy. In cases where local companies were acquired in the process of privatization, mixed ownership is more common. Establishment of joint ventures is the most common way of starting business in all three Baltic countries. In 1997 companies with foreign capital constituted 12 per cent of all businesses operating in Estonia, 16.3 per cent in Latvia, and 8.6 per cent in Lithuania, and most of these were joint ventures.

Conclusions

Due to their small markets and underdeveloped economies, the Baltic countries have limited possibilities to play key roles in the global strategies of multinational companies. Establishment of production companies for supplying specific national or regional markets does not seem to be a very high prospect. The biggest advantages of the Baltic States are related to their geographical position, and their proximity to a very large CIS market. However given the situation in Russia and the CIS, this should be considered more as an advantage for the future. The most important precondition for this region becoming an attractive place for investors is favorable trade conditions with Russia. Local understanding of both European management culture and the Eastern European way of doing business, as well as a well-developed infrastructure and relatively stable political environment, make the Baltic States an ideal choice for companies looking to expand to CIS markets.

The Baltic countries remain a place mainly for offshore factories which can use low-paid and highly qualified staff to produce specific items at a low cost that can later be exported. Investment into technical and managerial resources is not yet significant. Development and engineering is expanding and has good prospects for the future. At the same time transforming offshore plants into source plants with a broader strategic role is an obvious development for the future. This will happen as confidence continues to grow in the local management. It is reasonable to grant local management greater authority in all spheres of business organization, but it still makes sense to keep management control, especially financial. Expanding the functions characteristic to outpost companies – development of business intelligence using local business networks and good contacts between local and Russian businessmen – looks promising.

Cost and government globalization drivers play a more significant role for MNCs in the region. Not only low labor cost, but also good logistics and overall business costs, allow companies to increase their global competitiveness. By establishing Free Economic Zones (FEZ) Lithuania is creating new opportunities for multinationals, and many of them were planning to use this to their advantage after 1999, when FEZ started to operate. They will benefit not only from access to FEZ, but also in overcoming bureaucracy and customs problems which are the main factors discouraging activity in the region.

Further privatization of the largest state-owned companies in telecommunication, energy, and shipping industries, taking place in 1998–2000, and further restructuring of industry, also provides a chance for MNCs to enter the market without making considerable investment. The process of integrating the Baltic countries into the European Union is also taking place, with Estonia in the lead. According to the CEOs of many MNCs, this makes the region more attractive for investment.

Methods of global marketing do not need any more adaptation in the Baltic States than in any other European country. The main differences are that consumers here are more trustful and less knowledgeable about different product types or service aspects. Companies should pay attention to educating local consumers, otherwise their reaction may be unpredictable.

The main obstacle for expanding MNCs' activities in Baltic countries is the Russian crisis and the ongoing political uncertainty in the CIS. The economies of the Baltic countries are still considerably dependent on raw materials, oil, and gas imported from neighboring, including CIS, countries. About 25 per cent of imports and exports in all three Baltic countries are with Russia. Economic setbacks in Russia also affect GDP growth in all three Baltic countries. However, after the initial shocks of crisis have been weathered, many companies still consider the situation in Russia advantageous, with plans to expand their activity in this region. Know-how business activity in this region and a developed ability to cope with risk and uncertainty, are all skills with which managers in the Baltic States are well-equipped and can provide as services to foreign multinationals.

Detailed Statistics for the Baltic States

Table 7.7a   Latvia

Structure of Production in % of Total Gross Value Added	1993	1994	1995	1996	1997
Agriculture	11.8	9.5	10.8	9.0	7.4
Industry	30.8	25.4	28.1	26.4	25.7
Construction	4.3	5.9	5.1	4.7	5.0
Services	53.2	59.2	56.0	59.9	61.9
Gross Domestic Product (% change over the previous year)	−14.9	0.6	−0.8	3.3	6.5
Inflation Rate (% change over the previous year)	109.2	35.9	25.0	17.6	7.0

Balance of Payments (in millions of $)					
Exports of Goods	1035.0	987.9	1202.9	1347.8	1540.9
Imports of Goods	1032.0	1277.7	1712.4	2070.0	2724.4
Trade Balance	3.5	−291.0	−509.5	−722.2	−949.9
Services, net	323.2	348.5	416.3	347.3	383.0
Income, net	6.9	9.2	17.3	36.8	46.0
Net Current Transfers	77.1	128.8	62.1	84.0	73.6
Of which government transfers	46.0	103.5	31.1	46.0	28.8
Current Account Balance	409.4	194.4	−13.8	−254.2	−447.4
Reserve Assets (incl. Gold)	542.8	599.2	526.7	684.3	808.5
Reserve Assets (excl. Gold)	445.1	509.5	442.8	600.3	733.7

Foreign trade (in millions of $)					
Imports	930.4	1199.5	1597.4	2101.1	2762.3
Exports	983.3	955.7	1146.6	1307.6	1643.4
Balance	−52.9	−243.8	−452.0	−793.5	−1066.1

Average employment (in % of total)					
Agriculture and Forestry	19.5	19.3	18.5	18.3	18.3
Industry	23.1	21	20.4	19.9	19.8
Construction	5.5	5.5	5.4	5.7	5.7
Services	51.9	54.2	55.7	56.1	56.2

Industry and Agriculture (previous year = 100)					
Industrial Production Volume Indices	67.9	90.1	96.3	105.5	106.1
Gross Agricultural Production Volume Indices	78.0	80.0	94.0	90.0	100.2

Table 7.7b   Lithuania

Structure of Production in % of Total Gross Value Added	1993	1994	1995	1996	1997
Agriculture	14.2	10.7	11.7	12.2	12.7
Industry	34.2	27.0	26.1	25.8	24.0
Construction	5.1	7.2	7.1	7.1	7.3
Services	46.5	55.1	55.0	54.9	56.0
Gross Domestic Product (% change over the previous year)	−16.2	−9.8	3.3	4.7	5.7
Inflation rate (% change over the previous year)	410.2	72.2	39.6	24.6	8.4
Balance of Payments (in millions of $)					
Exports of Goods	1989.5	1961.9	2379.4	3091.2	4251.6
Imports of Goods	2141.3	2159.7	2992.3	3903.1	5415.4
Trade Balance	−151.8	−197.8	−614.1	−811.9	−1163.8
Services, net	−54.1	−52.9	−11.5	109.3	136.9
Income, net	8.1	8.1	−11.5	−82.8	−201.3
Net Current Transfers	113.9	151.8	96.6	130.0	233.5
Of which government transfers	93.2	113.9	54.1	65.6	102.4
Current Account Balance	−83.95	−90.85	−540.5	−654.35	−994.8
Reserve Assets (incl. Gold)	435.85	557.75	725.65	771.65	1108.6
Reserve Assets (excl. Gold)	361.1	491.05	662.4	708.4	1052.3
Foreign trade (in millions of $)					
Imports	2214.9	2274.7	3207.35	4128.5	5678.7
Exports	1968.8	1964.2	2378.2	3038.3	3889.3
Balance	−246.1	−310.5	−829.15	−1090.2	−1784.8
Average employment (in % of total)					
Agriculture and Forestry	22.5	23.4	23.8	24.2	21.9
Industry	25.7	22.4	21.2	20.2	20.1
Construction	7.1	6.6	7.0	7.2	7.1
Services	44.7	47.6	48.0	48.4	50.9
Industry and Agriculture (previous year = 100)					
Industrial Production Volume Indices	n.a.	7304.0	105.3	105.0	100.7
Gross Agricultural Production Volume Indices	95.0	80.0	106.0	110.0	106.0

Table 7.7c *Estonia*

	1993	1994	1995	1996	1997
Gross Domestic Product (% change over the previous year)	−9.0	−2.0	4.3	4.0	11.4
Inflation rate (% change over the previous year)	89.8	47.7	29.0	23.1	12.5
Balance of Payments (in millions of $)					
Exports of Goods	797.0	1283.4	1631.9	1619.2	2119.5
Imports of Goods	939.6	1626.1	2224.1	2564.5	3318.9
Trade Balance	−142.6	−343.9	−592.3	−945.3	−1199.5
Services, net	73.6	101.2	333.5	469.2	616.4
Income, net	−13.8	−28.8	2.3	1.2	−150.7
Net Current Transfers	103.5	110.4	111.6	90.9	117.3
Of which government transfers	103.5	104.7	88.6	75.9	96.6
Current Account Balance	20.7	−159.9	−144.9	−383.0	−616.4
Reserve Assets (incl. Gold)	431.3	417.5	510.6	587.7	791.2
Reserve Assets (excl. Gold)	397.9	414.0	507.2	584.2	788.9
Foreign trade (in millions of $)					
Imports	878.6	1599.7	2233.3	2926.8	4479.3
Exports	790.1	1260.4	1615.8	1883.7	2952.1
Balance	−88.6	−339.3	−617.6	−1043.1	−1528.4
Average employment (in % of total)					
Agriculture and Forestry	16.6	14.6	10.5	10	9.9
Industry	25.6	25.1	28.6	27.8	28.1
Construction	7.4	7.2	5.4	5.7	5.3
Services	50.4	53.1	55.5	56.5	56.7
Industry and Agriculture (previous year = 100)					
Industrial Production Volume Indices	81.3	97	101.9	102.9	113.4
Gross Agricultural Production Volume Indices	90.5	87.1	100.2	93.7	97.8

Bibliography

Business Central Europe. The Annual 1997/1998.

Central European Economic Review.

Comparative Statistics: The Baltic States (Riga, 1997).

EBRD Transition Report, 1996.

Estonian Investment Agency Review.

Lithuanian Development Agency News.

Lithuanian Economics Review. Dept. of Statistics, Lithuania.

Main Economic Indicators of Lithuania. Dept. of Statistics, Lithuania.

Regular Report from the Commission on Progress towards Accession: Estonia, 1998.

Regular Report from the Commission on Progress towards Accession: Latvia, 1998.

Regular Report from the Commission on Progress towards Accession: Lithuania, 1998.

Social and Economic Development in Lithuania, Latvia and Estonia. Dept. of Statistics, Lithuania.

Social and Economic Development in Lithuania. Dept. of Statistics, Lithuania.

Statistical Yearbook of Lithuania – 1996, 1997.

Verslo Zinios (Business News). The Annual 1996–1998.

8 Romania – Resources for the Region

CRISTIAN M. BALEANU

Overview

ROMANIA offers foreign investors the potential of a large domestic market that, with over 22 million consumers, is second in Central and Eastern Europe only to Poland. The country is rich in natural resources with fertile farmland, a large agricultural base and low labor costs. Romania was a relatively slow starter in the transition process and fell behind in terms of privatization of large state owned companies in core sectors. Since 1997, however, the Romanian government has become ever more committed to attracting foreign investment and offers a variety of incentives and tax breaks as well as very liberal investment rules. Although Romania still receives a low level of foreign investment compared to other C&EE countries, and is tackling foreign debt, it is now solidly on the road to establishing a market economy. It is a newly created market rich with potential that still needs to be developed and, in many ways, its success will depend upon the degree of foreign involvement.

History

Romania is a medium sized country, located in SE Central Europe. Its relief is almost equally distributed among mountains, hills, and plains; forests still cover 28 per cent of its surface of 283 391 square km. The Carpathian Mountains, the lower Danube (1075 km), and the Black Sea coast (234 km) are the main natural elements that define present day Romania. The geographical landscape influenced the history of the Romanians. The three principalities – Wallachia, Moldavia, and Transylvania – emerged in the thirteenth century, and succeeded in preserving their own cultural identity, the Latin language, and Christian faith, despite the expansionistic tendencies of three powerful surrounding empires (Ottoman, Russian, and Habsburg). The Romanian state was established in 1859, by the union of Wallachia and Moldova. In 1918, Bessarabia and Transylvania also acceded, forming Great Romania.

During World War II, Romania allied with Germany in the hope of recovering a part of the 36 000 square km of land and six million population, lost in 1939–1940 to Hungary (the Northern part of Transylvania) and the Soviet Union (Bessarabia and the Northern part of Bucovina). On 23 August 1944, King

Michael overthrew Marshall Antonescu's regime and cancelled all previous agreements with Germany. On 6 March 1945, a pro-Soviet Government was 'elected' under the pressure from Soviet troops, and Romania fell under the Soviet sphere of influence. On 30 December 1947, the King was forced to leave and Romania was declared a communist republic. The Northern part of Transylvania was recovered, but the rest of the territory lost in 1939 was given by Stalin to Ukraine and to the newly established Republic of Moldova. In the next half of the century, Romania suffered the bitter experience of the communist regime. In December 1989, the Romanian population rose up against Ceausescu's regime, opening a new page in its history.

More than half of Romania's people live in urban areas, including 25 towns whose population exceeds 100 000 inhabitants. The capital of the country, Bucharest, is situated in the South of Romania, and has two million inhabitants. A breakdown of the ethnic structure in 1997 was approximately 89 per cent Romanians, seven per cent Hungarians, two per cent Gypsies, and 0.5 per cent Germans and others. The main religious groupings include: Christian Orthodox (87 per cent), Roman Catholic (five per cent), Reformed (3.5 per cent), and Greek Catholic (one per cent) and others. From an administrative point of view Romania is divided into 41 counties. An additional regional structure also exists, comprising seven macro-regions plus Bucharest, and the National Agency for Regional Development was established, aiming to support decentralization and lay the groundwork for improved compatibility with European Union structures.

The official language is Romanian, the easternmost type of Latin language. English, French, and German are also widely spoken. Romania has strong cultural ties with France since both languages are Latin in origin and Romania has been strongly influenced by French culture.

Romania is a parliamentary democracy with a Constitution that was passed in Parliament and validated by referendum in 1991. The President and the two Houses of Parliament are elected by universal vote for a four-year term. Public administration is also based on the principles of local autonomy and decentralization. Local Councils of counties, towns and villages, and mayors are elected by direct vote. The Government appoints a prefect at the head of each county.

The major natural resources of Romania are: crude oil, natural gas, coal, non-ferrous ores, gold, silver, salt, and other minerals. Farmlands represent 40 per cent of the country's area, while pastures and hayfield cover about 20 per cent.[1] The main macro-economic indicators of Romania are illustrated in Table 8.1. In 1996 Romania had a GDP of US$35.5bn, and a GDP *per capita* of only US$1571, after Poland, the Czech Republic, and Hungary.

The Transition

Romania did not benefit from an early start in the transition process. In many respects Romania's starting point was more difficult than in other Central and

Table 8.1 Macro-Economic Indicators for Romania

	1989	1990	1991	1992	1993	1994	1995	1996	1997	1998
Real GDP change rate (%)	–5.8	–5.6	–12.9	–8.8	+1.5	+3.9	+7.1	+3.9	–6.6	–7.3
Monthly Inflation Rate (%)	n.a.	n.a.	10.3	9.6	12.1	4.1	2.1	3.8	8.0	2.9
Exports f.o.b. (US$m)	6343	3600	3538	4363	4892	6151	7910	8084	8431	8300
Imports f.o.b. (US$m)	3835	5436	4883	5784	6020	6562	9487	10555	10411	10911
Trade Balance (US$m)	2508	–1836	–1345	–1421	–1128	–411	–1577	–2471	–1980	–2611
Foreign Investment Dynamics (US$m)	n.a.	87.3	537.9	358.4	368.3	997.2	300.8	576.8	324.8	184.4
Unemployment Rate (%)	n.a.	n.a.	3.0	8.2	10.4	10.9	9.5	6.6	8.8	10.3

Source: National Commission for Statistics, Romanian Development Agency.

Eastern European countries. Pre-transition, the Romanian economy was focused on heavy industry and large infrastructure projects. This eventually led to a depletion of domestic energy sources and forced costly dependence on imports of energy and raw materials. After the collapse of the Soviet bloc in 1989–91, Romania was left with an obsolete industrial base and a pattern of industrial capacity wholly unsuited to its needs. Furthermore, unlike other transition economies, no attempts to reform had been attempted in the 1980s. Given this difficult legacy, and the real threat of massive social cost, the first transition government adopted a guarded and limited approach to reform. In the long term, however this strategy proved disastrous and the inability to introduce sufficient structural reform stunted economic growth and Romania's ability to compete within the C&EE region. While other C&EE countries steamed ahead with full-force reform, Romania lagged behind by only partly privatizing its industrial base, and relying on the agricultural sector to soak up displaced labor.[2]

From the privatization point of view, Romania is lagging behind other C&EE countries. The basic explanation lies in the structure of the Romanian economy, which prior to 1989 was based primarily on large state-owned companies. Privatizing these large companies proved to be a very complex process, often leading to social turmoil. Strong political and managerial leadership, which was not always available, was necessary to accomplish such a task. The left-wing coalition ruling the country before November 1996, was not convinced of the need for rapid privatization preferring to combine a slow privatization of small and medium companies, with the restructuring of the large companies. Privatization of the Romanian economy by sector is shown in Table 8.2.

Since 1997, the political landscape has undergone dramatic change, and with it economic policy and the attitude to foreign investment. The new government is highly committed to carrying out a variety of reforms felt necessary to bring Romania closer to a fully-fledged market economy including providing attractive

Table 8.2 Privatization by Sector

Sector	Valuation (ROLbn)	% privatized (by value)	% of companies privatized
Agriculture and fisheries	9 419	7	43
Metallurgy	8 766	58	11
Food processing	7 006	23	19
Machinery and equipment	5 736	6	23
Transport equipment	4 114	6	21
Chemicals	4 113	15	28
Transport and distribution	4 090	11	29
Non-metal minerals	2 520	26	47

Source: State Ownership Fund, as of 31 August 1997.

incentives to foreign investors. Although the newly elected right-wing coalition began an ambitious privatization process, lack of the necessary strategic know-how compounded political disputes, gradually reducing the pace of the reforms.

From November 1996, the Romanian Government pursued extensive economic liberalization, and harsh austerity policies. In 1997, economic policy concentrated on eliminating the large loss-making companies (seventeen of them were closed down in the first six months of that year), and the restructuring of the mining sector. Consequently unemployment increased from 6.6 per cent in 1996 up to 8.8 per cent in 1997, and ten per cent in 1998. Severance payments to workers reduced the urge to protest.

There has been a significant delay in the restructuring of the large state monopolies called 'Regie Autonoma' (RA), and the state-owned banks. Furthermore, in 1997 disputes concentrated more on political control of different members of the party coalition over these entities rather than on economic issues. By the end of 1998, the Government decided to speed up the privatization process and two important deals were concluded in that period: Hellenic Telecommunication Organization (OTE) acquired 35 per cent of Romtelecom for US$675m, and Société Générale acquired 41 per cent of the Romanian Development Bank for US$6.8m. The privatization of a second bank followed in 1999, when 45 per cent stock of Banc Post was bought by General Electric and Banco Portugues de Investimento.

Another major concern is that no major support has been provided to the development of the Small and Medium-sized Enterprises (SMEs), for which sector the foundations for the regulatory framework were laid in 1991–92 by the Government. Currently the SME sector is dominating the Romanian economy in a number of ways including the number of companies, contribution to total turnover, gross profit and investments. However, most jobs are still provided by the large (previously state-owned) companies. In 1996, large enterprises were the most important employers in Romania (64.6 per cent) although they contributed only 44.4 per cent to the turnover. This demonstrates their difficulties and even failures in overcoming the transition period. The SME sector contributes more than 50 per cent to total turnover and more than 65 per cent to the total gross profit.

Two phases of policies concerning SMEs can be identified in 1994-98. In 1993-96, the Romanian Development Agency, together with other government and nongovernmental organizations, increased awareness of the SME sector. This included new institutional support, basic financial and logistic support, and numerous assistance programs in favor of SMEs. The main drawbacks were the absence of a clear positioning of SME policy within the Government, and the lack of coordination regarding different measures (legal, financial, logistic). The initial policy led to an increase in the number of private SMEs in the Romanian economy. The second phase of SME policies, started in November 1996, contained more detailed objectives and measures. The need for strong government institutions, in charge of the design and implementation of a national strategy to support the development of the SME sector in Romania was satisfied only at the end of 1998, when the National Agency for Small and Medium Enterprises was established.

Foreign direct investment

Privatization has proceeded slowly in particular sectors, such as public utilities and finance, so that in practice the scope of large-scale foreign participation in the economy has remained limited. Economic instability and the lack of infrastructure has also discouraged long-term investment. Nevertheless, a variety of countries have chosen Romania in which to make relatively large scale investments. The top countries for foreign investment are illustrated in Table 8.3.

Although the current legal environment is favorable to foreign investors, Romania has acquired only a total of about US$5.17bn between December 1990 and 1998, a relatively small amount in comparison to Hungary, its smaller

Table 8.3 Largest Source Countries of Foreign Capital

Country	Capital (US$m)	% of Total Foreign Investments
Netherlands	566.8	11.0
Germany	481.3	9.3
France	475.4	9.2
Greece	438.5	8.5
Great Britain	432.6	8.4
United States	391.1	7.6
Italy	322.4	6.2
South Korea	278.0	5.4
Turkey	255.6	4.9
Austria	253.5	4.9

Note: As at December 1998.

Source: Romanian Chamber of Commerce and Industry, National Trade Register

neighbor. Of this, US$3.65m represents direct investment, mostly into former state-owned companies. A further US$1.52m represents future investments foreseen in the privatization contracts.

Additional attractive mechanisms are the Bucharest Stock Exchange (BSE), trading over 60 companies by December 1997, and the RASDAQ over-the-counter stock market trading over 2400 companies by December 1997. The peak for the Bucharest Stock Exchange was in 1997 when about ROL65bn and a volume of about 30 million shares were traded, and in July for RASDAQ when about ROL650bn and a volume of about 130 million share transactions were made. In August 1998, there were about US$360m invested in the Romanian capital market by foreign investment funds, the most important being Broadhurst Investment, with a US$100m investment.

Table 8.4 shows the distribution of foreign investment in Romania by sector of activity. New placements of foreign capital are foreseen in steel production, in telecommunication, in car manufacturing and in banking. Unlike other C&EE countries that are further along in the transition process, there is still plenty of room for foreign investors within the Romanian market, especially in respect to the size of the country.

Table 8.4 Foreign Investment in Romania by Sector of Activity

Sector	(%) of total
Heavy Industry	27
Professional Services	23
Wholesale	16
Food Industry	13
Retail	7
Light Industry	4
Agriculture	3
Transportation	3
Construction	3
Tourism	2

Note: December 1990–June 1999.

Source: Romanian Chamber of Commerce and Industry, National Trade Register

In October 1997, Euromoney magazine ranked Romania in 61st place on country risk. In May 1998 Standard & Poor's reported that it had decreased Romania's country risk rating reflecting Romania's prospects and its foreign debt on a long-term basis. The new ratings were B+ for the long-term foreign debt, BB for the long term debt in national currency and B for the short-term foreign debt. Romania's gross external debt increased from US$4.2bn in December 1993 to US$9.5bn in 1997. This situation has affected investors' confidence, despite the fact that Romania successfully met its financial obligations in 1999.

A number of multinational companies (MNCs) have chosen to invest in Romania and set up production facilities here. The largest companies investing in

Table 8.5 Largest Investors in Romania

Investor	Country of Origin	Capital (US$m)	Sector
Daewoo	South Korea	209.1	Cars, heavy Industry
Royal Dutch/Shell	Netherlands, UK, USA	91.0	Oil and gas
MobilRom	France	61.2	Mobile Telephony
MobiFon	Canada	57.5	Mobile Telephony
Coca-Cola	USA	32.8	Drinks
Unilever	Netherlands, UK	32.7	Consumer Toods
Nomura	Japan	30.0	Tires

Note: By July 1977.
Source: Romanian Development Agency

Romania are illustrated in Table 8.5. Some industries are more attractive to MNCs, for two main reasons: their export potential and their specific competitive advantages. Below are discussed some of the industries of most interest to MNCs.

Oil and Gas Sector

Romania has a tradition in the oil and gas industry that goes back to as early as 1857, when the first refinery in Europe began its operations at Rafov. At present, Romania ranks first in Eastern Europe and fourth in Europe in oil and gas production. These primary energy products represented more than half of Romania's production in 1996 (that is oil, 19.6 per cent, gas, 43.3 per cent).

At present, the internal oil production supplies 6.5m tons per year, while the refineries' total output is about 34m tons per year (the optimal size estimated for Romania being of 22–25m tons per year). Future reserves are estimated as being quite substantial. Oil production is provided mainly by seven state-owned refineries, the largest being Petromidia Constanta – with 21 per cent of Romania's total capacity. Three refineries are undergoing restructuring, while the other four units are in various stages of privatization. Several attempts have been made to privatize Petromidia, strategically located on the Black Sea coast. In 1998, an agreement was signed with the Turkish Akmaya, but this company withdrew in 1999, following the decision of the Romanian Government to suspend the facilities for foreign investors. On the other hand, the State Ownership Fund claimed unfulfilled obligations from the part of Akmaya. Currently, the Romanian authorities are looking for a new investor for Petromidia.

The production of natural gas is provided mainly by the state-owned ROMGAZ RA, providing a total volume of about 18m Nm^3 per year.

By the end of 1997, foreign investment in the oil sector amounted to US$130m, provided mainly by Shell and LUKoil. In 1998, Shell completed a US$4m liquefied petroleum gas terminal with annual handling capacity of 100 000 tons, and was planning to build another 150 million terminal at Constanta port. Other important international cooperation projects are planned for the future. These include development of an underground natural gas storing capacity; interconnection of the national gas transport system to the international network; rehabilitation and modernization of the oil transport system.

Metallurgical industry

The metallurgical industry has a significant tradition in Romania. In 1997 Romania was ranked the 21st producer of iron and steel in the world.

In 1998, the Romanian iron and steel industry comprised two main types of production flow: an integrated production flow, such as processing fluid cast iron and scrap iron, and a production flow based upon processing semi-finished products such as non-welded pipes, long rolled and flat rolled products.

Monthly production levels in April 1997, in thousand tons, were: raw steel, 548, hot-rolled solid finished steel products, 392, out of which steel pipes were 48, and cold-rolled steel plates and strips, 45. Total steel production was 6082 thousand tons in 1996, and labor productivity was 96.8 per cent of that recorded in 1990, as compared to 93.7 per cent in 1995.

Based on CEFTA provisions, Romanian iron and steel products will be allowed to freely enter the Hungarian and Slovenian markets from 1 January 2000, and the Polish market from 1 January 2002. The domestic demand for such products is estimated at 5700 thousand tons in 2002. More than 20 large companies in the sector are currently looking for foreign partners, but only one notable success has been recorded. This was the acquisition of OTELINOX Targoviste – the most modern Romanian stainless steel production unit – in January 1998 by Samsung, which paid US$37.4m for 51 per cent of the stock.

Electro-technical industry

Electro-technical production represents 2.2 per cent of the value of total industrial production in Romania. The sector employs 82 000 people (mid 1996). Exports represent 23.3 per cent of total industrial exports, and its added value is seven per cent higher than the industry average. Of total exports of US$179.7m in 1995, US$54m were represented by electric motors, US$21m by automatic equipment, US$14m by refrigerators, and US$6m by power transformers. In 1996, total exports increased by 142.4 per cent in comparison to 1989 and by 144 per cent in comparison to 1995. About 20 medium and large privatized companies are operating in this market. Important investors include ABB and Daewoo.

Cosmetics industry

By mid-1997 the cosmetics industry in Romania was represented by two state-owned companies (MIRAJ Bucharest and NORVEA Brasov) and five private manufacturers (Colgate-Palmolive Romania, GEROCOSSEN Bucharest, MIRALON Bucharest, FARMEC Cluj, and STEAUA Sibiu). In August 1997, Colgate-Palmolive Romania (a joint venture between Colgate-Palmolive

Overseas, Stella Bucharest, and Norvea Brasov), was the market leader for soap (58 per cent), hygienic oral care products (80 per cent), skin care products (35 per cent) and deodorants (14 per cent). Two years later, due to the increasing competition on the soap market in Romania, mainly on the part of the cheaper Turkish products, Colgate-Palmolive decided to close down its soap producing facility in Romania.

Domestic companies

Romania does not yet have any domestic companies that could effectively compete on a global scale with foreign MNCs. The top 25 Romanian companies are listed in Table 8.6. Most of them are either 'regie autonoma' or former state-owned companies, privatized by coupons and still under the control of the State Ownership Fund. During the period 1990–98 over 600 000 private start-up companies emerged in Romania. Only a few of them have succeeded in

Table 8.6 Largest Romanian companies

Rank	Company	Sector	1997 Turnover (ROLm)*
1	Renel	Electricity	19.2
2	Competrol	Fuel Distribution	9.6
3	Sidex	Siderurgy	8.3
4	Romgaz	Gas Industry	7.9
5	Sncfr	Rail Transport	4.8
6	Petromidia	Petrochemistry	3.4
7	Petrobrazi	Petrochemistry	3.3
8	Arpechim Pitesti	Petrochemistry	3.0
9	Automobile Dacia	Car Manufacturing	2.3
10	Cnc Dacia	Car Distribution	2.1
11	Rafo Onesti	Petrochemistry	2.1
12	Alro	Aluminium Production	2.0
13	Sn Tutunul	Tobacco Industry	1.9
14	Petrotel	Petrochemistry	1.6
15	Oltchim	Petrochemistry	1.6
16	Siderurgica	Siderurgy	1.4
17	Daewoo Automobile Romania	Car Manufacturing	1.3
18	Rafinaria Astra Romana	Petrochemistry	1.3
19	Azomures	Chemistry	1.2
20	Tofan Grup	Chemistry	1.1
21	Regia Autonoma A Huilei Romania	Coal Mining	1.1
22	Tarom	Air Transportation	1.1
23	Lafarge Romcim	Cement Industry	1.1
24	Regia Nationala A Padurilor	Forestry	1.0
25	Romtelecom	Telecommunications	1.0

Source: The Romanian Chamber of Commerce and Industry, The National Trade Register.

Note: *1 US$ = 7167.9 ROL (1997 average)

growing. The most prominent are Elvila – a major furniture producer and exporter, Ana Electronic – distributor and producer of Samsung household electric equipment, Ion Tiriac Bank Holding, and Tofan Group, the largest domestic tire producer.

Overall globalization drivers

In terms of globalization drivers, Romania is lagging behind other C&EE countries as it did not privatize the large state-owned enterprises as rapidly and efficiently. The purchasing power of its population is still relatively low and the infrastructure needs development. However the large population and abundance of natural resources means there is still a lot of potential for development. Furthermore, country costs, including labor and production, are still low when compared to other C&EE countries in the region. Marketing strategies are also easily transferable from other countries of C&EE and the West.

Market globalization drivers

Until 1998, most MNCs operating in Romania focused on the local market rather than establishing a strategic base for a regional European approach. A study among foreign investors, elaborated by Deloitte Touche Tohmatsu International in 1996, shows that market potential was the first reason given for investing in Romania (by 48 per cent), while the geographical position ranked fourth (by five per cent).[3]

Common customer needs

In Romania customer needs and tastes are very similar to those of Western Europe, adjusted only by the much lower *per capita* income. A middle class that can afford Western goods and services is not significant yet. Sergiu Brucan, a well known political analyst, estimated the monthly income for a Romanian middle class family as about US$1200, while the average monthly net income was in 1998 about US$100.

In 1993, the demand structure was made up of: food and drinks (48.3 per cent), transport and communication (12.5 per cent), education, culture and entertainment (11.4 per cent), rent and utilities (10.7 per cent), clothes and shoes (8 per cent), furniture and household appliances (7.5 per cent), and health (1.6 per cent). This structure does not differ greatly from neighboring countries, with a stronger emphasis on food and drinks (which is only 36.5 per cent in Hungary), and education, culture and entertainment (which is 7.0 per cent in Hungary). Consequently, agriculture, food production and distribution, gastronomy and fast food, and mass media and entertainment are very attractive sectors for foreign investors. Well known global names are Coca-Cola, Nestlé, Knorr, and Kraft Jacobs Suchard for the food industry; McDonald's, Pizza Hut, and KFC in fast food; and Ringier and MediaPro International in mass media. In January 1998, Time-Warner's pay-movie-channel Home Box Office registered its first 4000 subscribers, aiming at 20 per cent of the total 2.7m subscribers to the Romanian cable market.

The number of cars, as well as the number of radio sets, TV sets and refrigerators per 1000 inhabitants, is well below the European standard, although it increased substantially after 1989. Because of the prospect of entry into European markets, South Koreans are the most interested in these industries. Daewoo had invested US$150m into the Automobile Craiova factory by 1997, and Samsung, LG Electronics, and Hyundai are assembling TV sets locally. In February 1997, the main players on the Romanian TV set market were LG (18 per cent market share), NEI – a British-Romanian joint-venture, with 17 per cent, Samsung (16 per cent), Daewoo (13 per cent) and Philips (12 per cent).

Global/regional customers and channels

For the time being, only a few international companies are willing to assume the risk of relying on local companies as suppliers.

Daewoo had promised a 70 per cent integration of their cars produced at Craiova by the year 2000, but this target has proven difficult to achieve. The Koreans are mainly blaming Romanian companies' lack of capability to reach the necessary level of quality, and the delivery conditions.

At the same time, two major German companies have developed acceptable suppliers, relying mainly on traditionally strong ties between Germany and Romania, and even some people of German origin. For example Krupp-Bilstein set up a joint-venture with COMPA Sibiu to produce shock absorbers for cars, while Mercedes developed their own capacity to produce wooden-made parts for cars at Codlea, Brasov County.

Transferable marketing

As a rule, MNCs investing in Romania consider this country as part of the Central European region, and their marketing strategies and approaches in Romania are the same as those applied in Poland, Czech Republic or Hungary.

In 1992, when its operations started in Romania, Colgate-Palmolive kept some Romanian brand names produced by their partners such as STELA Bucharest and Norvea Brasov. However, soon they imposed their global products, such as Colgate toothpaste, Palmolive soap, or Mennen deodorant.

In 1995, Kraft Jacobs Suchard launched their first chocolate product called Poiana, but Milka imports have been important. Most packaging and branding looks very 'international', and often labels and instructions are multilingual. Advertisements and TV spots often show non-Romanians, but are translated into Romanian.

Cost globalization drivers

Cost globalization drivers have been one of the key factors attracting MNCs to Romania. The country provides a large, inexpensive and skilled labour force, a major advantage for MNCs interested in production. At the same time, many foreign investors are deterred by the poorly developed infrastructure, the complexities of the bureaucracy, currency exchange rules, and corruption.

Global/regional scale economies

Romania can contribute sales volumes to MNCs needing to achieve global or regional scale economies. At the same time, its geographical position, close to the CIS and Russia, favors additional exports to support a minimum efficient scale plant. That is why several MNCs (such as Daewoo, Siemens, Unilever, Procter and Gamble) located their production and distribution facilities in Romania.

Infrastructure

The overall quality of Romania's infrastructure is not very high according to Western standards. The railways need to be modernized and there are only 120 km of highway. There are four international airports, of which only Otopeni Airport, which extended its capacity in 1998, offers services at European standards.

Communications infrastructure is also in need of further development. In 1995 there were only 11.26 telephone lines per 100 inhabitants, compared to 11.49 in Poland, 25.91 in Slovenia, and 27.55 in Bulgaria. Consequently, Romtelecom was one of the first state-owned companies to be privatized. Telecommunications are still dominated by the recently privatized Romtelecom. Siemens and Alcatel are major providers of locally produced communication equipment. The mobile telephone industry is divided between MobiFon (50 per cent), MobilRom (45 per cent) and Telefonica Romania (five per cent), and is not accessible to ordinary people because of the relatively high prices.

Financial services is another area in need of restructuring. In 1997, six state banks and a cooperative bank that existed before 1989, dominated the sector. Two years later, the Romanian Development Bank and Banc Post, formerly state-owned banks, were acquired by foreign banks. The private banking sector, comprising about 20 banks, is by contrast, much smaller. Foreign banks, such as ING Barings, ABN-Amro, Citibank, Bank Austria, and Creditanstalt, became active players only in 1999.

Sourcing efficiencies

Romania is endowed with several power plants. Before 1989 it was common to have the electricity cut for 12 hours a day in villages and small towns, and for two to three hours per day in large cities and in Bucharest. This situation was due mainly to the high-consumption technology that existed in Romanian enterprises. After 1989, consumption decreased and the production of electricity increased. On the one hand, several big energy consuming companies were closed down, being unable to pay their debts to CONEL – the National Company for Electricity. On the other hand, the first nuclear power plant was inaugurated in 1996 at Cernavoda, using the modern and safe Canadian CANDU system. In 1998, Romania's production capacity was higher than its needs. Consequently, the construction of a second nuclear power plant has been delayed.

Retailing services in Romania are underdeveloped in comparison with its Western neighbors. Most of the retail business (95 per cent) is private but almost all trade areas are still owned by the former state-owned trade companies, now privatized by coupons and actually controlled by their management. Consequently, rents are high, especially in large cities. There is a lack of office space, and the construction and administration of business centers and offices is big business, especially in Bucharest and in other large cities. In the central area of Bucharest, the rent for offices varies from US$10–35 per square metre. In Bucharest there is only one supermarket chain, the local Mega Image, but future competition is expected from the Austrian Billa, which opened its first outlet in February 1999 and was intending to open three new stores in the capital city in the following five months, and other stores in Constanta, Brasov and Ploiesti.

Favorable country costs

Cost-related factors are considered to be crucial for a MNC deciding to invest in Romania. In 1997, the labor cost was ranked third (scored 12 per cent) after market potential and qualification of personnel. The poorly developed infrastructure was considered as being one of the main barriers after legislation, bureaucracy, currency exchange rules, and corruption.

Labor costs

The minimum gross monthly salary is about US$32, while the average gross salary amounts to US$136, exclusive of employers' contribution. Taxes on salaries are progressive and range from five per cent to 45 per cent (reduced from 60 per cent in 1998). Employers are legally obliged to pay additional 31 per cent contributions: 23 per cent to the social security fund, two per cent to health insurance, five per cent to the unemployment fund, and one per cent to the risk and accident fund.

Employees must pay a three per cent pension contribution, and one per cent to the unemployment fund. Salary levels vary, depending on the type of company and industry. Salaries tend to be the highest at MNCs subsidiaries and 'regia autonoma', 20–30 per cent lower at joint ventures, 40–50 per cent lower at Romanian private companies, and 60–70 per cent lower at state-owned companies.

The net monthly salary for a Romanian general manager of a foreign or joint venture company varies between US$1000–2300 to be paid in Romanian lei (ROL), while a secretary is paid the equivalent of US$150–300. Salaries tend to be lower in the food and textile industry and higher in electronics and telecommunications. A survey ran by KPMG Romania among 955 MNCs operating in Romania in the first quarter 1997 identified a significant difference between management salaries depending on the economic sector.

Table 8.7   Gross Monthly Base Salaries for Management

Sector	Minimum US$	Maximum US$	Average US$	Median US$
Manufacturing	306	3343	2130	2536
Agriculture	753	3343	1674	1299
Distribution	1468	2573	1980	1937
Import	393	1468	823	609
Consulting	225	999	545	478
Health & Care	900	3083	2023	2055

Source: Romanian Business Journal, 14–20 February, 1998, p. 4.

Technology role

The overall technology level of most former state-owned companies is obsolete since there was hardly any investment in new equipment made in the 1980's. However, the technical endowment in some industrial sectors, such as: roll bearings, electronics, and nuclear equipment, proved to be quite acceptable. Generally, the technical skills of management and workforce might be more advanced than their marketing or financial know-how.

Government globalization drivers

In the first years of the transition the Romanian government was opposed to a high level of foreign involvement in the economy and proceeded only warily with privatization, keeping control of most key industries. Many of the larger companies remained government-owned and under state control and the degree of foreign involvement was limited. Therefore the transition to a fully fledged market economy proceeded only very slowly. The situation changed in 1997, when the newly elected government pledged to open up the Romanian economy to foreign investment, privatize the large state owned enterprises, and make major structural and legal changes making investment simpler and easier. Currently there is a wide range of possibilities for investment, no formal restrictions on foreign direct investment in any specific sectors, and no compulsory performance requirements.

Favorable trade policies

Romanian market regulations have been gradually liberalized and modeled on the European Union. The new import customs tariff regulations have also been in force since 1992, in line with European Union standards. A customs duty drawback system and leasing legislation facilities are in place. Legislation in force in Romania allows the establishment of free zones since 1992, containing provisions meeting international standards on exemption from custom duties and value-added taxes, unrestricted import and re-export of goods, and the possibility of foreign currency payments in free zones. For the activities performed in free zones, companies are also exempt from profit tax. The law also permits the leasing or renting of land and buildings in the free zones to natural or legal persons for a maximum term of 50 years, according to the value of their investment and the nature of activity. For the moment, Romanian free zones have not attracted much investment. For example, the Galati Free Zone – set up in April 1994 close to Galati Harbor, with direct access to Danube and to the Black Sea Channel – has a total surface of 137 hectares. By December 1997, only 22 ha. had been leased to 28 Romanian and foreign companies. A further law was passed in 1999, that offers special facilities for investments in certain less favored areas (such as mining).

Favorable foreign direct investment rules

FDI rules in Romania are established through the following specific laws and Government ordinances: Law no. 35/1991, Law no. 71/1994, Ordinance no. 31/1997 and Ordinance no. 92/1997. Investors in Romania benefit from various advantages, as follows:

- The possibility of investing in any field and under any juridical form provided by law (see Global Organization and Management);
- Equal treatment for Romanian or foreign investors, resident or non-resident in Romania;
- Guarantees against nationalization, expropriation or any other measures with similar effect;
- Fiscal and custom duties facilities;
- Assistance in following administrative procedures;
- The right of converting the amounts in ROL from the investment, into the investment currency, as well as this currency's transfer in the originating country;
- The right to choose the appropriate trial or arbitration court for solving any conflicts;
- The possibility of reporting losses registered during a financial period in the profit before tax of the next fiscal period;
- The possibility of using accelerated depreciation;
- The possibility of deducting advertising expenses from the profit before tax;

- The possibility of hiring foreign persons. After the payment of legal taxes non-resident investors enjoy the right to transfer abroad without restrictions, and in freely convertible currency, the following incomes:
  (1) the dividends or profits obtained by a company, in the event that they are shareholders or partners, or the profit from a branch of it;
  (2) the incomes obtained by a partnership type of association, as well as the incomes resulted from selling the shares or social parts;
  (3) the amounts obtained from company liquidation, and
  (4) the amounts obtained as compensation against expropriation or other any equivalent measures.

Investors also enjoy the following incentives:

- Import of mobile assets, tangible or/and intangible, that represent contribution in kind to the social capital of a company, or represent contribution in a partnership or family association and that is needed for achieving the object of activity, is exempted from payment of custom duties and VAT;
- Import of technological equipment and machinery, made as direct investment, is exempted from payment of custom duties;
- In the case of making new investment, it is possible to choose one of the next two fiscal incentives, without cumulating them:
  (1) to deduct from the profit before tax the expenses regarding depreciation (including accelerated depreciation), or
  (2) to deduct from the profit before tax of one fiscal period a 20 per cent quota from the equipment acquisition cost that represent depreciating assets purchased during the fiscal period.
- To deduct from the profit before tax the entire advertising expenses and
- To recover back of annual losses declared by the taxpayers, during the next five consecutive years.

In 1999 tax exemption for investors was suspended for an indefinite period of time by the Romanian Government, following the request of the International Monetary Fund, in its attempt to keep the state budget deficit under control and to assure Romania's capability to fulfill its financial obligations. However, for larger investments (over US$50m), the aforementioned rules might apply, only upon special approval from the Ministry of Finance.

Role of trade blocs

In April 1997, Romania joined the Central European Free Trade Agreement (CEFTA), a group of six countries comprising 90 million inhabitants and a total GDP of US$273bn.

Romania has been a contracting member of the WTO since 1971, has ratified most codes adopted at the Tokyo Round, and has also been subscribing to the results of the Uruguay Round. On 1 February 1993, Romania signed an Association Agreement with the European Union. In 1997, Romania was not accepted by the European Union as a candidate for admission in the first tier of applicants, in spite of major support provided by France and Italy. In 1999, Romania became a member of CEFTA.

Several bilateral conventions have been concluded with other states concerning the protection and mutual guarantee of investments. In its commercial relations with the United States, Romania enjoys 'Most Favored Nation' status.

Government intervention

So far, there has been no foreign dominance of key industries, so that intervention into foreign companies by the Romanian Government has not been necessary.

As far as most domestic companies are concerned, during 1989–98, most government subsidies to ailing companies were eliminated. However, some large companies are still reluctant to pay their utility debts (electricity, gas and heating). In such cases intervention on the part of the government is necessary but a difficult decision to make because of the social and political consequences. Closing down large debtors can have serious social repercussions, as they often employ most of the active population of a small town. This fact has led to a hidden but highly significant amount of subsidy towards these companies, most of them active in steel and oil sectors.

Legal protection of contracts, trademarks and intellectual property

The legal protection of contracts, trademarks, and intellectual property fully complies with European Union standards and international agreements. The government is committed to combat corruption and has sought cooperation with the OECD in this respect.

Compatible technical standards

After 1990, substantial efforts have been made by the newly established Romanian Institute of Standardization to introduce European technical standards. Although most of the work was done at the national level, additional efforts have been necessary to have these standards observed at the company level.

Common market regulations

Regulations concerning prices and markets in Romania are generally similar to those applied world-wide.

Competitive globalization drivers

Due to the size of its population, Romania is expected to play a major role in the Central European Region. At the same time Romania, as a Balkan country, could gain strategic importance, especially in oil processing and transportation, since a major pipeline, from the Caucasus to Western Europe, crosses its territory. Several foreign companies are competing to get a prominent position in this industry, among which is the Russian LUKoil, which in January 1998 acquired a controlling stake in the Petrotel oil refinery and chemical plant. Probably for the same reason, Daewoo acquired the second largest shipyard on the Black Sea coast, at Mangalia.

Global Strategy Levers

Those multinational companies that have chosen to enter Romania have pursued strategies that take into account Romania's labor cost advantages, geographical position and market potential.

Market participation strategies

Since market potential seems to be the first reason for MNCs to invest in Romania, it is clear that they are highly interested in acquiring a substantial market share. Colgate-Palmolive has the largest market share in cosmetics, Shell is the second important fuel retailer after the state-owned Competrol, Unilever is the largest producer of detergent, and McDonald's is still almost a sole player in the fast food market.

The attempt of Daewoo to become a major supplier of cars on the Romanian market has failed, as the company was unable to compete with the lower-priced, old-fashioned Dacia cars. During the first half of 1998, about 69 000 new cars were sold in Romania, of which 61 per cent were produced by Automobile Dacia Pitesti and 14 per cent by Daewoo Romania. The most successfully sold imported brand, was Volkswagen, with only a one per cent market share (669 cars). Major changes in the Romanian car market are expected following the conclusion of the contract between Renault and the State Ownership Fund, for the acquisition of 51 per cent of the Dacia's stock for US$135m in 1999.

The market entry of one major MNC has attracted its traditional rival: Unilever was followed closely by Procter & Gamble in detergent production and distribution, while the first KFC fast food restaurant was opened in Autumn 1997 in Romania's capital, following the announcement of the record number of customers per day in Central Europe in the McDonald's Unirii Square restaurant in Bucharest.

Product and service strategies

Most of the products offered by MNCs on the Romanian market are subject to global standards in quality and shape, and little localization and adaptation seems to be necessary. However, MNCs are preserving established local brands.

Unilever decided to keep the local DERO brand name when it acquired the company, simply because for most Romanians 'dero' was equivalent to 'detergent'. Consequently, the new company was called 'Dero Lever'. Colgate-Palmolive also chose to preserve the highly successful Norvea toothpaste, with a price and quality comparable to its Dual Fluoride global product. These two brands are coexisting on the Romanian market, and their producer has about 80 per cent market share. The Austrian group BBAG, owning a major share in four Romanian breweries and having 17 per cent market share in Romania, produces and distributes their premium brand 'Kaiser', but also 'Arbema', a local brand.

Some companies have decided to create special brand names for the Romanian market. The Daewoo 'Cielo' is sold only in Romania, as well as the 'Bergenbeer' beer produced in Romania by the Belgian Interbrew company. The first product made by Kraft Jacobs Suchard in Romania was called 'Poiana', from Poiana Brasov, a well known tourist site located near Brasov, where KJS production facilities have been located. Moreover, KJS marketing attempted to appeal to Romanian's patriotism and 'Poiana' was advertised on the television by showing a couple of Swiss custom officers discovering this delicious chocolate 'Made in Romania'!

Activity location strategies

The actual utilization of Romania as a location for various value-chain activities is focusing mainly on raw material sourcing, manufacturing, local marketing, local distribution, and local customer service. MNCs are not taking advantage of Romanian R&D potential, and are still rather reluctant to involve indigenous suppliers in their businesses.

The main reason for an MNC to avoid local suppliers may have to do with the lack of confidence in their capability to meet the necessary standards related to quality and delivery conditions. The Romanian companies are putting much effort into finding acceptance as internationally recognized suppliers: more and more local companies are qualifying for certification according to ISO 9000 standards. As most MNCs have their own system to assess the capability of their suppliers to meet their standards, this certification is only the first step towards becoming a local supplier.

Under the pressure to reduce costs, MNCs acting in Romania prefer to establish their own greenfield suppliers, or to form a joint-venture under their control, as McDonald's did when they decided in 1997 to produce their hamburgers locally.

For some MNCs, Romania might have an ideal position in Central Europe and Eastern Europe. In the early 1990s, most MNCs were establishing facilities in Hungary, Czechoslovakia, Poland, and Russia, attracted by the economic advancement of the former three countries, and by the huge market potential of the latter. In between, there was a quasi-void area and consequently, some companies were attracted by this 'vacuum'. This was exactly the case of Procter

& Gamble (P&G) who established their Bucharest office as a regional headquarters with overall responsibility for direction, planning, coordination, hiring, and training for an area containing more than 58 million people.

Similarly, Coca-Cola focused on Romania as a region. The overall business is led by a Regional Director, while the actual production and distribution is managed by three independent companies, originating from Australia, Ireland and Turkey. Coca-Cola Amatil, with more than 50 per cent of sales, covers Bucharest and the south of Romania, Coca-Cola Molino covers north-east and north-west Romania, and Coca-Cola IMBAT operates in the eastern part of Romania. The consolidated Coca-Cola investment in Romania was more than US$200m in 1997, and it employed about 4000 people, of which 35 were expatriates.

As part of the Coca-Cola strategy in Romania, Coca-Cola Molino invested US$15m in a new company, Frigorex, located in Timisoara, producing commercial refrigerators. In 1997, Frigorex produced 25 000 refrigerators, out of which 50 per cent were exported in Eastern Europe. The total sales for 1997 were US$11.5m. A new investment program for this company was launched in 1998.

ABB also focuses on Romania as a region. The ABB group in Romania is represented by three production companies – ABB Energo, ABB Power T&D Romania and ABB Rometrics – and ABB Romania, a holding company which deals with implementing ABB strategy in Romania and coordinating the activity of all the ABB companies activating in Romania. In 1998, ABB bought 51 per cent of the stock in Automatica company in Bucharest, a major Romanian producer of automation equipment, for US$9m. Automatica has been a major supplier for the Ukrainean iron ore complex at Krivoi Rog.

Marketing strategies

Products and services are in most cases marketed in Romania similarly to other European countries. Most of the elements of the marketing mix can be transferred directly to the Romanian market: positioning, brand names, packaging, advertising, promotion, distribution and selling methods. The products should be labeled in Romanian. Television advertisements were quickly adapted and often translated. Western marketing elements and techniques were rapidly absorbed by Romanians and the marketing department was the first sector of a MNC operated by locals. By 1998, most of the market surveys were done by Romanians, usually guided by a Western specialist. Although this sector is dominated by market research companies like Lintas or GFK, local companies are very active, offering good quality marketing services. Prices of locally produced consumer goods are usually lower than in the rest of Europe, reflecting lower cost structures, but also lower purchasing power.

A typical case is that of McDonald's, which started its expansion in Central Europe in April 1988, when it opened its first restaurant in Belgrade. The decision to invest in Romania came in 1993, when the inflation rate amounted to 300 per cent. The first restaurant opened two years later, in June 1995, based on an initial investment of US$5.7m. Three years later, 28 outlets were operating in Romania, selling 50 000 burgers per day and generating a minimum gross income of ROL1bn (about US$110 000). Total investment by McDonald's in Romania up to the year 2000 is estimated at a minimum of US$100m. The total number of employees will have increased from 1600 in November 1996 to about 6000 by year 2000.

Naturally, McDonald's Romania is focusing purely on the domestic market. They started by offering their basic 'international menu', completely produced on the basis of imported raw materials. By the end of 1996, only buns, dairy products and lettuce were produced locally. In 1997, an additional US$1.8m was invested to produce locally ingredients for hamburgers, leading to an increased number of items in their menu. Following the introduction of a fishburger in March 1998, McDonald's launched a special World Cup soccer sandwich during the championship (soccer is highly popular in Romania), as well as Chicken McNuggets by the end of the year. The company's turnover in 1997 was about US$30m.

Before 1995, most of the production of Coca-Cola Romania consisted of Coca-Cola drinks in 0.2 litre bottles. In 1995, following the significant diversification of its product range, Coca-Cola Romania changed its sales system: from the direct sales system, when the products were sold directly from the truck, in the quantity and assortment available at that moment, to a pre-sell system. Each client was asked to conclude a framework contract with the distributor and the products were distributed on an order basis, in the quantity and assortment required by the customer. This decision led to a boost in Coca-Cola's sales. As a consequence, it was necessary to restructure the sales departments of the three companies, to hire new staff and to purchase new logistics.

MobiFon company was particularly successful in its market approach. A North American and Romanian consortium (Telesystem International Wireless from Canada, Air Touch Communications from the United States, Ana Electronic, LOGIC Telecom, RA Posta Romana, ISAF and the Romanian Investment Fund from Romania), MobiFon aims at offering state-of-the-art digital mobile telephony. In November 1996, MobiFon was awarded the first GSM license in Romania. The delivery and implementation of a nationwide network in Romania, called CONNEX GSM, is based on a framework agreement with Ericsson, including deliveries and installation of switching and radio equipment.

MobiFon launched its services in nine cities in April 1997. By February 1998, it covered 85 localities, towns and resorts, targeting 90 per cent of the Romanian population. A chain of corporate stores was set up in the main cities. Additional services were offered during 1997, such as electronic voice mail, and international roaming in 41 countries.

As part of its marketing strategy, MobiFon actively supports cultural, sports, and entertainment events, such as ATP Open Tennis '97, Romania Investment Summit, Ad Maniacs Gala, and the Jean-Louis Ferré Fashion Show, as well as the first international black humor festival 'Humhorror'. MobiFon's approach in public relations and advertising is really special. Its main characteristic is humour. Everything in Romania is promoting CONNEX GSM's image: all media, even walls and buses. For its TV and radio adverts, MobiFon employed Ioan Gury Pascu, a renowned young Romanian actor and singer, well known for his performances with the 'Divertis' group.

MobiFon sponsorships are addressing specific target groups by specific means. For example, in April 1998 MobiFon decided to support the 'Rising Star Award' scheme, a contest initiated by TDA, a British consultant, and supported, among others, by Lady Margaret Thatcher. This scheme is aimed at identifying, awarding and promoting the best Romanian companies for the current year.

Competitive move strategies

Due to its market potential, Romania should be considered by all MNCs which decide to enter the Central European market. This approach might be facilitated by the fact that all former communist regimes aimed at developing most of their economic sectors, in order to create a 'multilaterally developed communist society'. On the other hand, this region is too small for an MNC to make independent competitive moves of relevance in these countries.

Romania is often approached by an MNC via its western neighbor, Hungary. For example, PipeLife Romania, a US 2.5m investment, was set up by the Hungarian branch of PipeLife International, a joint venture of Solvay from Belgium and Wienerberger from Austria. The company started operations in 1997 and production of plastic pipes in October 1998. The headquarters and warehouses are established in the Bucharest area, while the production facilities are located in Harghita county, employing about 60 people.

One option for an MNC might be to establish production facilities in the largest countries, like Poland and Romania, and distribution facilities in the others. Typical examples are Colgate-Palmolive, in consumer goods, and Daewoo, in the automotive industry.

A relevant case is that of Unilever, which in 1993 acquired the Romanian detergent factory, DERO Ploiesti, the leading producer in the country. The principal brands at that time were the Romanian Bona and Dero detergent brands. After investing more than US$50m in a three year period, the company produces detergent powders (Dero, Bona, Omo and Domestos), home care products (Domestos and Coccolino), hair shampoos (Organics, Timotei), Ponds skin care range, the Close-up tooth paste, deodorants (Impulse and Rexona), after-shaves (Axe and Denim), toilet soaps (Lux, Lifebuoy, Dove and Amo). In 1998, the company moved in a new direction, by launching the Rama margarine and the Algida ice-cream. Both local and international brands are heavily advertised. In some cases advertising is basically international, with some adaptations, depending on the needs of each particular brand. For the local brands and some international brands, like Omo, they developed local advertising.

Unilever had to face many challenges in Romania. The company had to move quite fast to catch up with the Dero business and that meant reformulating and modernizing their brands, supporting them with strong advertising, upgrading the manufacturing process, and developing most other aspects of the whole business system. From the beginning it was necessary to establish a system of proactive distribution, run by business people motivated to succeed and to develop their activities, and knowing very well their marketplaces in their regions. In 1998, the retail system of Unilever was still dominated by small outlets but the first international players, such as Metro, were expanding their operations in Romania.

Organization and Management Approaches

The Romanian legislative framework offers numerous possibilities to foreign investors to organize their business activities. Companies established by foreign investors must have the usual characteristics, such as their own names, registered capital, registered office, employees, management and bank account. Special regulations exist for mere representative offices, which cannot perform commercial activities, but are entitled to carry out advertising, promotional, and other activities for their parent organization.

A foreign company can also develop business in Romania through a branch or a subsidiary. In Romania, a branch office is not a legal person, and the parent company will be held liable for all actions and debts contracted by the branch. Unlike branches, a subsidiary is a Romanian legal entity. In practice, MNCs' subsidiaries are normally established as limited liability or joint stock companies.

Joint ventures are not separately regulated by Romanian law, the term being commonly used to describe any of the number of forms of economic activity carried on with the contribution of foreign capital. A number of joint-ventures have been set up in Romania and these are illustrated in Table 8.8.

Most organization and management tools and practices are applicable by MNCs in Romania. As a consequence, most management positions might be held by locals. Favorite are those positions requiring a deeper understanding of the local environment, such as human resources, marketing, and accounting. Such a solution is more appropriate for MNCs employing expatriates from a Latin language country, as for those it is easier to understand and speak Romanian. After two to three years, the top management positions might be handed out to Romanians, as it happened in 1998, when Marian Alecu was appointed the General Manager of McDonald's Romania. An alternative solution was to establish mixed teams, consisting of an expatriate manager and a Romanian assistant manager. A typical example is Daewoo, where all divisions are led by such teams.

Although salaries paid to local managers are lower than those in the West, their level increased significantly during 1997, when the two telecommunications consortia, MobiFon and Mobil Rom, were pushing the salary level up by offering top executives' salaries that were 125 per cent higher than the average wages paid

Table 8.8 Major Joint Ventures in Romania by Sector

Sector	Company	Foreign Partner	Capital (US$m)
Oil and gas	Shell Romania	Shell Overseas Hold.	47.0
	Shell Petroleum N.V.	Shell Overseas Hold.	44.0
Heavy Industry	Daewoo Automobile Romania	Daewoo Heavy Industries	156.1
	Daewoo Mangalia Heavy	Daewoo Heavy Industries	53.0
Electronics	Emcom	Siemens AG	13.6
	Elcaro	Siemens AG	6.5
Electrotechnics	ABB Power Romania	ABB	4.4
	ABB Energoreparatii Romania	ABB	4.4
Chemistry	Unilever Romania	Marga BV	32.8
	Tofan Grup	Nomura	30.1
Wood Processing	Tenneco Packaging Romania	Tenneco Holdings	14.0
	Romcarton	Sikor France	7.4
Light Industry	Ric Pro SA	Conad Ltd	5.3
	Heim Milcov	Hohentz Trikotindustrie	4.4
Constructions	Mondial	Villeroi Boch AG	10.9
	Bouygues Romania	Bouygues	4.5
Banking	Banca Daewoo Romania	Daewoo Securities	22.8
	Banca Ion Tiriac	EBRD	18.7
Commerce	Metro Rom Invest	MSB Deelnemingen BV	28.7
	Lever Com Serv	Marga BV	12.1
Tourism	CCIB	Credit Lyonnais	19.2
	World Trade Center Bucuresti	Bouygues SA	15.3
Telecommunications	MobilRom	France Telecom Mobiles	61.2
	MobiFon	Telesystem International	57.5
Food Industry	Coca-Cola Bucuresti	Coca-Cola Amatil	32.8
	Compania de Imbuteliere Coca	Molino Holding	24.5
Services	Media Pro International	CME Media Enterprises BV	17.0
	Balli Metal	Balli Metals Ltd	11.6

Note: As of February 1998.
Source: Romanian Development Agency

by MNCs operating in Bucharest and 200 per cent higher than the market average for executives in similar positions. Consequently, about 20 executive search companies, foreign and local, are flourishing, and the high turnover of key people has become a major problem. It is not uncommon for a 35 year-old MBA graduate to change their position every six months, leading to additional costs for staff recruitment and selection on the part of the employer.

The first MNCs investing in Romania soon discovered that Romanians are also able to work in neighboring countries, in spite of the language barrier (Romanian and Hungarian are the only non-Slavic languages in Central and Eastern Europe). For example, in 1992 Colgate-Palmolive Romania developed the strongest marketing team in Romania. One year later, these specialists started to offer services to other Central and Eastern European subsidiaries. Colgate-Palmolive Romania's top management initially consisted of expatriates. The first Romanian Director of Human Resources was appointed in 1993. In 1996, she had already moved to another subsidiary in Central Europe.

A typical problem for MNCs entering joint ventures with former state-owned companies is over-staffing. To alleviate this burden and, at the same time, to create a positive climate in the organization, has proven to be a difficult task. An illustrative example was provided by the attempt of New Holland, a subsidiary of FIAT, to acquire Semanatoarea, the most important producer of agricultural equipment in Romania, in 1996. The rumor that most of the workers would be released led to a riot, and finally New Holland decided to cancel the agreement.

Another problem occurs when the foreign partner creates too high expectations among Romanians. One year after buying a 70 per cent stake for US$37m at Rulmenti Grei, a producer of bearings for industrial equipment located in Ploiesti, the USA based Timken had to face a major strike of the entire workforce of 1000 employees. The dispute arose when automatic salary increases promised by its management did not materialize. Fearing that a strike would affect the company's world-wide image, and hence its chances of purchasing a similar facility in Slovakia for which it was currently negotiating, Timken management accepted a 70 per cent increase in monthly wages in June 1998, followed by a further 20 per cent in September. Prior to the increase, the average wage at Timken Ploiesti was US$90 per month.

For Colgate-Palmolive Romania, the most difficult task seemed to be how to change the mentality of about 2000 employees without producing a major turmoil in the company. The expatriate managers soon discovered that the Romanians were more interested in production than in marketing and sales activities, they were not as concerned with client relationships or solving their problems. Romanian employees were used to working in a stable and well-structured hierarchy, dealing with only one task at a time, rather than in a complex, multi-tasking environment. The team spirit and the initiative were almost entirely missing, and everybody looked up to their boss, waited for orders and avoided responsibility wherever possible. The expatriate managers pushed to introduce new and modern management systems and skills (for example, computer knowledge and use has become obligatory). On the other hand, a lot of effort was made to understand existing local culture.

For Unilever, the biggest challenge in Romania was the restructuring of its workforce. In a three year period, the number of people was reduced from 1000 to 600, reflecting also the recruitment of additional people, particularly in areas like sales and marketing. In order to smooth this process as much as possible, a retraining and outplacement center was operated for a six month period. All the people involved were assessed, were offered and received retraining in whatever skills they required. There was also a process of bringing local employers into contact with the people to maximize their chances in an employment offer.

In 1997, the French group Lafarge bought a 51 per cent stake in the Romanian holding company Romcim, the largest cement producer in Central and Eastern Europe, which owned six plants in three Romanian regions, produced 7.5m tons, and employed 8000 people. Lafarge had to sign a contract with the Romanian Government which obliged it to keep the existing staff for a two year period. Besides the restructuring of the existing activity, in 1998 Lafarge Romcim launched a major voluntary leave program, supported by a compensation scheme, as well as training and outplacement measures, careful to avoid social turmoil, which might affect the company's image among the local communities. Additionally, Lafarge Romcim intends to finance local development projects, such as a business incubator which is expected to become operational in the year 2000 in Medgidia, where the company's largest plant is located.

A special case is Daewoo, the largest investor in Romania. As a part of their master plan to rapidly establish a Central European manufacturing base, by February 1998 Daewoo had invested more than US$300m in Romania. Of this, US$156m was invested into the Craiova car plant (the second largest Romanian car manufacturers, after Dacia Pitesti), US$53m in the Mangalia shipyard, US$25m in banking by Daewoo Securities, US$70m in real estate projects to develop apartments, offices and shopping malls in Bucharest, and US$10m in assembly sites for consumer electronics.

The relationships between Daewoo and its counterparts did not initially go so smoothly. After obtaining preferential conditions, provided by so-called 'Daewoo law', allowing Daewoo to pay part of their investment by incomes resulting from selling tax-free imported cars in Romania, and a big launch of their operations in 1995, Daewoo encountered several problems. Firstly, productivity did not increase as expected, leading to a much higher cost than foreseen. Secondly, cooperation with local suppliers did not developed as planned. The Koreans blamed Romanian suppliers for a lack of capability, necessary to meet required quality and delivery conditions, while the Romanians invoked the low prices, the lack of technical assistance, and the unreliable contracting conditions offered by Daewoo. In 1997, the first reliable local suppliers have been identified: Tofan Grup–Floresti, for tires, and Roti Auto Dragasani for wheels. The percentage of components manufactured in Romania is expected to increase from 35 per cent in 1997 up to 70 per cent by the year 2000.

Daewoo's activities in Craiova were affected most by the slow increase in sales, partially due to the stagnant purchasing power of the population and the powerful competition which came from Dacia Pitesti, the major Romanian car maker. Dacia is producing low quality cars, but at half the price (US$4000 instead of a minimum US$8000 for a Daewoo 'Cielo') and is very familiar to Romanian drivers and mechanics. At the same time, the image-conscious upper-class Romanians clearly prefer Western cars. In 1996, Daewoo sold in Romania only 28 000 cars, less than half its forecast of 71 000, and 3400 cars only in the first seven months of 1997. The main hope of boosting sales now lies in the introduction of a new leasing scheme, based on a three year term, with up to only 20 per cent advance payment, and with no additional guarantees from solvent customers. The support of the newly established Daewoo Bank proved essential for the success of this scheme. Daewoo management blames also the Romanian legal system for not sufficiently protecting local production against imports. As an example, nongovernmental organizations are allowed by law to bring cars into the country almost entirely duty-free. And more than 100 000 cars were imported by Romanian NGOs during 1995–98.

Despite this situation, Daewoo has decided to invest US$450m in a new engine and transaxle manufacturing unit at Craiova, for both Romanian and overseas needs. Furthermore, in May 1998, Daewoo has acquired 51 per cent of the shares owned by the State Ownership Fund in Mecatim Timisoara – the smallest of the three car manufacturers existing in Romania before 1989 – and has intended to invest US$100m in this company during 1998–2000. Daewoo plans to boost the company annual sales up to US$120m sales on the domestic market and US$30m on exports.

By the year 2000, Daewoo plans to increase its investment in Romania to about US$1bn. This strategy may be seen as evidence that Romania remains an important part of Daewoo's globalization strategy. However, the recent financial problems at the Group level, as well as the expected penetration of Renault on the Romanian car market, might affect Daewoo's plans in Romania.

Nomura International's approach to Romania was different from that of the other MNCs. Instead of acquiring local companies or building up greenfield facilities, they paid US$60m for a minority stake of 49 per cent in a newly established private company – Tofan Group - and arranged a further US$40m in long term debt financing. This was the first Japanese major investment in Romania, followed in March 1998 by Koyo Seiko, which bought a 51 per cent stake of Rulmenti Alexandria – the most modern Romanian ball bearing producer – for US$25m.

The case of the Tofan group is a particularly interesting story of successful entrepreneurship in Romania. In 1991, Gelu Tofan, a thirty-year old engineer employed by Danubiana Bucharest, then the major state-owned tire enterprise in Romania, succeeded in obtaining a bank loan with his family home as collateral. He bought a truck and started a transport and delivering company. His small enterprise grew rapidly and by 1994, Tofan established Tofan Recap, the country's single modern tire retreading factory. In 1995, his business success rocketed, and Gelu Tofan became a lei-billionaire. He continued to invest in new opportunities acquiring his former employer, Danubiana, in December 1995. With a 50 per cent market share, Danubiana's gross profit doubled in 1996, and the turnover grew by 10 per cent. An additional two factories, Victoria and Silvania were integrated into Tofan's empire in 1997, when the Tofan Group had a 70 per cent market share for production, and 85 per cent for distribution.

The Tofan Group's deal with Nomura was beneficial for the company's proficiency. A state-of-the-art promotional campaign supported the launch of a new and modern tire, MONTANA. Based on heavy investment in production facilities, the Tofan Group has decided to focus on export markets. The target for 1998 is US$100m. In the medium term, the Tofan Group wants to be in the top 20 of tire manufacturers world-wide.

The international financial institutions are increasingly interested in supporting the activities of MNCs in Romania. To finance the rapid expansion of CONNEX GSM, the European Bank for Reconstruction and Development (EBRD) has arranged and partially financed a US$190m loan facility to MobiFon in October 1997. At that time, this was the largest long-term loan granted to a private company in Romania that was not secured by the Romanian Government. In January 1998, the EBRD took a 38 per cent stake in Lafarge Romania, the subsidiary which holds 51 per cent of Lafarge Romcim. The EBRD will provide a total of US$160m for the modernization of Lafarge's Central and Eastern European plants.

Some MNCs operating in the oil and gas industry in Romania have a truly European vision. Typical examples are Royal Dutch Shell and LUKoil. Royal Dutch Shell set up a joint venture in 1995 in association with three state-owned distribution companies. Shell Gas Romania is operating on the local butane and propane gas market, along with ButanGas (a Romanian-Italian joint venture) and the state-owned public utility Romgaz. The company has invested US$4m in the construction of a liquefied petroleum gas (LPG) terminal expected to become operational in January 1998. The terminal is located in the Northern part of Romania, near the border with Ukraine. The terminal has a capacity of 100 000 tons per year and will supply a growing Romanian LPG market with imported gas from Russia. The LPG will be transported from Russia, through Ukraine, by railcars. It will be then pumped out from Russian railcars and transported via the Romanian railroad to the company's filling plants in three Romanian cities (Ploiesti, Constanta and Timisoara) and also to the Shell Gas subsidiaries in Hungary. For this project, Shell Gas Romania has formed a transnational company – Trans Gas Services - having as partners ButanGas Romania, Agip, and Total Gas Hungary. The third phase of this project is the construction of an import terminal for LPG on the Black Sea port of Constanta. This will require another US$50–100m investment effort from Shell.

In the beginning of 1998, LUKoil – the second largest Russian oil company – bought a 51 per cent stake in Petrotel, one of the four largest Romanian refineries, for US$300m (debts and future investments for the next four years included). LUKoil started its operations in Romania in August 1997, through a cooperation in oil processing with Rafo Onesti, another Romanian refinery. LUKoil took over Petrotel because of the reputation of Petrotel's well-trained and experienced workforce and its strategic position in the middle of the Romanian industrial belt. At the same time, the company is interested in developing joint ventures and other development programs with other Romanian refineries. LUKoil's most ambitious project is to extract oil from the Caspian Sea area and to distribute it directly to Central and Southern Europe through Romania. For this purpose, LUKoil holds an 18 per cent stake in an international consortium created to build a pipeline under the Black Sea. This consortium took 30m barrels per year from the Caspian (in May 1998), but it expects this to rise nearer to 45m barrels over the next five years. LUKoil has considered two directions for oil transport through Romania. The first is from Odessa to Italy via the Constanta Terminal. The second, the Northern alternative, through the Russian and Romanian systems to anywhere in Europe. In order to merge the efforts of the oil companies in the Black Sea area, infrastructural and financial projects are being planned, including the creation of a bank. LUKoil Romania is intended to be a center for sales in the Balkans, which will develop through low prices.

Conclusion

Our analysis shows that Romania is a good example of a country with great potential and also a great need for investment capital. Some of Romania's strengths are common to other Central European countries, especially those related to low labor cost coupled with people's high technical skills. Others are specific to the country, such as the relatively large size of the population, significant natural resources (including tourism potential) and an attractive geographical position.

Due to the serious delay in implementation of political and economic reforms, Romania is lagging behind its neighboring countries in Central Europe in terms of globalization drivers. The purchasing power of its population is comparatively low, thus holding back its market potential. Business support structures, including framework conditions specifically for small and medium-sized enterprises, are underdeveloped and bureaucratic barriers are still blocking economic initiatives. Inflation is on the decline, but foreign investors are still deterred by the lack of legal and political stability. Local companies are often not considered reliable business partners for MNCs.

Despite this reality, many important MNCs have become active in Romania, attracted by its large market potential, its natural resources, its low cost but highly skilled labor, its European background and planned accession to the European Union. Part of the MNCs are still considering Romania as a low cost site, developing labor intensive industries, while others are becoming interested in the local capacity to design and develop new products and to provide high quality technical services.

From a strategic point of view, Romania seen as part of Central Europe, although some MNCs are addressing Romania as a region. Only a few of them so far have established regional headquarters in Romania. Most MNCs are attracted by the higher capability of business support structures provided by the neighboring countries such as Hungary. Consequently, MNCs are applying regional rather than global strategies in Romania.

At the same time, newly created companies are emerging in Central Europe, which are becoming more and more international. Their competitive advantages over well-established MNCs lie mainly in their geographical proximity, regional network of people, and deeper understanding of the existing problems common to all countries in the region. Many Hungarian-based companies, for example, are expanding on the Romanian market. Some of them have already established production facilities in Romania, especially in those countries with a significant Hungarian minority. In the next five years, such regional companies will probably be the most important competitors of the MNCs operating in Romania.

When Unilever's Fergus Cass was asked 'What sort of model should a company follow when entering a country like Romania?', he replied 'It's difficult to use a model. I think a feature of our approach is that every new situation is unique. We have the ability to draw on many aspects of Unilever's experience. Clearly there was a considerable experience of operating in other Eastern European countries such as Poland, Hungary and the Czech Republic, but also the experience drawn from Western European countries and Latin America. There was no model but rather an approach to find solutions for this market and, of course, very much building on the significant local business we had acquired'.[4]

Bibliography

Adevarul, 8, 10 December 1998, 15 September 1999.

Adevarul Economic, 13–19 March 1998.

Bucharest Business Week, 1–7 June 1998, 22-28 June 1998.

Business Central Europe, The Annual, 1997/1998.

The Business Review, 10–25 January 1998, 9–15 February 1998, 13–19 April 1998, 4–10 May 1998, 25–31 May 1998, 1–7 June 1998, 22–28 June 1998, 9 February 1999.

Capital, 16 February 1995, 23 March 1995, 6 April 1995, 4 May 1995, 11 May 1995, 7 November 1996, 24 April 1997, 8 May 1997, 26 June 1997, 26 February 1998, 6 August 1998, 27 August 1998, 19 November 1998, 10 December 1998.

Central European Economic Review, September 1998.

Coopers & Lybrand, *Romania, Yes! – An Investment Guide*, Romanian Development Agency, 1995.

Curentul, 17 July 1998.

Deloitte & Touche, *Romanian General Business and Tax Overview*, March 1997.

Economist Intelligence Unit, *Country Report: Romania*, 1997–1998.

In *Review Romania*, July/August 1997, September 1997, October 1997, December 1997/January 1998.

Invest Romania, Spring 1997.

National Commission for Statistics, *Economia mondiala in cifre*, September 1997.

National Commission for Statistics, *Quarterly Statistical Bulletin*, 1998.

National Commission for Statistics, *Romania in Figures*, 1998.

Nine O'Clock, 1 March 1998, 24 May 1998.

Romanian Business Journal, February 1998, March 1998, May 1998, November 1998.
Romanian Development Agency, *Romanian Investment Review, 1996–1998*.
Romanian Development Agency, *Annual Report, 1994–1995*.
Romanian Economic Observer, March 1998.
Romania Libera, 10 December 1998, September 1999.

Notes

1 The Electronic Embassy, Washington DC, 1999.
2 OECD Transition Report, Survey of Romania, 1997.
3 Deloitte Touche Tohmatsu, *Succesul în coodonarea investitiilor straine in Romania*, 1996.
4 *The Business Review*, 4–10 May, 1998.

Acknowledgements

The Romanian author would like to address special thanks for the support provided during the preparation and elaboration of this material to Ms Adriana Ionescu, Adviser to the International Management Foundation Library, and to Dr. Ralf Rössing, Senior Adviser, GOPA Consultants.

9 Bulgaria – Building on the Black Sea

BOJIL DOBREV and STEFAN DOBREV

Overview

Surveys of foreign investors have shown a keen potential interest in Bulgaria. The primary attraction is the Bulgarian labor force, which is perceived as well-qualified, well-motivated and very inexpensive, even relative to other transition economies. Foreign investors also cite relatively easy access to the domestic market as an advantage for business in Bulgaria. A legacy of preferential access to the large Russian market, as well as a direct link between Europe and Central Asia through Bulgaria, are also strong advantages the country offers foreign investors and multinational companies.

But Bulgaria has received comparatively little foreign investment relative to many other C&EE countries. One of the deterrents is that Bulgaria represents a relatively small and poor domestic market that is both culturally and geographically farther away from most OECD countries than many competing transition economies. Bulgaria has a limited endowment of natural resources and relatively poor infrastructure and communications. The restructuring of manufacturing towards international competitiveness requires significant overhead investment in an uncertain and unstable environment.

With a population of just over eight million, Bulgaria lies in the center of the Balkan Peninsula. Located at the crossroads of southeastern Europe, Bulgarian culture is a unique blend of Slavic, Mediterranean, and Eastern European. Geographically, most of the country is mountainous ($110\,000\,km^2$), with the added feature of the Black Sea coast (350 km). Bulgaria borders Greece (494 km), the former Yugoslav Republic of Macedonia (148 km), Romania (608 km), Serbia (318 km) and Turkey (240 km). About a sixth of its 8.4 million inhabitants live in the capital, Sofia, with the rest of the country relatively sparsely populated. Unusually for the Balkans, ethnic relations among the Bulgarian, Greek-Orthodox majority and the Turkish Muslim minority, are largely unproblematic. Currently emerging from times of isolation and economic troubles, the country displays unique potential for development.

History

An early state with the name 'Bulgaria' existed in the Balkans between the seventh and tenth centuries and emerged again in the thirteenth and fourteenth centuries. This state was founded by a small union of Slavs and Bulgars from the Caucasus and Thracians, who lived in the area since before Roman times. It was the Bulgarian state that facilitated the spread of the Cyrillic alphabet, first in the Balkans and then to Russia. The interaction of Bulgaria with its mighty neighbor, the Byzantine Empire, remained strong over the centuries, and by 900 AD the Bulgarian state had adopted the Empire's religion and much of its culture. By the end of the fourteenth century, the Balkans were absorbed by the Ottoman Empire, although many of the region's inhabitants, including the Bulgarians, preserved their local language and some local traditions.

The modern Bulgarian State was initially founded as a duchy of the Ottoman Empire in 1878 after an Ottoman-Russian war, and only became completely independent in 1908. A democratic constitution and a successful economic policy led to a growth in economic prosperity for what was previously a relatively backward, agricultural country. In 1945, backed by the Red Army, the Communists established their government and in the subsequent decades, artificially forced industrialization and urbanization that was often economically unsustainable. By the end of the 1980s, the ruling regime was in trouble and crumbled with the fall of the Berlin Wall in 1989. Throughout Central and Eastern Europe such shifts in the balance of power provided opportunities to establish liberal market economies and democratic governments. In Bulgaria, however, the governments that followed, both the reformed socialists and 'democratic forces', failed to introduce social and economic policies that would facilitate a smooth transition to a free market. In 1996, the Bulgarian economy faced a major crisis, resulting in a dramatic fall in the average standard of living. The economic crisis was marked by a banking system in turmoil, a depreciating currency and contracting production and foreign trade. Since 1997, however, a number of much needed reforms have been passed, with the government ever more committed to structural reform, creating a rapidly changing environment.

The Transition

All of the transition economies have suffered, in one way or another, from inherited conditions associated with low competitiveness, the lack of developed financial and fiscal institutions, low confidence in economic policy and the accumulation of bad debt. Initial conditions in Bulgaria were more unfavorable, on average, in all these areas and the fact that Bulgaria had lagged behind many other C&EE countries in its reforms, reflects these initially adverse conditions. The financial crisis that Bulgaria underwent in 1996 highlighted the fragility of the economic environment and the need for solid structural reform.[1]

The 1996 crisis, however, has also provided the painful but necessary context to accelerate reform and deal decisively with loss-making banks and enterprises, to complete privatization, and to improve the overall banking system. While economic growth was dramatically interrupted in 1996, with a fall of GDP of about 11 per cent, by 1998 and 1999 a gradual recovery had begun. The real question is whether growth is sustainable, and whether planned reform will be implemented sufficiently to translate into long-term stabilization. The recent changes to the Bulgarian economy are illustrated in Table 9.1.

Table 9.1 Macro–Economic Indicators for Bulgaria

Bulgaria	1990	1991	1992	1993	1994	1995	1996	1997	1998
Nominal GDP ($bn)	19.2	7.5	8.6	10.8	9.7	13.1	9.9	10.1	12.2
GDP per capita PPP ($1000)	4.5	4.1	4.1	4.2	5.0	5.3	5.0	4.8	5.0
GDP (% change)	−9.1	−11.7	−7.3	−1.5	1.8	2.9	−10.1	−7.0	3.5
Industrial production (% change)	−16.7	−20.2	−18.4	−9.8	10.6	4.5	5.1	−10.0	−12.7
Budget balance (% of GDP)	−4.9	−3.7	−5.2	−10.9	−6.2	−6.6	−10.9	−3.9	1.3
Unemployment (%)	1.7	11.1	15.3	16.4	12.8	11.1	12.5	13.7	12.2
Average monthly wage ($)	157.5	55.0	87.7	116.9	91.5	113.1	79.4	76.3	106.5
Inflation (%)	23.8	338.5	91.2	72.8	96.0	62.1	123.0	1082.3	22.3
Exports ($bn)	2.5	2.7	4.0	3.7	4.0	5.4	4.9	4.9	4.3
Imports ($bn)	3.3	2.3	4.2	4.6	4.2	5.7	5.1	4.9	4.8
Trade balance ($bn)	−0.8	0.4	−0.2	−0.9	−0.2	−0.3	−0.2	0.0	−0.5
Current account Balance ($bn)	−1.2	−0.4	−0.8	−1.4	−0.2	−0.1	0.1	0.4	0.0
Foreign Direct Investment Flow ($m)	13.0	56.0	42.0	40.0	105.0	82.0	100.0	497.0	270.0
Foreign Exchange Reserves ($bn)	n.a.	0.3	0.9	0.7	1.0	1.2	0.5	2.1	2.7
Foreign Debt ($bn)	10.9	11.8	12.5	13.9	11.3	10.1	9.5	9.7	10.1
Discount Rate (%)	n.a.	54.0	47.8	63.1	93.9	38.6	342.1	6.8	5.2
Exchange Rate (/$)	0.8	16.7	23.3	27.7	54.2	67.2	175.8	1676.5	1760.4
Population (m)	8.7	8.6	8.5	8.5	8.4	8.4	8.3	8.3	8.2

Source: Business Central Europe, The Economist Group, http://www.bcemag.com

In the 1980s, Bulgaria enjoyed privileged access to Soviet oil and gas, an unlimited market for its wine and other agricultural products and electronic goods, and generated a GDP of around US$40bn.[2] With the disintegration of COMECON, Bulgarian industry lost both its traditional markets and its source of cheap natural resources. The large state-owned enterprises were not competitive in the new environment. Widespread mismanagement and the absence of central control, combined with the subsequent failure to push through the structural reforms required to adjust to post-Soviet conditions, led to decay and falling industrial output. This had a direct negative effect on GDP. The post-Soviet governments were unable to privatize or close down a sufficient number of ailing state enterprises. Combined with attempts to maintain nominal incomes, this led to accelerating inflation and a rapid depreciation of the national currency, the lev. As the same time, poor banking legislation made it possible to finance losses, in part through bad loans, which were unsustainable and eventually led to the collapse of the immature banking system.

Faced with the prospect of default on the external debt, in November 1996, the IMF proposed a currency board, as Bulgaria's best chance to restore confidence in the lev, eliminate unnecessary spending, and avoid hyperinflation. In 1997, depreciation was halted by pegging the lev to the German mark. The establishment of the currency board was followed by a reduction in inflation and interest rates and by a rise in foreign investment. The government has promoted new strategies for privatization and the recent agreement with the IMF will probably lead to the long outstanding closures of obsolete industrial plants. At the same time, branches and subsidiaries of foreign banks, like ING and BNP-Dresdner, could form the core of a new, healthier financial system and promote growth. Many issues concerning future policy and national priorities are still open, with both the opportunities and the risks attached.

Reflecting the recent changes, the population fell from 8.76 million in 1989, to 8.47 million in 1998, as a result of both emigration of a large portion of the Turkish minority (9 per cent of the population in 1993), low birth rates, and high mortality rates. In spite of the return of property and land to many pre-1945 owners, the urban/rural residency ratio is expected to stay strongly weighted towards urban areas since they are more developed and provide more employment opportunities. With priority given to the development of the private sector, the portion of the labor force employed in the public sector should fall. The unemployment rate has been relatively stable for some years, mostly due to the fact that loss-making state-owned companies, employing a large portion of the workforce, are not being closed down.

Foreign direct investment

Bulgaria was actually among the first of the C&EE countries to allow foreign direct investment, passing a joint venture law in 1980. Initially, freedom of intervention from the government was limited, but nevertheless, by 1989 there were 31 joint ventures operating in Bulgaria, mostly in machine building and electronics. At the start of the transition period in 1989, Bulgaria was again among the first of the C&EE region to adopt legislation to attract foreign investors. The Law on Foreign Investment, passed in 1991, was among the most liberal in the region, providing national status to all foreign investors, allowing 100 per cent foreign ownership, setting very low entry barriers and offering easy registration. In addition, the law provided for unrestricted profit reparation, supported also by foreign exchange legislation. Tax incentives were offered for large investment opportunities. In the next five years further legislation was introduced to support the favorable investment environment and assist foreigners. Foreign investment into Bulgaria, however, did not reflect the strong legislative initiatives and, so far, investment and multinational commitment in the region have remained relatively small.

The foreign direct investment (FDI) that has occurred, has become an important element of the development of the Bulgarian economy. Leading investor countries include Germany (with US$262m), and Belgium (with US$252m). Industrial facilities have attracted the bulk of the investment, a total of US$704m, and investment in trade comes second (with US$217m). After peaking in 1994, foreign investment has fallen, again as a result of deteriorating economic conditions and continuing political instability. However, with the introduction of the currency board in 1997, the interest of investors is on the rise.

Table 9.2 Main Foreign Investors in Bulgaria

Company	Industry	Country	Starting Year
Cable & Wireless	UK	Telecoms	1992
Delta	Greece	Food	1993
Clarina (Coca Cola)	Luxembourg	Beverages	1993
American Standard	USA	Machinery	1993
Ytong Holding	Germany	Construction	1994
McDonald's	USA	Fast Food	1994
Shell	UK/Netherlands	Gas stations	1994
Reiffeisen Bank	Austria	Finance	1994
ING Barings	Netherlands	Finance	1994
Siemens	Germany	Telecoms	1994
Danone	France	Diary	1994
K. Jacobs Suchard	Switzerland	Food	1994
Nestlé	Switzerland	Food	1995
BNP/Dresdner	France/Germany	Finance	1995
DHL	USA	Couriers	1995
ABB	Switzerland	Machinery	1996
WAZ	Germany	Publishing	1996
Daewoo	Korea	Hotels, WTC	1996
Solvay	Belgium	Chemicals	1997
Union Miniére	Belgium	Copper Refinery	1997
Marvex	Spain	Cement	1997

Source: Based on author's research.

Some of the best business opportunities are to be found in tourism, in both summer and winter, facilitated by the mild climate and the presence of both the coastline and mountain ranges.[3] Agriculture is also facilitated by the mild climate. Light industry, telecommunications, and trade, are also gradually growing. Building and construction, especially of airports, roads, and pipelines, are an important part of the ongoing improvement in infrastructure, which is so key to the country given its geographic position. The chemical industry remains an important sector, given the existence of better than average conditions and the existence of some niche-market natural resources. In 1997, the chemical industry still accounted for the largest share of industrial output (at 26 per cent).[4]

The greatest barriers to investment in the country are an unstable legal system, complicated bureaucracy, the limited size and purchasing power of the home market, excessive taxation, and a lack of infrastructure.[5] To this, most MNCs' local managers would add corruption. All of this is, of course, subject to change, although it will take some time. The risks involved, as well as opportunities, are the consequence of the government's inability to be clear about its development priorities. Unlike other countries, however, in the Balkans, there is comparatively little risk of political turmoil or violent outbursts.

History of foreign involvement

Before 1945, there was some presence of multinational companies, mostly as retail operations, notably IBM and Shell Oil. The closed market place and the structure of the economy, however, would not allow for more market-driven entrants and competition. In the communist era, the government granted concessions to several Western enterprises on a monopoly basis. IBM imported mainframes and trained support staff, Coca-Cola, PepsiCo and Schweppes franchised a state-owned bottling company, and Xerox had a joint venture production site. Two hotel chains, New Otani and ITT Sheraton, were given the opportunity to operate one hotel each, although ownership remained with the state.

Table 9.3 Leading Foreign Companies in Bulgaria

Company	Industry	Investment (US$m)
Union Miniére	Copper Refinery	300
Marvex	Cement	254
Solvay	Chemicals	227
Daewoo	Hotels, WTC	45
Shell	Gas Stations	45
McDonald's	Fast Food	25

After 1990, a number of MNCs (including Royal Dutch/Shell, L.M. Ericsson, and Cable & Wireless) made initial investments on a relatively small scale. They opened branches, acquired subsidiaries or formed joint ventures, in part through privatization deals. The unfavorable economic situation, however, prevented further expansion and most multinationals adopted a 'wait and see' policy. The exceptions, who continued to expand operations, included the Cable & Wireless joint venture 'Morifon' (mobile telephony), the Coca-Cola bottler Clarina, and the dairy products company, Danone. Only since 1996, have there been further expansion plans and an intensified activity on the part of other big companies.

Peculiarly, there has been little activity on the part of MNC telecommunication companies and retailers. While the former suffer from the still overly-regulated telecoms market, the latter face a strong decentralized retail market that does not presuppose the existence of chains and larger retail companies of any kind. An unsuccessful attempt to enter the market was made by the German group, Asko. On the other hand, the Greek joint venture, Stambouli, has done remarkably well.

Domestic companies

The largest local companies are still state-owned monopolies in regulated industries. These include the National Electric Company, the State Railways, Balkan Bulgarian Airlines, and the Bulgarian Telecommunication Company. The oil refinery 'Neftochim' on the Black Sea coast, is practically a monopoly. The large state-owned enterprises, mostly single plants in machinery and steel-processing, are also still important. International activities are often limited to exports of some niche products of the chemical and pharmaceutical industry (for instance, the now foreign owned, Sody-Devnja Plant), and transportation

Table 9.4 Leading Bulgarian Companies

Company	Industry
Neftochim, Bourgeois	Petrol Refinery
National Electric Company (NEC)	Electricity
Bulgarian Telecommunication Company	Telecoms
Bulgargas	Gas Supply
Kremikovtzi	Steel Plant
Stomana, Pernik	Steel Plant
Balkancar Holding	Forklifts
Bulgartabak Holding	Tobacco
Chirnko, Vratza	Urea
Agropolychim	Fertilizers
Elprom, Varna	Power Equipment
Sodi, Devnya	Chemicals
Vasov Machinery Plant	Defense
Varna Shipyard	Shipbuilding
Pharma Holding	Pharmaceuticals
Opticoelectron	Electronics
Vinprom	Wine & Spirits
Bulbank	Finance
United Bulgarian Bank	Finance
Biochim	Finance
Multigroup Holding (Private)	Trade

services, road haulage, and air freight. To date, only the private company, 'Multigroup Holding', has large scale international activity.

Country Globalization Drivers

There are a number of motivating factors for foreign MNCs to participate in Bulgaria. From the beginning, Bulgaria has been committed to attracting foreign investors and offers a variety of incentives. Although the population and its purchasing power are relatively small, and natural resources are limited, Bulgaria has an inexpensive and well-educated labor force. Bulgaria also offers foreign companies the geographic advantages of having access to the much larger neighboring Russian and Central Asian markets.

Market globalization drivers

Bulgaria's small size and economic history result in some special aspects in terms of market globalization drivers.

Global customers

The size of the domestic market is limited by the population size, and the small number of companies in the country. As in most holiday destination countries, however, the flow of foreign visitors (on average two million a year, 1996–98) in the summer results in a seasonal rise in market demand concentrated on the Black Sea coast. At the same time, markets of the neighboring countries are within easy reach from Bulgaria, provided that the infrastructure improves as planned – around 60 million people in ten countries live within 500 km of Sofia.

Domestic market demand in Bulgaria is strongly concentrated in the capital and in the four or five other large cities. Customer needs are common to the rest of the region and have been relatively stable for the past few decades. Unusual for a post-communist country is the housing market. It is customary to own one's house (or flat in the cities) and to pay for it in full at the time of purchase. Therefore more than 70 per cent of the housing is owner-occupied and without outstanding mortgages. This is reflected in a rapid growth of the local construction industry, which was 90 per cent privately owned by 1998.

The mass product market is led by consumer electronics and automobiles. The demand for services is highest in education, tourism and travel and food and catering. The changes in the economy during transition brought about a rapid shift in the social structure and hence in the distribution of income-related consumption habits. As of 1998, 80 per cent of the population were classified as low income, 15 per cent as middle income and five per cent as high income. Thus the middle class consumption patterns typical for Western countries apply to only 15–20 per cent of Bulgarian consumers.

Few companies use Bulgaria as a source of outputs for their global operations, although a few commodities are available on a sufficient scale or at sufficiently favorable conditions. But the relative isolation from world markets during communism means that there is no tradition of out-sourcing from Bulgaria. There are, however, important exceptions. The country has a long tradition as a significant tobacco exporter, and Bulgarian tobacco is used in some global brands, such as Philip Morris and Camel. Bulgaria is also a source of some niche-market minerals and plants used in the chemical and pharmaceutical industries. Until the beginning of the 1990s, it was a big supplier of rose oil for the global cosmetics industry. Nevertheless, global customers did not usually have significant operations of their own or headquarters in the country. Only recently have British Gas and Total started drilling for oil on the Black Sea shore.

Transferable marketing

Marketing in Bulgaria, although generally similar to marketing elsewhere, has to consider some peculiar features of the local mentality. First, excessive advertising in the normal way tends to raise suspicion and, hence, put people off. Instead, consumers rely on immediate impressions (their own, or their friends') of the product or service for their purchasing decisions. This makes direct sales, as Xerox recently introduced, and network marketing extremely effective. Second, foreign brands are valued extremely highly at the expense of local brands and no-brand products – even if, as Nestlé found out, the foreign-brand product is produced locally.

Other than this, all standard marketing tools and techniques can be applied in the same way as in developed Western countries. Television advertising, product presentations, lotteries and games, and so on, are used as in most countries and sometimes transferred directly from another market with very little adaptation. There are numerous media channels, mostly private, that cover all consumer groups. As of 1998, the only nationwide television channel was state-owned, but private stations broadcasting locally exist, and some of them have national licenses pending.

Lead countries

Before 1990, within the closed system of the communist bloc, each country had a specialization. Therefore, Bulgaria had a lead in computers and peripherals, in chemicals and pharmaceuticals, in foods and beverages, in forklifts and in some branches of the defense industry. In the 1990s, as COMECON fell apart, competitiveness became crucial and most exports were severely hit by political turbulence abroad, especially embargoes. And as the economic environment worsened, innovation slowed down. Nevertheless, the country remains a major manufacturer of some niche products in the chemical, light and defense industry.

Cost Globalization Drivers

Cost globalization drivers in Bulgaria have four favorable aspects: currently, the well-qualified, low cost labor force, and – potentially – physical infrastructure and technology.

Favorable country costs

The cost of labor remains low, the average monthly salary (in most industries) being about US$100, with little additional cost. Also Bulgaria is one of the cheapest countries in Europe in which to live and work. This quality/price ratio is consistently pointed out as one of the main incentives for doing business in Bulgaria. It should be noted, however, that the cost situation may change. First of all, wages will probably rise in the future. Second, top specialists working for foreign companies, those people MNCs are talking about, are customarily paid up to ten times the average wage.

Education

Good education is traditionally highly valued in the Bulgarian's popular perception. As a result, there is a well developed educational system at all levels – from primary school to higher education – producing technically highly-qualified individuals for all sectors. The recent decline in the quality of education offered by the large state universities has largely been compensated for by the emergence of new, private institutions applying international teaching standards. Many, like the American University, the International University, and the New Bulgarian University, successfully use franchised programs from the United States, Britain, the Netherlands, and Germany. Bulgaria has a total of forty universities with about 223 000 students and 22 000 lecturers. MNCs are eager to stress the good technical knowledge and skills of their local staff, and some (KPMG, IBM Bulgaria) consider it a key success factor. The necessary additional training has usually been confined to company-specific issues and introducing a Western business culture and working practices.

Infrastructure

In the future a very strong cost globalization driver in Bulgaria could be the transport and communication infrastructure. The development of a good transport infrastructure is facilitated by the very short distances within the country – the longest commonly traveled route, Sofia to the Black Sea Coast, is only 380 km. In addition, the geographic position of the country leads to a very large volume of transit traffic – the most logical routes from Europe to the Middle East and Eastern Mediterranean, from the Mediterranean to Russia and to Central Asia, from the Caucasus oil fields to Europe, all pass through Bulgaria. This transit traffic is probably the only area where global scale economies can be generated in the otherwise not very large Bulgarian economy. Naturally, such a geographic position also favors logistics of both production and distribution.

The utilization of this position, however, requires good infrastructure. Some infra-structural elements are already in place – a fairly extensive road network (34 000 km) reaching all parts of the country, the main railway lines, four international airports (Sofia, Bourgeois, Vama, and Plovdiv), two large ports on the Black Sea (Bourgeois and Vama) and two ports on the Danube (Russe and Vidin), a gas pipeline from the Ukraine and another one planned between Bourgeois and Alexandropolis in Greece. All this, however, needs upgrading and extending if it is to play the role of a globalization driver.

Similar conclusions apply to the communication infrastructure. Putting it in place is eased by, again, the small size of the country, and in part by the existing facilities – Bulgaria leads in Eastern Europe by the number of phone lines *per capita* (33.5 per 100 inhabitants). Most of the equipment is, however, desperately outdated and in poor repair. Several projects funded by the EU Phare program and carried out by major telecommunication companies (Siemens, Ericsson, Cable & Wireless) are expected to improve the situation dramatically, and by 2000 the country should have a completely state-of-the-art network. The projects include digitalization, increasing the number of gateways to international networks, the provision of more value added services and, at a later stage, deregulation. In the meantime, new technology markets have seen rapid growth. In 1997, the mobile phone operators Mobifon (NMT450, a Cable & Wireless N) and MobilTel (GSM, an Alcatel N) have seen the number of their subscribers grow by 215 per cent, to 26 900 and 12 000 respectively. At the same time, there are 15–20 Internet providers and two national networks (Bulpack and Sprint, providing telecoms services via the X400 Network Sprint).

Technology

The Bulgarian information infrastructure is relatively well developed, as a result of the traditionally leading position of the country in IT in the former communist bloc. About 550 firms, about half in Sofia, retail hardware and software, and some are involved in software development. In 1997, 30 000 computer systems were purchased for US$47m. The technology market is growing very fast and an increasing number of companies offer 'total solutions'. All major international producers of hardware and software are present in the country either directly, like IBM, Xerox or Hewlett-Packard, or through a dealer network. For thirty years there has been a network of regional computing centers located in the large cities and providing various EPP services. A major problem, however, is that a national policy on IT development still does not exist.

The development of business information databases started in 1990. In spite of the very small information market, by 1998 ten to fifteen companies (public and private) were offering company profiles, legal information, products and services information and statistics. The information market is expected to increase and international cooperation to expand as a result of the process of restructuring of the economy and the process of privatization. At the same time, low labor cost and high employee qualification apply to software development as much as to other sectors. Thus, IT development in joint-venture companies is expected to expand. In short, the information infrastructure is potentially on its way to become a cost globalization factor in Bulgaria.

Government globalization drivers

Foreign trade and capital inflow is essential to Bulgarian economic restructuring and development. Thus the attitude of both the government and the legislature towards foreign investment and trade is extremely favorable. In the long run, this should result in a helpful legal environment for foreign investors.

Foreign trade policies

The foreign trade regime, although conforming to WTO requirements and subject to continuous liberalization, is still somewhat complicated. At the end of 1989, the state monopoly in foreign trade was abolished, and all companies were given free access to international markets. There are no export taxes or unilateral export quotas. However, in 1998, export of some types of goods (mainly food and beverages), were subject to registration. International obligations and domestic legislation prohibited the export of a limited number of commodities.

Import regulations depend on the type of goods and still have a more or less temporary character. Again, a limited range of goods is subject to registration. There has been a universally applied import tax of two per cent (as at July 1998), although energy resources, pharmaceuticals, and some other goods are exempt.

At present, six duty-free zones operate in Bulgaria. The state provides the land and the infrastructure, and considers their development a priority of its economic policy. Their advantages have attracted a number of MNCs, including Hyndai, Daewoo, and Groupe Schneider.

Role of regional trade blocs

The country is a full member of WTO and an associated member of the EU. Free trade agreements have been signed with EFTA, CEFTA and Turkey. Bulgaria is a party to conventions on general trade (the Vienna Convention) and on trade of specific goods (the Wassenar Agreement on dual use goods). International practices for customs procedures and cross-border transactions, have been adopted. Investment protection agreements exist with 26 countries, and double taxation treaties with 36 countries. Bulgaria has applied to join the European Union and is currently being considered as part of the second wave of European Union entrants from C&EE.

Bulgaria is one of the initiators of a Black Sea Economic Cooperation Zone, but the project is still at a preliminary stage. There are signs that a *de facto* trade bloc is about to be formed under the umbrella of Greece. This trend is reflected by the large number of Greek investments in the country.

Foreign direct investment rules

There is a Foreign Investment Law, enforced by the Foreign Investment Agency and the Advisory Council on Foreign Investment and Financing. There are no legal restrictions on foreign participation in businesses, or on the repatriation of profits and capital. There is protection against expropriation and adverse changes in legislation. Foreign nationals and organizations may own buildings, but not arable land. However, Bulgarian businesses, even foreign-owned in part, are not subject to restrictions in this respect.

Reliable legal protection

The Bulgarian 'Law for the Protection of Competition' applies equally to all business entities, foreign and domestic, regardless of their legal form. It is enforced by a government agency – the Commission for Protection of Competition.

Intellectual property, including software, is legally protected. Initially, this legislation was hardly enforced, and illegal copies of audio (CD) and video records and software were widely available. After strong political pressure on part of the US, however, such breaches of copyright were curbed down, practically to zero by 1998. Technical standards in the country are completely harmonized, metric, and conform to those used in most of the EU Enforcement is still strict.

State-owned competitors and government intervention

A good entry point for foreign capital can be the opportunity provided by the privatization of large, state-owned enterprises. The process has been a government priority for several years, although with mixed results up to 1998. Privatization took three forms: direct sale of assets, voucher privatization and return of assets, mainly land and property, to the pre-1945 owners expropriated by the communists.

However it should be noted that most utilities are still a regulated monopoly. This includes telecommunications (except mobile telephony) and all energy related industries, including electricity, coal mining, oil refineries. Plans for partial privatization still exist.

Direct sale of businesses (cash privatization) began in 1993, with the declared aim to give long-run efficiency improvement priority over short run revenues for the Treasury. Thus the various auctions, management buy-outs, and so on, are only part of the process of determining the future buyer. Negotiations on outstanding liabilities, the extent of future investment, and employment are at least as important. All large deals are conducted through the government privatization agency, the rest through ministries, municipalities, and so on. The main incentive to buyers is the provision for state debt instruments, available at a discount in the open market, to be used as a means of payment at face value. A profit tax holiday, but no other tax breaks, are also given.

These fairly complicated procedures gave scope for an over-bureaucratic system and, allegedly, corruption in the past. As a result, only 30 per cent of all formerly state and municipality owned assets were privatized as of April 1998, leaving a lot of opportunities open. However, the process is currently under review, and is expected to accelerate in the future.

The voucher (mass) privatization, in two waves, has resulted in about two million individuals becoming minority shareholders in former state-owned enterprises, mainly through investment funds. However, since this has hardly had an effect on the operations and organization of these previously state-owned companies, the results are less than satisfactory.

Urban property was successfully returned to its pre-communist era owners in a very short period of time (1994–95 restitution). However, the corresponding process with arable land and forests is still going on. By 1997, about 60 per cent of the land was back in private hands.

Although important cornerstones of modern legislation are in place, they still do not amount to a comprehensive investment-friendly legal environment. While setting up a presence in Bulgaria is relatively unproblematic for MNCs, complications arise at the operational stage. Regulations are incomplete, incoherent, and sometimes directly contradictory. Changes at short notice are usual. Enforcement is often weak or biased. The civil service is slow and paperwork often very cumbersome. MNC managers frequently complain about corruption. At the same time, the process of radically reforming the legal framework and the administration gives the chance to build on the experience of other countries in the region to develop a framework which is far-reaching.

Common marketing regulations

Marketing regulations are recent, and follow those accepted internationally. Still an exception are the often unclear and permanently changing rules for television advertisements.

Competitive globalization drivers

Bulgaria could, potentially, have a global strategic importance in the tourism industry. However, achieving such a position presupposes a number of factors that still have to materialize: complete privatization, opening of the transport market, improvement of the infrastructure and the level of services. Other than this, as a reasonably developed country, the role Bulgaria will play in the strategies of global companies is similar to a number of other countries in the region – mainly as a still unsaturated new marketplace. This is reflected by the type of activities located in the country – mostly retail, customer support, and training of staff, and only to some extent production. At the same time, MNCs of most industries are likely to meet their global competitors in Bulgaria. Overall, with the quality, modern management, and business culture, they bring to the country, MNCs are more than welcomed by a society currently in transition.

Overall Global Strategies

Because of its limited market size, many MNCs have chosen larger neighboring markets in the C&EE region. Nonetheless, MNCs can be found in most industry sectors and have pursued a variety of strategies. There is still much room to make use of the variety of incentives offered to foreign investors and continued opportunities to capitalize on inexpensive labor costs and high levels of education for production activities.

Market participation strategies

The market participation of MNCs differs widely across industries. Four groups of industries can be identified: (1) those where MNCs hold the high end of the market in their sector, (2) those where there are only MNCs in the market, (3) those where they share the market with local companies as peers, and (4) those where local companies dominate the market.

Where MNCs hold the high end of market share in their sector, they offer a high-priced, high quality range, with the traditional customer service, while local companies or less-well-known firms from other countries offer the cheaper, no-frills products and services. This applies to two major areas, fast food, and information technology and office equipment.

In fast food, major chains with operations in Bulgaria are McDonald's, Pizza Hut, KFC, and Dunkin Donuts. They are well known and popular brands in spite of the relatively small number of outlets they have. So far, all domestic chains and individual restaurants following a similar concept have failed to attract enough customers and have closed down. Only fast food outlets offering local specialties at a lower price could stand the multinationals' competition.

In IT and office equipment, IBM, Hewlett-Packard and other well known names sell mostly to other foreign companies' offices and businesses where quality and customer service are crucial. The rest of the market, including home equipment, is served by local or East Asian assemblers at considerably lower prices.

Where there are only MNCs in the market, there are practically no local companies in the market, and MNCs compete with each other. This applies, with a few exceptions, to business services, to large scale office equipment, and to automobiles.

Most business services, such as audit by international standards, tax advice, IT and management consultancy, are only offered by the Big Five firms. Customers are mostly other MNCs, the civil service and regulated monopolies. Hill International is leading in the local HR market. The only World Trade Center is owned by Daewoo. In a way, they had to create the market by explaining to local managers the need for their services. The exception is legal services, which are almost entirely in local hands.

As far as industrial and large-scale office equipment, communication equipment and cars are concerned, there is practically no local production, so the entire market is held by MNCs. Xerox, Canon, Minolta and others have been established since the early 1990s, because of their privileged position in communist times. The entire upgrade of the telecoms network is being carried out by L.M. Ericsson, Siemens and Alcatel. All cars are imported, with Russian brands and Daewoo having significant market shares.

The third group comprises MNCs who compete with local companies on the same level. Industries in this group include oil and gas retail, much of food and beverages, courier services, household chemicals, construction, and to an extent, banking. In fuels, Shell Bulgaria is expanding, but still lacks an all encompassing network of gas stations, and, given the regulated prices, is just one of many players in the market. In food and beverages some notable companies are Nestlé, Kraft Jacobs Suchard and Danone. In detergents and cosmetics, Johnson & Johnson, Procter & Gamble, and others have retail operations. On paper, foreign banks compete with about sixty local banks. However, the quality of their service and their lack of affiliation with local business groupings set them apart in a market of their own, and thus are in a way in the second group.

Finally, there are the sectors where there is practically no MNC presence. In some, this is due to regulation, notably energy, but also railways and telecoms (the latter being subject to deregulation in 2001). An exception is the gas industry, which is entirely in the hands of Gazprom of Russia. In other such sectors, there is simply very strong and cost-efficient local presence – for instance clothing, hotels or road haulage.

The small size and purchasing power of the Bulgarian economy limits somewhat the importance it has as a potential market for most industries. However, the trends in the country make it significant for several sectors. First, the accelerating restructuring of the economy, on both a macro and micro level, and an increased opening to the world markets carry with them a rise in the demand for business services, including operational consulting. (The market is, however, currently dominated by the Big Five.) Second, the ongoing upgrading and improvement in the country's physical infrastructure carries with it large and sophisticated construction project financed by the government and the European Union. Third, the country's telecoms will be deregulated soon, creating a demand for state-of-the-art providers. And finally, the transit traffic passing the country makes it important for all types of passenger and freight transport.

In most cases not all MNCs active worldwide in a given industry have entered the country. For example, global companies such as Exxon or Burger King are still absent. This leaves room for foreign investment in many (other) local markets.

Global products and services

In Bulgaria, global products and services can be divided into three categories depending on their type and customers: corporate services, mass-market services, and mass-market products. Although all of them are received well in their global form, a few small differences exist in the Bulgarian market.

The market for corporate services: (accounting and advisory, equipment solutions, and so on) was created with global brands only in the beginning of the 1990s. Usually no adaptation beyond that required in other countries was necessary. Mass-market global services (such as Shell gas stations and McDonald's fast food) entered the Bulgarian market in a form familiar to local consumers, and enjoy an enormous popularity with practically no adaptation whatsoever. Finally, in the group of mass market products some minor changes were needed, such as in design and packaging, to adapt to local habits. For instance, Danone took over some local brands, while slowly introducing its international range. IT and software companies had to stress rather more their lower-price, less-features ranges. Coca-Cola and Pepsi had to use bottles that fit local standards in order to participate in the country's recycling chain.

KPMG opened its Sofia office in 1992 and, from the very beginning, offered the full range of services to local businesses and government institutions as well as foreign investors. Frequently, the firm would give a company legal and tax advice on a potential investment project in Bulgaria, go on to provide corporate finance services in connection with the actual transaction, and then advice on the restructuring of the company in Bulgaria and its IT systems, and later audit its finances.

Shell Bulgaria was first established in 1994. The brand name was already well known in the country, due to the fact that Shell Motor Oils had been on the Bulgarian market for seventy-five years. By 1997, the company had invested US$35m, and is planning for a further US$45–65m by 2003. Much of this investment has been in the construction of gas stations. There are twenty in operation, with a further three due to be completed. In addition, Shell Bulgaria is involved in chemicals and lubricants, and is looking at LPG and aviation fuels. Shell has introduced a concept new to Bulgaria – the 'Select' shop as an integral part of the gas station. The concept has been so successful, that Shell's local competitors are copying this approach.

However, not all product or services introductions have been successful. So far, *bona fide* management and strategy consultancy services could not be feasibly introduced, due to the small scale of most businesses' activity. Rover, now owned by BMW, failed to introduce its Maestro model into the market. Even though local production lowered the price, the car was too outdated in terms of technology and marketing was poor.

Activity location

Bulgaria's place in the value chain of MNCs takes one of three forms: (1) production-type activities (including R&D and general management) in the country for sale abroad, (2) sale-type activities (including customer support) of goods and services originating abroad, and (3) production-type activities in the country for sale in the country.

Activity location strategies of the first type are chosen by Daewoo, which located its Eastern European headquarters in Sofia, although no production has taken place so far. Potentially, Xerox may also locate its regional headquarters in Bulgaria. Also included in this type of strategy should be the cases of Union Miniére and Solvay – two Belgian companies who bought a copper refinery and a chemical plant respectively, and produce niche products for their other operations and their customers worldwide, although there is no research located in Bulgaria. Such a value chain organization is possible in two other industries in Bulgaria: light industry, especially food and beverages or textiles, and tourism, the only service that is, in a way, exportable.

The second type of activity location is the most common. Products of all types, from computers to cars to detergents, are produced in locations in Central Europe or Greece and sold in Bulgaria. Given the limited scope for efficient scale production in Bulgaria in most industries, this is likely to remain the most popular choice.

Finally, production, mostly for the local market, is seen in two areas: most services, from courier to business advisory to fast food to gas stations – are by definition produced locally. In addition, some light industry MNCs also produce locally for the Bulgarian market – Nestlé and Kraft Jacobs Suchard produce chocolates (while importing the rest of their product ranges through distributors), and Danone produces a whole range of diary products only for the Bulgarian market. At the same time, IT companies like IBM conduct R&D, such as software development, for products destined for the Bulgarian market in Bulgaria. R&D in this area for the world market will, however, possibly be introduced in the near future.

Global marketing

The marketing activities of MNCs in Bulgaria have had a twofold effect: (1) MNCs promote their products and services and, at the same time, (2) introduce a marketing culture in the local business environment. In this sense, a fundamental adaptation was hardly necessary.

Market positioning follows the backbone approach of most MNCs for the country – slow expansion after careful observation of developments in the country. Practically all companies make strong use of their international brand names, a powerful tool in Bulgaria. Consumer products are usually labeled according to the strategies used in Greece or Central Europe, sometimes even without translation of the text, and this is accepted as normal by consumers. Advertising for the mass market uses traditional channels including billboards, newspaper TV ads, and (less so) radio, creating a strong presence for names like Shell, McDonald's, and MobilTel. Advertisements must be in Bulgarian, and since mistranslations of advertising texts can be a problem in Bulgaria as much as anywhere else, practically all MNCs use local advertising agencies. As in other countries, companies aiming at the corporate market reach their clients through personal contacts, while advertisements are mostly aimed at recruiting.

Product promotion uses the normal range of workshops, product presentations, promotional games (Shell, Danone, Coca-Cola), active presence on fairs and exhibitions (Xerox, IBM, 3-Com), both on newly established, industry specific events and on the long-standing Plovdiv Universal Fair.

Sales for the mass market usually take place through a network of independent distributors (Nestlé, Kraft Jacobs Suchard, Danone) or company owned, locally staffed distributors (Daewoo). Distributors usually purchase products on a wholesale basis directly from the parent company abroad or from the local production facility. The local offices of MNCs usually keep out of the process and provide customer support and sales to corporate customers. Direct sales for the mass market are seen only from companies with a worldwide tradition in this method (Zepter). For the corporate market, Xerox introduced direct sales for the first time in 1997, with remarkable success.

Given the low average incomes in the country, prices of MNCs' products for the mass market have to be equal, or at most marginally higher, than those of the local equivalents, something achieved mainly through the cheaper cost of local production (Nestlé, Danone, for example). Imported products (automobiles, cosmetics, consumer electronics, and so on), offered at world-market prices, usually take only an upmarket niche.

The corporate market for products and services is less price-sensitive than this, and yet more spending-constrained than its counterpart in most other countries. Pricing of MNCs tries to take this into account.

Foreigners identify three success factors for marketing in Bulgaria: a good quality of the products that justifies a higher price, a strong corporate presence through lobbying and being active in the business community, and getting to know, without adopting, the local habits and customs.

Global competitive moves

As with most countries, MNCs include Bulgaria in most of their global competitive moves. For instance, Xerox's change in company philosophy and organization worldwide fully affected the subsidiary in Sofia. There was an exception when the pace of reforms in Bulgaria slowed down in the beginning of the 1990s. At this time, many MNCs had entered the country, as elsewhere in Eastern Europe, and decided not to enter some other countries. However, since the pace of reforms justified neither expansion, as in the Czech Republic, Poland or Hungary, nor pulling out as in war zones of the former Yugoslavia, competitive moves in Bulgaria became specific to the country – mostly along the lines of a 'wait and see' policy. In all cases, investment in Bulgaria should be seen as long-term engagement, displaying little immediate returns, but a lot of future opportunities.

Global Organization and Management

More than 87 per cent of the foreign companies doing business in Bulgaria are connected to MNCs operating in more than three countries.[6] Most companies who have production operations in the country (Nestlé, Danone, KJS, Rover) or provide services (Shell, McDonald's, the Big Five accountancy firms), have established separate independent companies according to the Bulgarian Law of Commerce. Most of those who retail products and provide customer support have either a wholly-owned subsidiary (IBM, ICL) or just a branch (Unisys, Daewoo, and ING Bank). Many IT companies (Apple, Compaq, Epson, Intel, Microsoft and Borland), have a network of local dealers, but no actual operations of the company in the country. Finally, some MNCs just maintain an office (Bull, Philips, L.M. Ericsson) or have a local company as a designated business partner (Silicon Graphics). There are also a few joint-ventures (Siemens, Cable & Wireless and Sprint.)

The frequently changing legal environment, tax and finance regulations, and the complex local accounting practices necessitate an extensive adjustment of financial management, which can be an entry deterrent or make organization difficult.

Similarly, HR management has to be adapted to the local customs and mentality, placing a lot of emphasis on retraining. Recruiting takes place mostly through personal contacts, and, to an extent, press advertisement. The services of HR consultancies are mostly used by foreign companies (only five per cent of Hay International's clients are Bulgarian). Careers Services and Careers Fairs at schools and Universities are only at the start of their development.

Employment contracts are usually signed for a limited period of time and include a three to six month long test period. The strong influence of the trade unions in larger enterprises and traditional industries should be taken into account. In addition to wages, compensation may include food vouchers, a clothing allowance, free transport, and a discount on the company's products. Medical care provided by the state is free and available to everyone. Top management may also be given a company car, a mobile phone allowance or club memberships.7

Currently, most MNCs employ only a limited number of expatriates in the top positions (never more than 20 per cent of all employees), and thus employ mainly local staff. In the long run, there is a policy of replacing all expatriates. In a few cases, such as Hill International, IBM, and Arthur Andersen, this has been successfully achieved. Bulgarians not only head these companies in the country, but also participate in their global management. The former Bulgarian Country Manager of IBM Bulgaria is now Regional Manager for C&EE, located in Moscow. Even though there is a large majority in the workforce unsuitable for employment by MNCs, there is also a number of exceptionally well-educated individuals in the country qualified to fill positions of high responsibility.

In most cases, MNCs find that local staff, especially in Sofia, are, at least after some training, well qualified in their particular field, but lack broader management skills. In a few cases (ING Bank, Shell Bulgaria), employing local managers led to problems. In general, they often lack the necessary business culture, display lack of familiarity with the principles of a free marketplace, do not understand the role of information or of business advice. The reasons obviously lie in the recent past, and in spite of the numerous management training programs, the emergence of a new, better generation of managers and a major change in management culture will take time.

The main problems MNCs face in the country are the bureaucratic civil service, corruption, and the unfair competition from local companies, where tax evasion and behind-the-scenes dealing is widespread. On the other hand, as a key success factor MNCs indicate a good team, patience and a long-term view, and an understanding for the local business environment and culture and the changes that take place in them.

Most companies plan to expand their operations in Bulgaria. Thirty-five per cent of the foreign companies in Bulgaria reported an annual turnover between US$1m and US$5m for 1997. The percentage was expected to rise to 40 in 1998, as more companies pass the US$1m mark. About half of the companies employ fewer than 50 people, but ten per cent of them expect to go over this number within two years. Eight per cent of the companies employ more than one thousand people.[8]

Most MNCs own their Bulgarian operation, 64 per cent have an over-90 per cent stake, 30 per cent between 50 and 90 per cent and only three per cent own less than 30 per cent.[9] Local companies are usually given almost complete autonomy over their operations by the parent company, but do not yet participate extensively in global management. More participation in the global network can be expected, however, as activities expand.

Conclusions

Multinational corporations entered Bulgaria with four main types of activities: services, production, retail and business advisory. Companies offering services (fast food chains like McDonald's and KFC, hotel chains like Sheraton and Intercontinental) and Royal Dutch/Shell were among the first large foreign investors to make use of the existing market. With their brand names and quality they are still successfully operating in the country. In 1998, Metro of Germany announced the planned opening of three supermarkets, and in 1999, a Hilton hotel was being built in Sofia. Foreign investors have expressed interest in the

Bulgarian Telecommunication Company and in the flag carrier Balkan Bulgarian Airlines, whose privatization is due in 1999. Foreign banks like ING Barings, BNP Dresdner, Reiffeisen and some Greek institutions set foot in the country. In 1997 Daewoo purchased the Sofia World Trade Center. DHL is the main foreign provider of postal services, with offices in fifty cities.

Production by MNCs in Bulgaria started as a consequence of the privatization of leading local plants – to begin with in the food industry (Kraft Jacobs Suchard, Nestlé, Danone), and later in chemicals (Solvay), non-ferrous metals (Union Miniére), cement manufacturing (Marvex) and power plant equipment (ABB). More foreign investment is expected with the planned privatization of the oil refinery, Neftochim.

Retail selling of brands produced by MNCs is seen in practically all sectors! Automobiles (Daewoo, Peugeot, Renault, Ford, and so on), household chemicals (Johnson & Johnson, Procter & Gamble), consumer electronics (Sony, Philips), computers (IBM, HP, ICL), software (Microsoft, Borland), office equipment (IBM, Xerox), communications (L.M. Ericsson). MNCs in retail either use representative offices or the networks of local distribution companies.

In business advisory services, the Big Five accounting firms were among the first MNCs to enter the country. They still serve mainly foreign companies in the country, the large state-owned enterprises and the administration.

In spite of the differences between the strategies of MNCs in Bulgaria, two things stand out in common. First, most companies see their work in Bulgaria as a long-term investment, carefully watching the development of the country and waiting for new opportunities to emerge. Secondly, MNCs apply all the lessons they learned, four to five years before, under similar conditions in Central Europe in the Czech Republic, Hungary and Poland.

When approaching the Bulgarian market, MNCs use their international brand names and offer the entire range of their products and services, as they do in the other countries of C&EE. All of them extensively employ local staff, although top management positions are initially held by expatriates, since good local managers need to be trained. MNCs provide a wide scope of training programs in the country. Almost all foreign companies are members of the Bulgarian International Business Association, which provides them with information and contacts. All MNCs have to adapt to the local culture and habits. This is especially true for their marketing and human resource management.

The main problems MNCs face are an incoherent and unstable legal system, cumbersome bureaucracy, limited purchasing power of the population, excessive taxation, and lack of developed infrastructure. Nevertheless, 57 per cent of foreign investors in a survey conducted by KPMG indicated a fulfilment of their primary expectations and optimism about the future, nine out of ten enjoyed living in the country, and 88 per cent were considering further investment.

At the same time, the Bulgarian government has an ambitious program aimed at accelerating structural reform, privatization, and development of the private sector, and, in the long run, meeting the criteria for joining the European Union.[10]

Located in the center of the Balkans, Bulgaria holds a strategic position for trade and communication between Eastern Europe, Russia, Western Europe and the Middle East. It can have a highly efficient infrastructure. It has beautiful scenery and a mild climate. A well-educated labor force is readily available. Most of all, it has a dynamic business environment with huge development opportunities – giving it the potential to play a key economic role in the southern tier of Eastern Europe.

Bibliography

EBRD, *Transition Report*, 1996.

EITO, *European IT Observatory*, 1997.

Foreign Investment Agency, *Bulgaria Business Guide*, June 1998.

Foreign Investment Agency, *Current Foreign Investment Climate in Bulgaria*, 1997.

High Tech Forum, June 1998, Sofia, *High-Tech Market Review*, June 1998.

KPMG Bulgaria, *Foreign Investors in Bulgaria – Survey*, July 1998.

National Statistical Institute, *Statistical Yearbook*, 1997.

OECD, *Investment Guide for* Bulgaria, 1996.

Notes

1 OECD, Investment Guide for Bulgaria, 1996.
2 *Financial Times*, EU Expansion: Ambitions of Bulgaria and Romania undimmed, Thursday September 18, 1997.
3 KPMG Bulgaria, *Foreign Investors Guide*, June 1998.
4 Foreign Investment Agency, *Bulgaria Business Guide*, June 1998.
5 KPMG Bulgaria, *Foreign Investors in Bulgaria – Survey*, July 1998.
6 KPMG Bulgaria, *Foreign Investors in Bulgaria – Survey*, July 1998.
7 Bulgarian International Business Association, *Salary Survey*, November 1997.
8 KPMG Bulgaria, *Foreign Investors in Bulgaria – Survey*, July 1998.
9 KPMG Bulgaria, *Foreign Investors in Bulgaria – Survey*, July 1998.
10 KPMG Bulgaria, *Foreign Investors in Bulgaria – Survey*, July 1998.

Acknowledgements

The authors would like to thank: A. Alexandrov, Managing Director, Hill International, M. Alexieva, Marketing Manager, Xerox, W. Drysdale, Managing Partner, KPMG Bulgaria, R. Engman, President, Ericsson Bulgaria, J. Gaston, Senior TelTitorial Partner, Price Waterhouse Coopers, B. Genovski, Partner, Arthur Andersen & Co., D. Koronakis,

General Manager, Nestlé Sofia, A. D. and T. G. Munozy Oribe, Country Manager, ING Bank, for the time they spent with us discussing their experiences in Bulgaria and providing us with invaluable information about their companies' activities. KPMG Bulgaria, for the wealth of data in their survey. The Library of the Bulgarian Chamber of Commerce and The Bulgarian International Business Association for providing a variety of information sources, and International University, Sofia, for providing the facilities for this project to be carried out.

10 Ukraine – Europe's Frontier

BOHDAN HAWRYLYSHYN and PAVLO SHEREMETA

Overview

Ukraine is the second largest country in Europe after Russia. Despite the fact that Ukraine reappeared on the European map only in 1991, and despite its current economic problems, many recognize its huge potential and strategic importance for Europe's economic and political stability and envision its role as a bridge from Western and Central Europe to Russia and the Middle East.

Historical background

Ukraine is located in Eastern Europe and borders Belarus in the north, Russia in the east, Poland, Slovakia, Hungary, Romania and Moldova in the west, and has a direct access to the Black and Azov seas in the south. The territory of Ukraine approximately equals those of Great Britain, Belgium, the Netherlands and Italy combined. The population of Ukraine is approximately 50 million people, the fifth largest in Europe. There are four cities with over one million inhabitants. Ethnic Ukrainians constitute about three quarters of the population, while ethnic Russians, about one fifth.

Many historians believe that the name 'Ukraine' literally means 'frontier land' and appeared in the sixteenth century. Another theory is that this name has much older origins, dating from the twelfth century and is derived from the Ukrainian word 'kraïna', which means 'country'. The history of Ukraine as a European state begins with the medieval princedom of Kiev (or Kyiv in Ukrainian). An ancient city on the Dnipro (Dnieper) River, Kiev stands at the crossroads of the old north-south trade routes from Scandinavia to Greece, and east-west trade routes between Europe and Central Asia. This strategic location ensured the country's intensive diplomatic affairs and its prominent role in Europe. In 1362, the Kiev princedom was absorbed by the Grand Duchy of Lithuania and later all the Ukrainian lands were put under the direct control of the Polish crown. At the same time, numerous merchants, notably Italians from Venice and Genoa, settled on the Black Sea coast, especially in Crimea. But it was not until 1648 that an uprising, led by Bohdan Khmelnytskyi, led to the re-establishment of the Ukrainian Cossack republic. The way in which Cossacks conducted their internal affairs is now seen by many in Ukraine as a historic attempt to develop democracy. The Cossack state, however, badly needed allies in order to survive and was compelled to sign a military treaty with the Moscow Tsar in 1654. The treaty guaranteed a great degree of autonomy for Ukraine, which, however, gradually dwindled away, as Eastern Ukraine was forced to enter the Russian Empire. After an unsuccessful attempt to assert Ukrainian independence from Moscow, led by Hetman Mazepa (an elected state and military leader) who joined King Charles XII of Sweden in his war against Peter the Great, many of the original Cossack institutions were disbanded.

Since the larger western part of the country remained within Poland and Austria, Ukraine was incorporated into European politics, economy and culture. This not only helped to preserve, but even develop the sense of its national identity. This acted as a catalyst for two bids for Ukrainian independence in this century. The first one was unsuccessful in 1917–1921, when the short-lived Ukrainian People's Republic was occupied by the Bolsheviks' army. The communist rule established thereafter was devastating for Ukraine. Until 1990, 95 per cent of Ukraine's economy was controlled by Moscow. The consequences for those few who overtly promoted the use of the Ukrainian language and culture in Soviet Ukraine were grave. The Ukrainian-speaking rural population in the eastern and central regions was devastated by the collectivization campaign and artificially-induced famine in the 1930s. Ukrainian-speaking western Ukraine, annexed by Stalin only after World War II, was brutally marginalized. Being exasperated by all of this, 90 per cent of Ukrainians voted for independence from the USSR in December 1991.[1]

There are three important historical conclusions, which have influenced and will continue to influence the role of Ukraine in Europe. Firstly, Ukraine is not a newcomer on the map of Europe. Foreign investors may still benefit from the strategic location of Ukraine as a traditional crossroads between Europe and Central Asia. With or without national independence, Ukrainians have been active participants in European affairs in the past. Similarly, Ukraine is not a new democracy. Finally, Ukrainian politics and culture is, and will continue to be, influenced by the historical fact that the nation was divided for a long time. This is the major reason why Ukraine has not yet chosen a common clear vision for itself. But based on its most recent history we predict that Ukraine will remain an independent state and will move gradually towards membership within the European Union.

Politics

With the adoption of independent Ukraine's constitution in 1996 and the recent signing of friendship treaties which settled border issues with all seven neighboring countries, Ukraine is one of the more politically stable countries among the newly independent states of the former USSR

In parliamentary elections in 1998 the Communists emerged as by far the largest single party, winning more seats than the three runners-up together, though the leftward swing was not big enough to give the Communists and their allies a clear majority. The new parliament was better structured thanks to a new election law passed in 1997, according to which half of the deputies were elected according to the lists proposed by parties and half by simple majority in each constituency. This ensured a stronger role of political parties in political decisions. Entrepreneurs and managers of different companies also got quite a large number of seats in the new parliament. Although this new legislation was a positive step towards a well-managed 'democracy' in the Western sense, factors such as an unstable economy and growing social discontent with the pace and results of reforms, led to a rise in popularity of the radical left-wing and right-wing political parties. In late 1999, the government of President Kuchma was re-elected in further elections.

Economy

The most encouraging economic trend in Ukraine is its macroeconomic stabilization from 1994 to 1996 resulting in a low inflation rate of 16 per cent in 1997 versus almost 5000 per cent in 1993 (see Table 10.1) which allowed the introduction of a new and more stable currency, the hryvnia. As a result of the 1998 financial crisis the hryvnia was devalued by 80 per cent, while the Russian rouble suffered a 333 per cent devaluation as of December 1998. However, Ukraine is slower than its Central European neighbors, such as Poland and Hungary, in privatizing and restructuring its economy. This has resulted in a continuous decline in Ukraine's GDP.

Governmental data indicate that the official GDP in 1998 was just 37 per cent of that of 1990. It has been estimated, however, that approximately 50–55 per cent of Ukraine's GDP is produced by the unofficial 'shadow', or unregistered economy, which is not shown in the official statistics. This assumption is based on the rate of energy consumption. Many Ukrainians are involved in such unregistered production and trade, including the trade in foreign goods. The appearance and growth of this 'shadow' economy is the result of high taxes,

Table 10.1 Macro-Economic Indicators for Ukraine

Indicator	1990	1991	1992	1993	1994	1995	1996	1997
Population (million)	51.8	51.9	52.0	52.1	51.9	51.7	51.3	50.9
Nominal GDP (US$bn)	n.a.	n.a.	8.0	13.5	37.6	37.1	44.0	49.7
Real GDP (% change)	–2.6	–11.6	–13.7	–14.2	–23.0	–12.2	–10.0	–3.2
GDP per capita PPP (US$)	4490.0	4069.0	3720.0	3299.0	2620.0	2352.0	2184.0	2175.0
Inflation (%)	5	91	1210	4735	891	376	80	16
Exports (US$bn)	n.a.	50.0	11.3	12.8	13.9	14.2	15.5	15.4
Imports (US$bn)	n.a.	53.4	11.9	15.3	16.5	16.9	19.8	19.6
Current–account balance (US$bn)	n.a.	–2.9	–0.6	–0.8	–1.2	–1.2	–1.2	–1.3
Foreign direct investment flow (US$m)	n.a.	n.a.	200	200	100	400	500	600

Source: Business Central Europe, http://www.bcemag.com/_bcedb/history.idc

difficult registration processes for new ventures and widespread corruption. This unofficial economy softens the negative impacts of economic decline in the 'official' economy, but does not bring any revenue to the State budget and facilitates the flight of capital.

There are sectors of Ukrainian industry that are growing on the 'officially recognized' market (see Table 10.2). Several multinational companies, such as Coca-Cola, Philip Morris, Procter & Gamble, operating in the fast moving consumer goods industries have been lured by the size of the Ukrainian market (fifty million people) and are running fairly successful operations in Ukraine.

While many multinationals may perceive Ukraine as a potentially attractive market, due to the difficulties discussed above they are reluctant to expand their activities past those of distribution.

Foreign trade

Ukraine's imports and exports have increased in recent years. The former Soviet Republics, primarily Russia, remain Ukraine's main trading partners. In 1996, export data show that Ukraine's most exported products were ferrous and non-

Table 10.2 Largest Ukrainian Companies

Company	Industry	1996 Revenues US$m	1996 Exports US$m	1996 Export as % of revenues
Kryvorizhstal	Iron and Steel	1988	166	8.3
Illyich Iron & Steel	Iron and Steel	1779	550	30.9
Azovstal	Iron and Steel	1580	442	28.0
Zaporizhstal	Iron and Steel	1200	261	21.7
Donbasenergo	Electricity	908	n.a.	n.a.
Zaporizhska Nuclear Power Station	Electricity	898	123	0.0
Centrenergo	Electricity	756	n/a	n/a
Dzerzhynsky Iron & Steel	Iron and Steel	681	142	20.8
Dniproenergo	Electricity	613	n.a.	n.a.
Nyzhniodniprovsky Pipe Rolling	Iron and Steel	610	95	15.6
Alchevsky Iron & Steel	Iron and Steel	585	102	17.4
Dniproshyna	Tires	539	n/a	n/a
Balakliya Ukrgasprom	Gas production	538	6	1.0
Nikopol Ferroalloy	Non-ferrous metals	530	209	39.3
Zahidenergo	Electricity	517	n.a.	n.a.
South-Ukrainian Nuclear Power Station	Electricity	507	1	0.3
Rivnenska Nuclear Power Station	Electricity	504	n.a	n.a
Donetsky Iron & Steel	Iron and Steel	503	80	15.9
Kremenchucknaftoorgsyntez	Chemistry	463	32	6.9
Odessky Pryportovy Zavod	Chemistry	458	181	39.6

Source: Kompanion, Annual, 1996.

ferrous metals (steel and aluminum), which accounted for 32.3 per cent of its exported goods. Chemicals contributed to 11.9 per cent of Ukraine's overall exports, and unprocessed food (grain, and so on) another 9.9 per cent. Ukraine's biggest import items are mineral products (52.4 per cent), mainly oil and gas from Russia and Turkmenistan, and machinery (13.3 per cent).

Foreign direct investment

There are three clear stages in Ukraine's foreign investment history. From 1991 to 1993, foreign investment was low due to perceived political instability after the break-up of the former Soviet Union, and a lack of knowledge about the market opportunities and business legislation within the newly independent country. In 1994–96, after the new Ukrainian president proposed plans for radical economic reforms, foreign investments rose significantly. In 1997, however, growth in foreign investment and the activities of multinational corporations in Ukraine slowed down due to uncertainties regarding the upcoming elections, parliamentary in 1998 and presidential in 1999. Difficulties in navigating in the emerging and often confusing landscape of Ukrainian business law also slowed investment. Both of these situations hinder the speed at which multinational corporations can penetrate the Ukrainian market. Interestingly, foreign investors have sharply increased their commitment to Ukraine over the nine months of 1998 amid the crisis. Foreigners invested a total of US$705m in Ukraine during this period, which is up 40 per cent over 1998. This figure includes two big investments: Daewoo had to invest US$150m to keep its import tax exemptions and the European Bank for Reconstruction and Development has pledged some US$250m in 1998.[2] The United States, the Netherlands, Germany, Britain and South Korea are the top investors into Ukraine as of September 1998, together accounting for more than half of foreign direct investments. Nevertheless, Ukraine's foreign investment figure *per capita* lags behind those of most former Soviet bloc countries, and Central and Eastern Europe.

Multinational companies consider the following factors favorable for investments in Ukraine: inexpensive and qualified labor, a stabilized currency, privatization that allowed the creation of joint-ventures, psychological acceptance of the need for foreign investors, market need for high-quality products produced by MNCs. As unfavorable factors MNCs identify frequently amended legislation on foreign investment and taxes, which increases business risks; loss of tax privileges for joint-ventures; government corruption; and poor infrastructure.

Social

Public enthusiasm created high political and social hopes and expectations towards a new Ukrainian state at the beginning of the 1990s. The social climate is less optimistic now. Among the factors, mostly caused by the poor state of the economy, are the decline in population (–0.8 per cent in 1997), a breakdown in the unreformed welfare and pension systems, and the decline in the quality of Ukraine's healthcare and education systems, all resulting in a mounting dissatisfaction with the country's leadership, especially among poorly paid public sector workers. While there are now better quality, 'foreign' or private domestic health care and education institutions, their services are priced out of reach for the majority of the Ukrainian population.

A study commissioned by the Ukrainian Trade Union Federation has shown that if US$50 per month is taken as a poverty line, 74 per cent of the population lives in poverty. Amid this widespread poverty, some five to ten per cent of the population, mostly located in Kiev and other large cities, prospers, as shown by the ever-increasing number of expensive Western stores and cars.

In 1998, President Kuchma proposed a program which aims to speed up economic reforms. Pivotal points in this plan were: to improve the general business environment by reducing the level of red-tape and increasing government efficiency, to deregulate the economy and increase competition, to reduce the tax burden and undertake fiscal and financial sector reform, to further the course of privatization and restructuring, to develop an investment policy with special emphasis on boosting the development of the capital market, and finally, to undertake social reforms, including revision of the pension system, health care, and welfare provisions, with an emphasis on private insurance and better targeting of state support for the needy. The speed of this plan's implementation depends greatly upon the legislation adopted by the newly elected parliament and the outcome of the future presidential elections.

Country Globalization Drivers

The Ukrainian consumer market consists of fifty million people whose needs are widely segmented and continue to be influenced by foreign advertisements. Consumer needs in cities such as Kiev, Odessa, Lviv, (Luor) and Kharkiv (Kharkyv) are more sophisticated due to easier accessibility to Western products through close proximity to airports and more developed means of transport. There are, however, several barriers involved in penetrating the entire Ukrainian market. over 30 per cent of Ukraine's population is rural and, thus, difficult to target. An underdeveloped road system in many areas, the poor state of repair of existing roads, and little access to public transportation in rural areas hinder distribution to this segment of the population. This rural segment has less purchasing power, and less information about available goods. There are many regional and city newspapers, but no real 'national' newspaper. Few national TV channels are accessible from all areas of Ukraine, and some of the more popular TV programs, especially those shown in Eastern Ukraine, are produced in Russia.

Regional

Ukrainian consumer markets can be divided into two regions, Western and Eastern Ukraine, with the Dnipro River as the geographical divider. Generally speaking, the western Ukrainian consumer market is more open to Western ideas than the eastern Ukrainian one for obvious reasons. Although most of Ukraine was isolated from the West for seventy years, western Ukraine belonged in earlier times to the Austro-Hungarian empire and later to Poland, Czechoslovakia and Romania. The architectural influence of these countries can be seen today in Western Ukraine, in striking contrast to Soviet-designed cities and towns in Eastern Ukraine. Furthermore, the role of religion in strengthening ties between Western Ukraine with its neighboring countries must be emphasized. The Greek Catholic (or Uniate) church, which is subordinate to the Pope in Rome, is still strong in the region, and this further strengthens Western influence. In contrast, the Eastern Ukrainian Orthodox church has close ties with Moscow.

While Ukrainian, the official language, is used almost exclusively in Western Ukraine, Russian is more predominant in eastern Ukrainian cities. The Eastern Ukrainian consumer market is therefore more easily influenced by Russia and the Russian market.

Common customer needs

Ukrainian consumers behave differently in many ways from those in Western countries. One way to demonstrate such differences is to compare Ukrainian shopping habits to the shopping habits in developed economies. In Westernized countries, consumers may go to a supermarket or a discount store for consumer goods about once a week. In Ukraine, although supermarkets or discount stores are slowly appearing in the larger cities, much of the shopping for consumer goods takes place in small private shops, kiosks and bazaars. Because of a lack of 'one-stop-shopping' facilities, Ukrainians shop in these outlets several times a week. Products can be purchased in smaller quantities and are priced lower than at supermarkets. If more resources were invested in the distribution network and facilities such as warehouses, railway stations and river ports, products would be accessible to a larger portion of potential consumers.

Transferable marketing

Since the Ukrainian language is based on the Cyrillic alphabet (as is Russian), Western companies should change their product labels but also their overall marketing messages to tailor them to the special needs of the population. More and more brands already adapt their messages to the Ukrainian environment, for example using Ukrainian actors set in Ukrainian homes, rather than dubbing foreign commercials into Ukrainian as was commonly done several years ago.

The general belief among professionals is that Ukrainians have been receptive to the various marketing 'gimmicks' they have been exposed to. Promotions campaigns can be quite successful here. But advertisers say that how well promotion is received depends upon the persuasion skills of those conducting it and the brand's reputation. Ukrainians tend to be suspicious of large-scale promotion, which they perceive as too aggressive. Such suspicion may stem from Ukrainians' short history with advertising. Sometimes they wonder what is wrong with the product if someone gives something away for free or at a discount. Therefore, it is important to build credibility around the product to overcome this in-built distrust. Coca-Cola's first promotion in 1996 was new and novel. Since then 'Coke' has done five promotions and progress in the public perception and growing acceptance of such methods, is evident.

Cost globalisation drivers

Global/regional scale economies

The Ukrainian market is potentially large enough to support efficiently sized plants, but currently there are two factors which hinder this. The first issue is the low purchasing power of the bulk of the population. The second problem is that taxes and registration fees of any start up significantly increase its cost, which compounds the problem of the attractiveness of the market.

In terms of factors of production, Ukraine can provide inexpensive and fairly-skilled labor and renown fertile agricultural land, the 'black earth'. Better motivation of workers and newer technology, however, are needed in order to make Ukrainian labor and land more productive.

Ukrainian labor is inexpensive, in terms of the cost per hour. Real labor cost, however, as estimated by unit manufacturing cost, is high because of low productivity. Ukrainian GDP roughly equals that of Hungary, although Hungary's population is 20 per cent that of Ukraine. A Ukrainian miner produces an average of 112 tons of coal a year,[3] compared with 420 tons in Poland, 2000 tons in the United Kingdom, and 4000 to 6000 tons in the United States.[4] Taking into account the unit manufacturing cost, Ukrainian labor, within nonrestructured companies, is fairly expensive. High taxes and overhead costs of these companies, as well as official and unofficial fees for government 'services', further increase the real cost of operation.

The issue of land as a production factor is similarly connected to the labor productivity problem. Although Ukrainian agricultural land is among the most fertile in the world, currently it is not a comparative advantage of Ukraine. In addition to low labor productivity, antiquated machinery and the inability to purchase agrochemicals in sufficient quantity, and the fact that full private ownership of land, as practised in the West, is still not possible, converge to maintain the output of the agricultural products much below the real potential.

Capital, however, is the most problematic factor in Ukraine. Domestic savings are low in absolute and relative terms. People still do not trust the Ukrainian banking system, and prefer to keep their capital abroad or 'under the mattress' in foreign currency. Cumulative foreign direct investment is only some US$2bn. The Ukrainian capital market is also very small in scale, and knowledge and skills on how to tap into the global capital market are as yet undeveloped.

Logistics

Although Ukraine has a relatively well developed internal transportation infrastructure, such as railroads, highways, local roads, metros, buses, and trams, the collapse of the Soviet Union left these assets in a poor state of repair. Railroad transport in Ukraine, although having extensive coverage, is time-consuming and inefficient. A 600 km trip from Lviv to Kiev by car takes on average seven hours, but by train it takes eleven hours. The problem of crossing Ukraine's western borders is exacerbated by the fact that the Ukrainian rail-system was designed with a wider wheel gauge than that in Western Europe. Trade with Europe via rail is, thus, hampered by the task of changing the wheels under the carriages, which takes several hours.

Infrastructure

Ukraine's power and energy sector face many challenges. First, is the challenge of effective privatization. Energy used to be supplied to customers almost free of charge. Now, in theory, private consumers pay 80 per cent of the cost of electricity. In reality many do not pay their bills. The agreement with Ukraine, according to which Ukraine should shut down all of Chernobyl's reactors by 1999, creates the problem of developing substitute sources of electricity. The second challenge is that of efficient use of energy. Many factories in Ukraine are energy intensive and cannot settle their electricity debts. Given low income levels, pricing energy for commercial or private use constitutes a political issue. Because of this, power plants lack sufficient funds to ensure a regular supply of coal, and thus the means to invest in new, more efficient, power generation technologies.

The Ukrainian financial system is only emerging and thus targeting mostly large Ukrainian exporters and multinational companies. There are several foreign-owned banks, such as France's Crédit Lyonnais, which service these multinational clients. In comparison to other governmental organs, the National Bank of Ukraine (NBU) is the most effective institution. This gives hope that the Central Bank system will improve regulations, which will serve as a stimulus for developing an effective private banking system.

Telecommunications

Ukraine's telecommunications system is slowly improving. On 1 January, 1997, the average number of telephone lines per 100 people was just 18.3, which already represented an increase from only 14.7 per 100 in 1992. While Ukraine uses global technical standards in energy and frequency, their antiquated phone system is in need of upgrading. Three Global System for Mobile (GSM) licenses have been allocated by the government, but a dispute has emerged over the frequency distribution. Utel which arrived in 1992, acquired the long-distance sector of the telecommunications industry. Local phone services are, for the most part, still owned by the government. However, the number of western telecommunications companies is increasing, even though vague or non-existent legislation creates difficulties for such companies. Due to such difficulties, Motorola, in a highly publicized move, has greatly scaled down its operations in Ukraine. Lucent Technologies recently started a large project to install a fiber optic network system to upgrade and in some cases replace local networks and long distance channels.

In 1997, Ukraine showed one of the highest growth rates in internet users in all of Eastern Europe. Several internet service providers, such as LuckyNet and Global Ukraine, have sophisticated channels to the Internet backbone via satellite and radio modems and continue to increase their customer base, although the price of personal computers and Internet services puts them out of reach for the majority of the Ukrainian consumer market. 'Standard' Internet service provider services are similar to those found in the West and prices are similar as well. Recently, the first Ukrainian Internet cafe opened in Kiev, charging customers approximately US$2 per hour for access to the Internet.

Technology role

Given its high-standard education system, some world-class industrial research capabilities and relatively inexpensive labor, Ukraine may be an excellent place to host corporate R&D departments carrying out sophisticated activities. This opportunity has already been spotted by Korea's Samsung, which is discussing with Ukraine's Science Ministry the possibility for joint scientific projects, and the participation of Ukrainian scientists in Samsung's scientific research work.

Government globalization drivers

Ukraine has legislation that welcomes foreign investment but other government policies and the still weak legal system cause significant problems for MNCs.

Trade policies

In general, Ukrainian trade policies may seem to be less favorable than those of Central European countries for both importers and exporters, particularly because of non-tariff barriers. For example, all trade companies need to get trade licenses for contracts worth over one million dollars. For this, they must travel to Kiev, apply to the Ministry of Foreign Economic Relations and Trade, and wait for between five and 30 days for a response. In addition, there is a 90 day limit for returning foreign currency from foreign trade operations to Ukraine's banks. Furthermore, the Cabinet of Ministers is discussing reducing this period to 30 days, which would make many foreign trade operations impossible.

FDI rules

Foreign direct investment legislation, including provisions on the repatriation of capital, was adopted in 1992 and was quite liberal in comparison to other East European countries. This legislation provides a five year tax break for foreign investors and joint ventures. The only problem is that the legislation has been amended three times since its inception. Now Ukraine has several conflicting legal provisions on foreign investments, which upsets and confuses multinational companies. The lack of clear legislation regarding private land ownership, particularly by foreigners, exacerbates this problem. In 1995 and 1996, currency regulations were stopped as a result of macro-economic stabilization and at the request of the IMF. However, in response to recent events in the global capital markets, the National Bank decided to tighten currency exchange operations in Ukraine, making currency arbitrage transactions much more difficult.

Ukraine established the Presidential Committee on Foreign Investment made up of high-level Ukrainian officials and most significant foreign investors. The panel advises the President on solutions to Ukraine's thorniest commercial disputes, usually between foreign investors and local bureaucrats. Foreign investors see the presence of this Committee as a positive factor in attracting foreign investments to Ukraine.

Regional trade blocs

Ukraine received associate membership status in CEFTA (the Central European Free Trade Association) in 1996. There are some external and internal pressures for Ukraine to join the Russian dominated CIS customs union, but the current political leadership believes that this partnership would not be beneficial to Ukraine. This refusal creates problems for Ukrainian exporters, especially in the food industry and, in particular, sugar and alcohol for the Russian market.

Government intervention

Despite the calls for industrial policy of some sorts, direct government intervention in most industries is limited.

Legal protection

Legal protection of contracts, trademarks, and intellectual property remains weak. The Ukrainian market is flooded with cheap, unauthorized copies of movies and computer software. In one rather well-known local dispute, a Kiev-based radio station, Gala Radio, discovered that another station was using its name on the airwaves. Both parties have been navigating Ukraine's legal maze for close to a year in hopes of a resolution, both continuing to call themselves 'Gala Radio'.

Compatible technical standards

European technical standards are gradually being introduced at the national level and by some advanced Ukrainian companies and joint ventures. But more substantial efforts are needed to have companies adhere to these standards.

Common market regulations

As state-owned companies are mostly not in good economic condition and are in serious need of restructuring, government's intervention includes creating the 'rules of the game'. Both foreign and local managers complain that the economy is seriously over-regulated. Procter & Gamble had significant difficulties in Ukraine over the certification of its products, when bureaucrats alleged that the company was dumping sub-standard soap detergent and other products. Under pressure from the IMF, Ukraine passed several laws in the summer of 1998 to ease the regulatory business environment. One law slashed the registration period from 60 to seven days and subjected bureaucrats who delayed the process further to fines. Another law simplified the tax scheme for small business. Whereas previously it was not uncommon for audits, performed by tax authorities, to occur several times monthly, another new law limits the maximum number of such audits.

Competitive globalization drivers

Ukraine's competitive globalization drivers are relatively weak at present and await the strengthening of the economy.

Global/regional strategic importance

Multinational companies recognize Ukraine's strong regional strategic potential. The real value of such a potential, however, hinges upon successful economic reforms in Ukraine. Ukraine is, as mentioned before, a large potential market for the region but is in great need of legislative and governmental stability to strengthen its attractiveness. Lack of successful economic transformation in Ukraine could cause region-wide political instability, thus stunting economic growth of the entire region.

Presence of foreign competitors

Multinational companies operating in Ukraine face serious competition in almost all industries, especially in fast-moving consumer goods from companies in other countries, particularly from neighboring Poland and Turkey. The price/quality combination of the imported products is still attractive, in spite of import tariffs levied by the government.

Interdependence via MNC value chain

Multinationals are only beginning to create strategic interdependence between Ukraine and other countries of Eastern Europe through sharing activities such as manufacturing and other components in the added-value chain.

Internationalized domestic competitors.

Ukrainian exports were very much oriented towards Russia and other CIS countries until late 1998. Managers of the Ukrainian companies still have good business contacts in these countries and excellent knowledge of their markets. However, the Russian crisis of 1998 shows a serious vulnerability of such one-directional orientation and we may expect an increasing interest by Ukrainian companies in the Western markets.

Global Strategy Levers

Multinational companies can be quite successful in Ukraine despite the difficulties of its business environment, as companies like McDonald's and Coca-Cola have proven. Although McDonald's entered Ukraine relatively late (1997), in comparison to other countries in Eastern Europe, its move was aggressive and turned out to be successful. They opened four restaurants within six months in Kiev, and one in Kharkiv. By December 1998 they had sixteen restaurants in Ukraine.

Coca-Cola Amatil built its first bottling plant in Lviv in 1994. By then Coke had already captured approximately half the market due to an excellent distribution system and aggressive marketing campaign. Still, there is a lot of market opportunity to tap into as Ukrainians get richer and become more used to carbonated soft drinks, especially in the countryside.[5]

Market participation strategies

We can distinguish four types of multinational companies with regard to their market participation strategies in Ukraine. First, at the lowest end of market participation, there are multinationals such as Viessmann that have their operations in Ukraine still administered from Russia. Secondly, most corporations have opened country representative offices perceiving Ukraine as a potentially attractive market, but due to the difficulties discussed above, are reluctant to expand their activities past that of distribution. Thirdly, there are corporations that have opened their production facilities in Ukraine. Among those are ABB, RJR Tobacco, Philip Morris, Kraft Jacobs Suchard, S.C. Johnson, and Tetra Pak. The fourth group, which at the moment includes only Coca-Cola, are the corporations who view Ukraine as part of their global corporate strategy and plan to create Ukrainian production operations to satisfy not only Ukrainian consumer demands but also those of the region.

Product/service strategies

Despite its low *per capita* income, those foreign MNCs who enter Ukraine can generally market globally standardized products and services with minimal adaptation, provided they target the top end of the market. Products and services targeted at the mass market require much more adaptation.

Other companies show that effective marketing strategy may include some adaptation of the product to the specifics of the Ukrainian market. Kraft Jacobs Suchard developed its Korona chocolate brand with an increased amount of cocoa specifically for the Ukrainian market, which got used to the sweet and black chocolate. Given the inefficiencies of the Ukrainian information infrastructure and difficulties in targeting a dispersed rural population, Korona, which includes different versions of chocolate products, also proves the success of the 'umbrella' type of brand.

Similarly, Reemtsma and a local marketing and advertising firm have teamed up to make a success story out of Prima cigarettes. In 1996, Prima's reputation in Ukraine as a down-market, non-filter cigarette was so firmly established that the thought of changing that image would have seemed absurd. Reemtsma purchased the Prima brand name two years ago and developed a filtered version of the cigarette. The company also added to its product line another brand, the Prima Luxe, which boasts a new type of charcoal filter and sells for much cheaper than its foreign-made counterparts. The results speak for themselves. Prima's share of its market segment has surged from 32 per cent to 50 per cent.[6] Largely because of Prima's growth, Reemtsma has jumped ahead of its biggest rival, Philip Morris, for the number-one spot in the Ukrainian cigarette market. It seems that the main reason for Prima's success was that Reemtsma created a demand rather than just satisfied an existing one.

Activity location strategies

Only a few MNCs have set up operations in Ukraine and many have had significant problems. We discuss here the notable successes.

ABB has been operating in Ukraine since 1992. It has started a Ukrainian company with headquarters in Kiev and branches in three Ukrainian cities. It has also created three joint-venture companies producing transformers and other electricity producing and distribution equipment. These companies employ altogether 2400 people. ABB's revenues in Ukraine will reach US$60m in 1997, which is 50 per cent more than in 1996. The joint ventures' exports have also grown significantly, reaching US$10m in 1996. ABB contemplates further investments of US$5–10m in Ukraine, subject to an improved business environment and growth in demand. ABB's advertisements are visible on big boards inside Boryspil airport and along Ukraine's major highways. The company sponsors international seminars on East European business development issues. Once a year ABB-Ukraine takes part in energy equipment exhibitions. The company distributes its brochures at research and development institutes, construction companies, and it works actively with government agencies, taking part in tenders.

McDonald's started its Ukrainian marketing research project in 1993. In 1995 it created McDonald's Ukraine Ltd., and in 1997 it opened its first restaurant in Kiev. The move is rather aggressive; the company plans to invest more than US$100m into the Ukrainian market, with plans to build 85 restaurants which would employ 6.5 thousand employees by 2001. By December 1998, McDonald's had already opened sixteen restaurants in Ukraine, capitalizing on its first-move advantage with every restaurant serving close to five thousand customers daily.

Global/regional product and services

Although many Western firms have come to grief in the highly-regulated Ukraine market place, Coca-Cola has flourished. Coca-Cola's campaign to dominate the Ukrainian market took eight years using a systematic development approach employed successfully in dozens of countries before Ukraine. First import the product, then the syrup, then bottle in-country. Coke's Ukrainian program began immediately when Ukraine became independent in 1991. The first step was to build on the existing trade name recognition, which in Ukraine's case was Fanta. Introduced to the Soviet Union for the 1980 Olympics, Fanta products not only possessed name recognition in newly-born Ukraine, but provided Coca-Cola with an existing distribution network upon which to build. Within two years, Coke had already developed a distribution network using products trucked in from neighboring countries like Hungary.

Coke began manufacturing in Ukraine in 1995, when Australian bottler Coca-Cola Amatil joined up with a Lviv bottler. This was followed in May of 1998, by the delayed opening of the much larger Kiev bottling plant, which Coca-Cola Amatil also financed. Now, Coca-Cola owns about half of Ukraine's estimated 300m litres ($148.5m) beverages industry, according to industry analysts' estimates. Compared to most countries of similar size in Europe, those figures are extremely low. Ukrainian sales represented only five per cent of the total of 4.48bn litres of carbonated and non-carbonated beverages sold by Coca-Cola Beverages in central and Eastern Europe where it does business. But Ukraine, with a population of 50 million, constitutes 25 per cent of Coca-Cola Beverages' 200 million potential customer pool.

As a result of the 1998 financial crisis, Coca-Cola's performance deteriorated somewhat, but it is highly unlikely they will ever be low enough for Coke to consider pulling out. By the middle of 1998, Coca-Cola had clearly accomplished in Ukraine what it has done in much of the rest of the world. After having taken the decision to enter the market, it systematically increased its investment to move from penetration, to manufacturing, to market dominance, in less than ten years.

South Korea's Daewoo Corporation has arrived in Ukraine in a big way, pouring unprecedented sums into local car production and taking on the task of changing lackadaisical Soviet work practices. The Daewoo project is the largest single foreign investment of post-Soviet times, with $1.3bn promised into the AvtoZAZ plant[7] in the central Ukrainian city of Zaporizhya considered a cradle of Ukrainian culture as the seventeenth-century meeting place for Cossack councils.

Founded in 1863, the plant produced ploughs until 1959, when the Communist Party leadership in Moscow decided to launch the production of the first Soviet 'people's' car. Since then, more than two million cars have been produced at AvtoZAZ, including the Zaporozhets and Tavria, widely compared to the East German primitive, but popular, Trabant. The factory ran into serious problems in the 1990s, producing and selling just a few thousand cars per year.

According to the current deal, signed in 1997, the venture will produce annually 80 000 modernized Tavrias, 180 000 new-model Daewoo Lanos, Nubira and Leganza models, and 60,000 Opel Astra and Opel Vectra cars by Daewoo's world-wide partner, US carmaker General Motors. The plan for 1998 is more modest for 40 000 Tavrias, 30,000 Daewoos and 8000 Opels, with Korean and U.S. cars mostly preassembled at the AvtoZAZ subsidiary unit in the southern Black Sea port city of Illichivsk. Half of future output is aimed for exports to Europe, the former Soviet Union, Middle East and Latin America.

The Ukrainian authorities, desperate to increase the flow of investment and invigorate the economy, offered significant concessions to Daewoo. Ukraine's parliament, in a law passed six months ago, scrapped import taxes on spare parts, gave a ten-year break in profit taxes and the future venture tax-free land status, and forgave AvtoZAZ an $80m debt. In April 1998, the government met the AvtoZAZ-Daewoo lobby's final demand, banning imports of cars more than five years old and slapping a $5000 duty on those of less than five years. But these decisions have run into opposition from the European Union, a camp the government can ill afford to alienate.

Thousands of Ukrainians have been bringing in second-hand cars from Germany, the Netherlands, Belgium and Switzerland because they are cheap and more reliable than Soviet-made cars. The EU's stand was backed by local car dealers and some media. The European Commission, the EU's executive body, said the privileges could complicate Ukraine's bid to join the World Trade Organization. The stakes are high as the EU is Ukraine's second largest trading partner after Russia, with trade turnover last year at $4bn, and is the biggest aid provider, having given Kiev $4.5bn since independence.

Ukrainian officials stress that the law makes no specific mention of AvtoZAZ or Daewoo, but was aimed at any investor who brings at least $150m investment and would get the same privileges.

Swiss Nestlé acquired 32 per cent of Lviv-based chocolate producer Svitoch in 1998. Amid increasingly bitter competition, many local companies are seeking foreign investors' capital to bolster their defenses in a hostile market, as well as seizing more market territory from their competitors. Svitoch's deal with Nestlé follows a lengthy and painful courtship period, during which more than one foreign firm pressed its suit. Over 1999, Svitoch expects to receive the bulk of Nestlé's $40 million investment, most of which will be used to modernize equipment and step up production. Nestlé will also give its partner access to the latest technology and marketing strategies. Plugging into the Swiss firm's huge distribution network will give Svitoch a chance to boost its sales both at home and abroad.

McDonald's plans to phase out foreign ingredients in several years. Dairy ingredients will be the first group given over to local producers, with McDonald's shake and sundae mix expected to be made in Ukraine by the Italian-Ukrainian joint venture Parmalat, which has established a joint venture with a local dairy farm. Then Ukrainian cattle will become a source of hamburger meat, with McDonald's German supplier having established a joint venture with a farm in the Luhansk region. The local partner received Western processing equipment and training in European meat-handling standards in 1997. Fresh Start Bakers, a division of Campbells Soups, will set up a bakery in Ukraine. A McDonald's distribution complex, consisting of freezers and coolers, will be established at around the same time. Ironically for a potato-rich country, potato products will be among the last products to be produced locally. This is due to peculiarities of the potato and the french fry-making process.

Procter & Gamble has been working in Ukraine since 1993. The company has invested more than $50m in Ukraine over 1996–97. Undeterred by the conflict with the State Standards Committee over product certification issues, the company announced plans in early 1998 to invest another $230m into Ukraine over the next five years. These investments will go toward manufacturing new products in Ukraine, developing the company's distribution infrastructure, stepping up marketing and upgrading the company's Tambrands factory in Boryspil, outside of Kiev. The Tambrands factory is the company's main producer of women's hygiene products in Europe. The route the company takes in Ukraine will depend on how it perceives long-term rewards. Ideally, the company is looking for a pay-back period of five–ten years. If the financial crisis of 1998 were to persist and this target no longer look feasible after three–six months, the company may have reconsidered its investment policy for Ukraine.

S.C. Johnson was a pioneer in the Ukrainian market when it decided to buy an 80 per cent share in a joint venture with a household chemicals association in the then Soviet Ukrainian republic in 1990. Later, Johnson watched production plunge along with the buying power of Ukraine's increasingly impoverished consumers. There was a time in 1993 when it also thought about quitting. But being the family owned business with a long-term commitment the company decided to stay. Company officials also mention the President's support as an important factor in the corporate decision not to leave the Ukrainian market. Johnson's sales started increasing and the profits went into refurbishing its aging plant. In 1995, it bought out the joint venture and brought out new products.

A leading German international cigarette maker Reemtsma is expanding further in Ukraine. Ukrainian Prime Minister Valery Pustovoitenko cut the ribbon at the official opening of Reemtsma's new, modern cigarette manufacturing plant, on the outskirts of Kiev, in November 1998. The German company said that its investment in the facility demonstrated its confidence that business prospects are bright and that eventually the plant will account for production of a third of the 90bn cigarettes predicted to be consumed each year in Ukraine early in the next century. The Kiev site, which also houses the company's Ukrainian headquarters, was Reemtsma's second plant in Ukraine. The company, which has another factory in Cherkasy, employs a total of 2000 people in the country. The manufacturing capacity at the Kiev site, presently 8bn cigarettes per year, will be increased in order to produce cigarettes for export to neighboring countries, including Russia. Other cigarette producers such as Philip Morris, R.J. Reynolds and BAT have operated production facilities in Ukraine since the mid-nineties.

Marketing strategies

Western or global marketing strategies, with appropriate modification, can be successfully applied in Ukraine. Coca-Cola provides an example of gradual intensification of marketing efforts. During the early 1990s marketing costs stayed down, as in those days Coca-Cola marketing was limited to signs for café and restaurant owners. Marketing in Ukraine has, therefore, focussed on raising product awareness while advertising was cheap, rather than stimulating demand among the country's predominantly poor residents. That was Coke's textbook corporate strategy, and its result can be clearly observed in cities throughout Ukrainian, where red Coke logos adorn a disproportionate amount of kiosks, taverns and billboards. Marketing intensified, including radio and television advertising by spring 1996. In addition, Coke has sponsored rock concerts in several Ukrainian cities, and donated $1m to the Ukrainian National Olympic Committee in early 1998. The multinational does not advertise so much in smaller towns, largely because much of the rural populace can only pay in what they get paid in: commodities like grain and tractor parts. But 'Coke' still manages to get its message out to Ukraine's rural areas.

McDonald's price policy represents an interesting case as the price of a Big Mac is somewhat higher in Ukraine than, for example, in Hungary. This may be the effect of at least three factors. Firstly, the Ukrainian currency might have been less undervalued than the forint, but more than the rouble. Secondly, the costs of doing business in Ukraine are higher, including transportation and administrative overhead. Finally, the marketing strategy of the company in this country targets a narrower, wealthier, consumer segment.

Global Organisation and Management

It is difficult to generalize the way in which MNCs typically organize their Ukrainian operations. Organization structures and management styles of MNCs tend to reflect the structural 'norms' of the country or area from which they originate. Asian companies like Daewoo tend to have tightly controlled branches, closely supervised from their corporate headquarters. With the project, AvtoZAZ-Daewoo, mentioned earlier, some 50 Korean experts have arrived in Zaporizhya, The Koreans' presence has had immediate and far-reaching consequences on the plant's 20 000 employees. The literary work of Bolshevik leader Vladimir Lenin, which placed Soviet 'red directors' prominently on office book-shelves during communist rule, have been supplanted by 'The Great World of Business' by Daewoo Group chairman Kim Woo-Choong.

On the other hand, the Ukrainian offices of European companies, such as Crédit Lyonnais Ukraine, tend to be more flexible and autonomous. The organizational design of ABB's Ukrainian companies followed the global matrix format. Ukrainian managers head the joint ventures and most departments at the Kiev headquarters, while expatriates still hold the positions of Country Manager and Business Area Managers. An interesting 'social' project taken on by ABB in 1996, and completed in 1997, was the construction of residential multi-apartment buildings, complete with all the necessary infrastructure in Khmelnytsky. The US-based ABB Susa constructed 17 residential buildings with a total of 620 apartments for military personnel and their families living in that city. The newest technologies, management, and planning systems were used in this project.

The structures and management styles of American companies located in Ukraine vary, but for the most part fall somewhere between the practises of the two examples mentioned above. There is no evidence that multinationals adapt such management processes as strategic planning, budgeting, and performance review to Ukrainian conditions. Compensation and human resource management are adapted in order to reflect national labor regulations.

When multinational companies first arrived in Ukraine in the early 1990s, expatriate personnel were employed in top and second-level management positions. There is a growing trend within these companies to replace higher-level, more expensive expatriate managers with Ukrainian managers particularly in such positions as directors of marketing, public relations, finance, and so on. Now Procter & Gamble has only 13 expatriates in a staff out of 525 employees. Five foreign managers in key positions are helping to launch McDonald's in Ukraine. But by the end of the decade the international fast-food chain expects that Ukrainian executives and managers will run an exclusively Ukrainian show. As far as local staffing goes, two key targets were achieved. First, by the end of 1997 the current local staff of about 1000 almost doubled in size. Second, in 1998 Ukrainian management trainees were trained in Ukraine, rather than in Romania or Poland as was the case in 1997. That led to relaxation of English-language requirements for prospective managers.

Several multinational companies have already appointed Ukrainians to the CEO positions at their Ukrainian subsidiaries. But this does not prove to be a guarantee of success. In general, however, Ukraine still suffers from a shortage of well qualified, experienced, globally-oriented managers, though the supply is gradually increasing. Many foreign managers emphasize the importance of management skills in their employees. Independent and critical thought was regarded as seditious by the Soviet education system and by some conservative managers. Therefore, an ability to use knowledge and initiative in unanticipated circumstances generally remains a weak aspect of many Ukrainian employees. Hence, many multinational companies hire training and consulting firms to conduct skill-enhancing seminars and courses. Strong corporate cultures can now be recognized in most Ukrainian offices of multinational companies.

Conclusions

A number of multinational companies have started activities in Ukraine, benefiting from its great potential based on a large population, favorable geographic location with access to the Black Sea, inexpensive but well-educated labor, and significant natural resources. At the same time, Ukraine is considered by many as 'one of the biggest disappointments in Eastern Europe,' with its lagging privatization, slow corporate restructuring, over-regulated economy and as a result, prolonged recession. Therefore, few multinational companies established their production or research facilities in Ukraine until 1998. Most corporations do not go further than selling their products or services here.

Yet as the success of several companies, for example, ABB, Tetra Pak, S.C. Johnson, Coca-Cola Beverages, Kraft Jacobs Suchard, and R.J.Reynolds, and Reemtsma shows, first mover advantages in starting production operations in the rather difficult Ukrainian business environment, as well as broader and deeper knowledge of the Ukrainian market, can bring significant returns. There is no doubt that sooner or later the Ukrainian economy will be deregulated and restructured. At that moment competition may heat up seriously, but those companies that have production operations already running may get a significant competitive advantage in having a profound knowledge of the Ukrainian market, well-trained personnel, and effective human resource management practices.

Notes

1. For more details see *The Times Guide to the Peoples of Europe*, edited by Felipe Fernandez-Armesto (Times Books, 1994).
2. Foreign investment surges amid turmoil. *Kyiv Post*, 20 November, 1998.
3. From Plan to Market. *World Development Report 1996*, The International Bank for Reconstruction and Development/ The World Bank.
4. Note however, that many of the mines in the USA are open-pit or shallow with wide seams, whereas in Ukraine they are mostly very deep with thin seams.
5. Source: *Financial Times*, 3 March, 1998.
6. Prima stubs out the competition. *Kyiv Post*, 24 November, 1998.
7. Rostislav Khotin, 'Daewoo Puts Money, New Work Practices in Ukraine', *Reuters*, 25 March, 1998.

11 Conclusion – The Sun Rises in the East

ANDRZEJ KOŹMIŃSKI and GEORGE S. YIP

This book has analyzed what the countries of Central and Eastern Europe (C&EE) can offer to multinational companies and foreign investors. Rather than giving the usual economic analysis of the region, each country chapter outlines the way in which the country has developed, the current situation as well as the potential opportunities to emerge. Since the collapse of communism in the early 1990s and the opening up of the region to business venture and foreign enterprises, many multinational companies (MNCs) have become increasingly involved in a range of activities in the area. They have found countries in C&EE to be highly attractive, both as a source of new markets and production sites, and also for other value added activities such as research and development. MNC involvement has in turn stimulated economic growth and promoted political and legal changes in the region. Because of generally low levels of savings and fiscal policies limiting enterprises' ability to invest, the national prosperity of individual countries will, to a large degree, depend on their continued ability to capture MNC investment in the future.

Common wisdom and experience indicate that when entering the post-communist countries of C&EE, MNCs look for the benefits that can be represented in the simplified form of the 'Eight S Formula'.[1]

- Size of the market and present and prospective rates of market growth. Larger countries with consistent patterns of faster economic growth are the winners in this respect, Poland being the best example;
- Strong margins resulting from the 'novelty appeal' of Western products and services, from government granted privileges, or lucrative exclusive contracts. Some less stable and less advanced in transition post-Soviet countries are perceived as attractive from this point of view by companies which do not avoid risk;
- Stability (political, legal and monetary) is a strong motivator attracting MNCs to the most institutionally mature countries of the region included in the first group of the EU candidates (Czech Republic, Estonia, Hungary, Poland and Slovenia);
- Skills available at a lower price (scientific, engineering, artisan, and so on). Relatively high educational levels combined with low wages and salaries make the region highly attractive from this point of view;

- Support covers both infrastructure and business services such as consulting, auditing, legal advice, financial services etc. Multinational providers of business services following their clients also targeted countries capable of attracting the highest inflow of foreign direct investment (FDI). Government-financed infrastructure (roads and railway systems) still remains to different degrees underdeveloped in the region due to the generalized crisis of public finance. The level of telecommunication services, however, is rapidly raising due to the attractiveness of these markets for international providers;
- Suppliers enabling low levels of manufacturing costs can be found more easily in the countries where 'linkage industries' (such as glass, plastics, metals, chemicals, wood etc.) have been already privatized and became more flexible and capable of satisfying high standards imposed by the MNCs;
- Safety of the operations resulting from a stable legal system and effective law enforcement is likely to be found in the most advanced countries already negotiating EU membership;
- Springboard effect can be obtained if one country (usually providing more comfortable business environment) can be used as a platform for penetration of other (usually less developed) markets. For example, Poland and Baltics offer such opportunities for penetrating post-Soviet markets and Slovenia for the Balkans.

The analytical methodology of this book has been designed to provide much more precise and focused insights into both the current and potential roles of the nine countries and their economies in the global and regional strategies of foreign MNCs. Furthermore, this book provides a comprehensive, easy to use, and effective framework for analyzing the many facets of what makes a particular country attractive to investment. At the same time, the methodology takes into account, and is sensitive to, potential pitfalls and blocks that MNCs may encounter in implementing their ideal strategies.

Traditional economic analyses all too often view countries in isolation, focusing on standard economic indicators (such as market, cost, and competitive conditions), viewing these on a stand alone basis, with little regard for cross-country synergies. In contrast, this book underlines the role that distinct countries can play within the region as a whole. While the region is highly diversified culturally and each country is unique in what it has to offer, the common Communist past and the heritage of the economic ties between Communist bloc countries (developed within the framework of COMECON) will inevitably encourage future cross-regional interaction.[2]

From a larger perspective, these countries can play an important part in the multi-country, regional and global strategies of MNCs and, in an increasingly intercultural world, should not be treated as isolated markets. Poland is centrally located in the region and still maintains economic ties with post-soviet countries such as Ukraine, Belarus, and the Baltics. Because of that, several MNCs use Poland as a regional hub or a platform for penetration of these markets. Similarly, the Baltic countries of Lithuania, Latvia and Estonia have attracted a variety of

MNC investment, despite the relatively low GDP growth rate compared to other countries, the relatively small domestic markets, and the economic decline caused by the links with the Soviet Union and the latter's collapse. Viewed independently, these relatively small countries cannot hope to compete for foreign investment. However, with their role in the region and in the scheme of regional trade, they play an extremely important part in uniting Eastern and Western markets. Particularly important is the access to the eastern coast of the Baltic Sea with Lithuania establishing itself as the regional transport hub, having the only ice-free port in the region. The European Union (EU) has recognized Lithuania as a transport center in the region linking the EU with Eastern Europe. The EU's Transportation Commission declared the two routes running through Lithuania among the ten priority transport routes connecting Scandinavia with Central Europe, and East–West routes linking the huge Eastern European markets with the rest of Europe. Large trading partners include Russia, Germany, and Poland.

From the point of view of foreign investment Estonia has been exemplary in the region, with many MNCs choosing to locate their headquarters there because of the favorable conditions created for investors, and on the first wave list for EU membership along with the larger countries of Central Europe. While Estonia set almost no limitations in its attempt to attract FDI and forge links with Western economies, Latvia and Lithuania have been more cautious and conservative in their privatization programs. The different approaches to transition process means that together the Baltic countries have managed to promote new links with the West while still keeping close contacts and good terms with their giant neighbor and traditional trading partner, Russia. Both are important considerations and facilitators for MNCs looking to trade in throughout Central and Eastern European region and in Russia.

The Czech Republic and Slovakia, in spite of their 'velvet divorce' in 1993, are still closely economically linked and have complementary orientations: Slovakia more to the East and the Czech Republic more to the West. Such a combination offers interesting opportunities to MNCs and can trigger considerable cross-country synergies.

Country Ratings

We finish this book by providing summary quantitative ratings of each economy in terms of its globalization drivers, global strategic levers, and organization and management approaches. These have been discussed in more detail in each country chapter. We also discuss how MNC managers can best use these ratings to develop strategies and approaches for their companies.

Quantitative analyses across each of the nine countries are provided in 'star ratings' tables that compare country economies according to individual globalization drivers, global strategic levers, and organization and management factors. The tables are based on a five-point scale (illustrated by the number of stars). The economies are scored relative to Germany, a developed and neighboring European country, and the dominating economic player in the region. Germany typically rates scores of five (the highest). The chapter authors have designated the individual country ratings and all final assessments are the collective evaluations of all co-authors. The star ratings evaluate the situation as of 1999, and attempt to give a future prognosis for the new millenium. It should be noted that in almost every case, the ratings will become more favorable in the future as globalization continues and as many C&EE countries forge closer links with Western Europe and other developed economies of the world. These likely changes were discussed in more detail in the individual country chapters. Additional stars in brackets indicate the likely near-term improvements.

Providing summary ratings of nine countries situated in, all too often, a volatile social, political and economic environment, invariably comes with its risks. The generalizations made, however, are for the sake of aiding readers, and making evaluations of past, present and future situations more accessible. Methodologically, the star ratings are based on the findings of a team of researchers (and their assistants), all of whom have significant expertise in the region. Using a quasi–Delphi technique, the star ratings have been reviewed through a number of rounds, including two face-to-face meetings of the group members.

Managers should use these ratings as initial references from which to start their own investigations. Used in such a way, they will provide the foundations on which managers can develop their own ratings for particular, more specific businesses and industries in the region. Undoubtedly, such ratings will also raise questions about exceptions in the region, and managers are good at finding the exception to the rule. And there will undoubtedly be plenty to chose from in this unpredictable and, as yet, relatively unexplored region. But in many cases it is here that the opportunities lie. The star rating system will, hopefully, provide a starting point as well as a spur to search for the reasons why ratings for a particular industry or business are out of the ordinary.

Market globalization drivers

Table 11.1 should help to understand the extent to which each economy has market globalization drivers that favor market participation, offering global products and services and using global marketing strategies.

Companies looking for markets where *customer needs* are most similar to those of developed Western countries should look first to the countries who are in the first wave of candidates to the EU. They are the most Westernized. Western consumption styles and tastes are generally accepted throughout the region, particularly by the younger generation. But purchasing power is a barrier. The formation of a middle class with higher disposable incomes is the most important factor contributing to the process of globalization of customer needs in the region. The middle classes are also linked to the private sector, so the most advanced

Table 11.1 Summary of Market Globalization Drivers

	Common Customer Needs	Global Customers	Global Channels	Regional Customers	Regional Channels	Transferable Marketing	Lead Countries
Germany	*****	*****	****	*****	*****	***	*****
Hungary	****	**	***	**	***	****	***
Poland	****	**	***	**	***	****	***
Czech Rep.	****	**	***	**	***	****	**
Slovakia	****	**	**	**	***	****	**
Slovenia	*****	**	**	***	***	****	**
Baltic States	****	*	*	*	**	****	**
Romania	***	*	**	**	**	****	*(*)
Bulgaria	***	*	*	**	**	****	*
Ukraine	**	*	*	*	**	***	*(*)

* = very low, ***** = very high

Source: Based on country analyses. See definitions in Appendix.

Notes: Ratings vary somewhat by industry.

countries in privatization have more Westernized consumer needs. Companies venturing to poorer and less advanced-in-transition countries can expect greater differences in consumer needs. These ratings have implications not just for product design and selection but also for the amount of market research needed. So the fewer the number of stars, the more time and money should be budgeted for market investigation.

Companies, particularly those in industrial or business-to-business sectors with large customer accounts, need to be very aware of where they can find *global customers*. Such global customers require not only more attention but cross-country global coordination. In the region of C&EE these global customers are MNCs operating simultaneously in several countries of the region. ABB, Coca-Cola, Nestlé or Daewoo are good examples of such companies present in all countries of the region. They also have regional headquarters or significant leading operations in one of the countries, most likely in one of the countries most advanced in transition and with high market potential. Building a presence in these or other countries with high scores for global customers will create opportunities to deepen relationships with these very important accounts.[3] The same arguments apply for the ratings on *regional customers*. In Central and Eastern Europe they are mainly Western European companies focusing on the region and not fully globalized yet (e.g. German Benckiser in detergents) or Russian oil and natural gas companies (such as Gasprom or Lukoil) supplying the whole region.

Countries with high ratings for *global channels* and *regional channels* pose greater opportunities and demands for customer service. In the last five years, the most advanced in transition countries of the region with the highest market potential (Czech Republic, Hungary and Poland) experienced an unprecedented expansion of global retailers such as Carrefour, Auchan, Leclerc, Metro and others. This dramatically upgraded the channels of distribution existing in these countries. Companies wishing to build extensive and sophisticated distribution systems in the region of Central and Eastern Europe should, clearly, select these locations in preference.

The entire region is highly attractive for companies seeking *transferable marketing*, that is, the ability to use marketing approaches from the United States and Western Europe. Central and Eastern Europeans are clearly aspiring to Western consumption and life styles. Because of that they readily accept Western marketing and enjoy it as contrasting with the communist past. Some relatively rare exceptions from this rule result from religious convictions, moral codes or national pride. For example, the Benetton poster, depicting a priest and a nun kissing, was banned in Catholic Poland.

Managers entering the region should also consider the importance of *lead countries*, with the attendant benefits of early exposure to markets and customer innovation. In the C&EE region, Hungary, Poland and, to a lesser extent, the Czech Republic can be considered as regional lead countries. Most products new in the region, such as life insurance, pension and investment funds, mortgage financing, shopping malls, leasing and direct selling were, and still are, first introduced in those countries having the largest sophisticated, urban middle class and leading in transition. After being tested in these countries and eventually adjusted to local conditions, new products are introduced in other markets of the region.

Cost globalization drivers

Managers can use Table 11.2 to help evaluate cost-related globalization drivers in the region. In particular, the country ratings indicate where best to locate which activities.

Companies seeking markets that can contribute sales volume to achieve *global scale economies* or *regional scale economies* should look first to the largest (measured by population size and purchasing power) markets of the region: Poland, Hungary, and the Czech Republic. If transition will spur their economic growth, then Ukraine, Romania and Slovakia should be the second choice.

The whole region looks attractive for its *sourcing efficiencies*. Managers can look to Poland, the Czech Republic and Ukraine as the sources of raw materials such as coal, copper, sulfur. Wood and agricultural products can be obtained cheaply all over the region. Supply industries such as steel, metal, paper, textile, glass, chemicals, and plastics were relatively well-developed under communism and can still provide a wide range of by products and components at relatively low prices. Old state-owned companies, however, are not very reliable. Nor are they customer-responsive suppliers capable of maintaining high quality and delivery standards. That is why sourcing efficiencies can be found much more easily in the countries where supply industries have already been privatized.

Table 11.2 Summary of Cost Globalization Drivers

	Global scale economies	Regional scale economies	Sourcing efficiencies	Favorable logistics	Good infrastructure (overall)	Financial infrastructure	Physical infrastructure	Business country	Favorable country costs	Technology role
Germany	*****	*****	***	*****	*****	*****	*****	*****	*	*****
Hungary	**	***	***	***	***	***	****	***	***	***
Poland	***(*)	****	****	****	**	***	***	***	***	***
Czech Republic	**	***	***	***(*)	***(*)	**(*)	***(*)	***	**(*)	**(**)
Baltic State	*	**	**	****	**	**	***	***	****	**
Slovak Republic	**	***	***	***	***	**	***	***	***	**
Slovenia	*(*)	**	***(*)	****	****	***	****	****	**	**(*)
Romania	**	***	**	***	**	**	**	**	***	**
Bulgaria	*	**	**	***	**	*	**	**	***	**(*)
Ukraine	**	***	****	**(*)	*	*	**	**	****	**

* = very low, ***** = very high

Source: Based on country analyses. See definitions in Appendix.

Notes: Ratings vary somewhat by industry.

Cheap and highly skilled labor is certainly one of the advantages of the region providing for *favorable country costs*. Several MNCs operating in industries, such as mechanical, clothing, automotive and construction, use companies located in C&EE as subcontractors. This is particularly attractive for companies with extensive manufacturing operations in high-cost countries such as Germany. Some words of caution, however, are appropriate here. First, in the most advanced countries with a highly privatized industry, wages are growing and rapidly approaching the levels of the poorest EU countries such as Greece and Portugal. Second, relatively high taxes (including social security and wage tax) considerably increase labor costs and this is not going to change because of the public finance crisis. Third, labor productivity remains considerably lower than in the EU countries, and even private or privatized companies are often not up to Western standards of technology, efficiency, flexibility, and responsiveness.

Because of such weaknesses in C&EE enterprises, several MNCs have started their own component manufacturing green-field operations in the region. Automotive and automotive parts industries provide many such examples, such as Volkswagen's engine factory or Toyota's gearboxes factory in Poland. Such foreign direct investments, however, are made almost exclusively in the most advanced and the most stable countries. Low levels of safety, stability and discipline combined with bureaucracy and corruption can increase the risk of such operations to unacceptable levels.

Nearly all producers of goods, and some of services also, do better in countries with *favorable logistics* either as a production site or market or both. Centrally located countries with good access to other larger markets receive high ratings. The whole region is characterized by relatively favorable logistics due to the central location in Europe on the crossroads between West and East, North and South. Countries with access to well-developed seaports (such as Poland, Baltics, Romania, Bulgaria and Ukraine) offer additional potential benefits. Leveraging of these benefits, however, requires *good infrastructure* covering *physical, financial* and *business infrastructure*. Physical infrastructure, particularly roads, railways, energy and communications has been neglected and outdated under communism. The regions democratic governments are not yet fully capable of closing the gap between C&EE and EU countries because of tight monetary policies and the crisis of public finance. Hungary and Slovenia are relatively better equipped. Poland, one of the transition leaders, is clearly lagging behind and Romania, Bulgaria and Ukraine receive poor ratings. Hungary, Poland and Slovenia have the most developed financial infrastructure (banking system, capital markets, financial services, insurance). Small Slovenia has the best business infrastructure, although other Central European countries are rapidly catching up, helped by a massive presence of multinational business service providers.

For companies seeking economies that can play a *technology role*, C&EE offers a mixed set of opportunities. In the Communist era, R&D activities were concentrated in 'institutes' or 'centers', which were relatively well funded by the state. Quite often, these organizations grouped world class scientists and engineers. R&D units, however, were almost completely cut-off from industry and from markets. As a result, they developed a bureaucratic and academic culture incompatible with business requirements. Because of that, most of the R&D units disintegrated or degenerated, particularly as the MNCs entering the region had no means properly to assess their potential. What has remained from the R&D potential of Central and Eastern Europe is being slowly and partially absorbed by foreign, rather than local, companies operating in the region because local companies have very limited R&D budgets.

Government globalization drivers

The ratings of government globalization drivers can help managers to evaluate the kind of roles that governments play in each country of the region, and the types of strategies that companies should pursue to deal with government intervention (Table 11.3).

Opening of markets is a fundamental principle of the International Monetary Fund, World Bank, European Union, World Trade Organization and other international organizations actively supporting the transition process. This pressure on C&EE governments resulted in general acceptance of this principle and adoption of *favorable trade policies* throughout the region. Hungary and Slovenia come out as the most open countries. Ukraine is lagging behind the others, probably because of hard line communists still maintaining considerable political influence. Some trade barriers still remain or, are even added to, as a result of pressure from unions trying to protect non-competitive industries and to save jobs. 'Sensitive' sectors such as steel, agriculture, automotive or textiles are the most often affected by protectionist trade policies. In the case of the countries negotiating accession to the EU such policies will have to disappear within the next couple of years or become compatible with the policies of the EU, member countries. Other countries of the C&EE are likely to follow.

Alternative to, or as well as, trade, companies may choose to invest directly in these economies. In terms of *favorable investment rules*, the ratings are similar to those for favorable trade policies. The difference for managers lies in the relationships they will have to build with governments, typically much more long term for investment. Countries maintaining consistently pro-Western orientations, in spite of political fluctuations and government changes, are more capable of such a long-term relationship with foreign investment. Hungary, Poland and the Czech Republic are examples of such countries. In Slovakia, Bulgaria and Romania pro-Western orientation prevailed after their last parliamentary elections, but it remains to be seen what the next political change bring. In the past, former communists, after winning elections, have tried to change or at least to 'adjust' government policies toward foreign investment. Ukraine is torn between pro-Western and pro-Russian orientations, the latter represented by a strong parliamentary faction of non-reformed Communists. The government's policies maintain a shaky equilibrium between these two orientations.

Conclusion – The Sun Rises in the East

Table 11.3 Summary of Government Globalization Drivers

	Favorable trade policies	Favorable FDI rules	Role in regional trade blocs	Freedom from government intervention	Absence of government owned competitors	Legal protection	Compatible technical standards	Common marketing regulations
Germany	*****	*****	*****	****	*****	*****	*****	******
Hungary	****	****	***	***	****	***	****	******
Poland	***	***	***	***	***	***	****	******
Czech Republic	***	***	***(*)	***	***	***	****	******
Slovak Republic	**(*)	***	***	**(*)	**(*)	***	*****	****
Slovenia	****	***	***	***	****	****	******	*****
Baltic States	***	***	*	***	****	***	****	****
Romania	***	***	***	****	**	**(*)	***	****
Bulgaria	***	**	***	**	*	**	***	***
Ukraine	**	*(*)	*(*)	*	*	*(*)	***	*****

* = very low, ***** = very high

Source: Based on country analyses. See definitions in Appendix.

Notes: Ratings vary somewhat by industry.

An economy's *participation in the trade blocs* can ease the paths of MNCs already active in other members of the bloc. The Central European Free Trade Association (CEFTA), with the Czech Republic, Hungary, Poland, Slovakia, Slovenia, Bulgaria and Romania participating, is such a bloc. All these countries, however, are in the process of negotiating accession to the EU, which gives a temporary character to the bloc, and all experience internal difficulties, when trying to open sensitive markets such as agricultural products, automotive, steel, coal or chemicals. Ukraine is a member of the Commonwealth of Independent States (CIS) but Russian economic problems and lack of political leadership make it rather questionable as a trade bloc.

Three government globalization drivers, *freedom from government intervention*, *absence of government-owned competitors* and *reliable legal protection*, have similar implications for MNCs. In particular, poor ratings on these dimensions require MNCs to have excellent skills, experience, and connections for coping with uneven and shifting playing fields. In this regard there are clear differences by national origin of MNCs. On average, German, Scandinavian and Italian firms are the most likely to succeed in difficult C&EE environments because of historical experiences and a past history of cooperating with communist regimes in the 1970s and the 1980s. Americans are rapidly catching up with them, leveraging the only super-power status and popularity of the 'American model' in that part of the world. Asians experience the most difficulties due to the large cultural distance, communication difficulties and negative stereotypes. There are also differences by industry. Companies in industries that rely on patents, proprietary knowledge, or intellectual capital need to think hard about participating in countries with poor legal protection and weak law enforcement. And if they do participate, they need clear strategies for minimizing their risks of loss. Clearly Romania, Bulgaria and Ukraine pose the greatest problems.

Most countries of the region have highly *compatible technical standards*, making it relatively easy for MNCs to market their products and technologies. The level of compatibility is likely to increase with the progress of EU accession negotiations.

Lastly, most C&EE countries have *common marketing regulations* that enable marketers to use international marketing standards, techniques and instruments such as advertising spots. The countries with strong religions, such as Poland, can impose some marketing restrictions.

Competitive globalization drivers

Understanding competitive dynamics is particularly important in the C&EE because the region has become a strategic battleground for MNCs operating in such key industries as automotive and automotive parts, tobacco, breweries, detergents, telecommunications, retailing. Important sales growth can be expected in C&EE catching up with Western European standards and several key

Conclusion – The Sun Rises in the East

countries of the region are likely to join the EU by 2004 or so. Prospects of EU membership strongly influence the perspectives and investment plans of MNCs with large operations in the region, such as Korean automaker Daewoo. Table 11.4 summarizes the region's competitive globalization drivers.

Table 11.4 Summary of Competitive Globalization Drivers

	Global Strategic Importance	Regional Strategic Importance	Presence of Foreign Competitors	Interdependence via MNC Value Chain	Internationalized Domestic Competitors
Germany	*****	*****	*****	****	*****
Hungary	**	***	****	****	***
Poland	***	***	****	****	***
Czech Republic	**	***	***	**	**
Slovak Republic	*(*)	**(*)	**	**	**
Slovenia	*(*)	**(*)	****	**	***
Baltic States	**	**	***	**	**
Romania	**	***	***	**	*
Bulgaria	*	**	***	*	*
Ukraine	**	***	**(*)	*	*

* = very low, ***** = very high

Source: Based on country analyses. See definitions in Appendix.

Notes: Ratings vary somewhat by industry

Global strategic importance and *regional strategic importance* of countries concern the extent to which an MNC's activities have strategic importance beyond those countries. While the region as a whole is gaining global strategic importance, individual countries have relatively small global importance because of the size, achieved level of development, and still-remaining uncertainties associated with future prospects. Precisely because of these reasons (size and growth), Poland is the only notable exception, with moderate global strategic importance partially due to its central geographic location and its emerging role as a key US ally in the region. The regional strategic importance of countries is more evenly distributed, mainly because each of the analyzed economies can be used as a platform for further expansion.

Presence of foreign competitors poses another set of challenges for MNC managers. Three regional leaders – Czech Republic, Hungary and Poland – have already attracted a very large number of MNCs, intensifying competition and increasing such costs as office rent, and compensation of local personnel. Cost-benefit analysis of alternative locations in the region is becoming an increasingly important task for corporate planners.

Lastly, MNC managers need to be concerned about each country's global and regional *interdependence via MNC value chains* (the extent to which MNCs have created strategic interdependence between a country and others through sharing of activities such as manufacturing, distribution, R&D or other parts of the value chain). Several MNCs looking for economies of scale have internationalized their value chains concentrating different activities in different countries of the region, particularly for production and distribution. In consequence, both activities and competitive struggles in these markets have regional repercussions. So managers need to extensively coordinate with other geographic units when developing and implementing plans for operations in every single country of the region.

Recommended MNC overall strategies

The ratings of country globalization drivers have direct implications for recommended MNC strategies. We discussed some of the implications in previous sections. The ratings in Table 11.5, presenting global strategic levers, make direct recommendations. Again, we need to caution that this book's recommendations can only be a starting point for MNC managers. They need to customize our overall recommendations for their individual situations.

Table 11.5 Summary of Recommended MNC Overall Strategies

	Market Participation	Global Products/ Services	Activity Location	Global Marketing	Global Competitive Moves
Germany	*****	*****	****	****	*****
Hungary	****	*****	****	****	***
Poland	***	****	**	****	****
Czech Republic	****	****	****	****	**(*)
Baltic States	***	****	***	***	**
Slovak Republic	***	****	***	***	**
Slovenia	***	*****	****	****	***
Romania	***	*****	***	***	**
Bulgaria	***	****	***	***	*
Ukraine	***	****	**	***	**

* = very low, ***** = very high

Source: Based on country analyses. See definitions in Appendix.

Notes: Ratings vary somewhat by industry.

Market participation constitutes a first, critical decision for MNCs. Based on all the various globalization drivers, no one C&EE economy receives a maximum five-star rating, which is reserved for leading world economies such as Germany. The region as a whole, however, would certainly receive an unambiguous five-star rating: all major MNCs have to be present 'somewhere' in the region. Furthermore an MNC should adopt a portfolio strategy, with participation in a number of regions (North America, Western Europe, Asia[4] and so on) and markets within each region. Standard portfolio theory says that firms should invest in a mix of high risk – high return and low risk – low return assets. The same applies to participation and investment in countries. Until very recently (late 1990s), C&EE, as a whole, was perceived by managers as a high-risk region. The time has come to make a distinction between stable democracies, accomplished transition economies, NATO member countries negotiating accession to the EU, and less advanced transition economies with still good prospects in the more distant future. Typically, an MNC should position itself in one of the mature market economies of the region and in one less advanced economy associated with somewhat higher risk. The Czech Republic, Hungary, Poland and Slovenia are gradually becoming low risk countries.

The extent to which MNCs should offer in each economy *global products/services* relates particularly to the *common customer needs* driver that we discussed earlier. C&EE is clearly culturally ready to accept global products. In less advanced countries, problems arise from disposable incomes and distribution networks.

Once an MNC has decided to participate in one or more of these countries, another portfolio decision concerns where to locate which value-adding activities. Applying global strategy means not reproducing every activity in every country. The *activity location* column in Table 11.5 indicates the overall attractiveness of each economy for locating activities in general. Poland scores somewhat lower than other transition leaders because of poor physical infrastructure (roads and railway system) not improving rapidly enough. In the next section we will discuss each individual activity.

The whole region is culturally and psychologically ready for *global marketing*. In less advanced countries scoring three stars, the degree of readiness is likely to increase with economic development.

Lastly, MNCs have to recognize when they need to make *global competitive moves* rather than just local ones. Because of its strategic importance, Poland scores the highest in this respect. Other countries, in the first group of candidates negotiating accession to the EU (Czech Republic, Hungary, Slovenia and Estonia), are likely to increase their role in the global strategies of the MNCs as their accession dates approach.

Recommended MNC activity location strategies

Table 11.6 shows detailed ratings for individual activities. Its clear message is that attractiveness of the region is moderately high and, also, moderately differentiated by countries. Moreover, it is likely to rise with transition progress, economic development and further internationalization of individual economies.

Recommend locations for *research and development* activities depend mostly on the globalization driver *technology role*, all countries of the region offer some prospects in this respect. But effective use of existing potential (mainly human capital) would call for creation of new research and development units by MNCs. General Electric in Hungary, Lucent Technologies and Motorola in Poland provide examples of companies doing exactly that.

Table 11.6 Summary of Recommended MNC Activity Location Strategies

	Research	Development	Procurement office	Production operations	Regional marketing team	Regional sales force	Regional distribution ceter	Regional customer service	Regional HQ
Germany	*****	*****	**	***	*****	***	*****	****	****
Hungary	***	****	***	****	****	****	****	***	****
Poland	***	***	****	****	***	****	****	***	****
Czech Rep	***	***	***	***(*)	***(*)	***(*)	***(*)	***	***(*)
Slovakia	***	***	**	***(*)	**	***	***	**	**
Slovenia	***	***	***	***	***(*)	***(*)	***(*)	***(*)	***
Baltics	***	***	**	***	***	***	***	***	***
Romania	**	**	***	***	***	***	**(*)	**	*(**)
Bulgaria	***	**	**	**	**	*	*	*	*
Ukraine	***	***	***	***	*	*	*	*	*

* = very low, ***** = very high

Source: Based on Country analyses. See definitions in Appendix.

Notes: Ratings vary somewhat by industry.

The location of regional *procurement offices* depends on access or proximity to important sources of supply. Poland is clearly the leader in this respect with its central location, natural resources and relatively well developed supply industries. Poor transportation infrastructure diminishes the attractiveness of these assets. Other countries of the region are also offering interesting prospects and some of them (for example Hungary or Slovenia) have superior transportation systems.

Many MNCs' first interest in C&EE is as a site for *production operations*. All the most advanced-in-transition countries offer attractive price/productivity ratios for labor, plus engineering skills and relatively predictable and orderly business environments. Other countries of the region can be attractive as production sites because of cheaper labor, idle production capacities, cheap energy or abundant raw materials.

Many MNCs may wish to set up *regional marketing teams*, instead of, or in addition to, national ones. Factors affecting the choice include communications and transportation infrastructure, and the ability to attract or hire talented and cosmopolitan marketing executives. The most 'exciting' places of the region, such as Budapest, Prague or Lubljana, offer the most attractive prospects in this regard.

In some cases, MNCs can also use a *regional sales force*. Cost becomes more of a factor than for, typically, smaller regional marketing teams. Language differentiation of the region imposes use of English as the *lingua franca* and somewhat limits the use of regional sales forces to larger business-to-business customers. The same language issue applies to *regional customer service* operations.

The location of a *regional distribution center* depends primarily on the logistics need of a company and on the infrastructure of a country. Several countries in the region can be used as hubs for different purposes, but no one country stands out clearly as the winner in this regard.

Lastly, most MNCs need a *regional headquarters* in the region. Major considerations include communications and transportation infrastructure, quality and availability of office space, proximity to key markets and the quality of life. Hungary, the Czech Republic and Poland seem to be the best suited for headquarters locations because of various sets of those reasons. Budapest and Prague are the most Westernized and cosmopolitan cities in the region with the best infrastructure. Warsaw is located in the center of the largest market and is emerging as the most important financial center of C&EE with the best developed capital market (to date almost immune to the repercussions of the Russian crisis). Slovenia and Romania can be perceived as gateways to the Balkans, where after a series of devastating conflicts, a reconstruction process is likely to start. Baltic countries can provide valuable links between Russia and Scandinavian markets.

Recommended MNC organization approaches

Managers should use Table 11.7 to help develop the appropriate organization and management approaches in each country. Despite similarities resulting mainly from the common Communist past and regional economic cooperation within

Table 11.7 Summary of Recommended MNC Organization Approaches

	Can go alone	Give significant autonomy	Use global management processes	Participate in global processes	Can use local managers	Can use foreign managers	Source of global and regional managers	Instill global culture
Germany	****	*****	*****	*****	*****	*****	****	****
Hungary	****	*****	****	*****	*****	*****	****	*****
Poland	****	*****	***	****	*****	*****	****	*****
Czech Rep	****	*****	****	****	****(*)	*****	****(*)	****
Slovakia	***(*)	****	***	***	***	****	**(*)	***
Slovenia	****	*****	****	***	*****	****	***(*)	*****
Baltics	****	****	***	***	****	****	****	*****
Romania	***	****	***	***	****	*****	****	*****
Bulgaria	***	***	**	**	***	****	**	****
Ukraine	***	***	**	*	***	****	**(**)	****

* = very low, ***** = very high

Source: Based on Country analyses. See definitions in Appendix.

Notes: Ratings vary somewhat by industry.

COMECON, cultural differences and differences in transition, as well as economic development should be taken into consideration as factors affecting MNC management.

MNCs can go alone and not need local partners if foreign investment laws allow and if operating conditions are easy in a country. From the legal point of view, 100 per cent foreign ownership is, in principle, allowed in all the countries analyzed in this book. Exceptions might be linked to the sectoral policies of governments and might be brought up in negotiations with foreign entrants. One should be also aware of the relative scarcity of local partners that are mature, reliable and compatible with Western business practices. Cultures and management processes of state owned enterprises (SOEs) make it hard for them to cooperate with Western partners. SOEs are also quite often highly politicized, with powerful and militant trade unions, and political appointees occupying key managerial positions. These characteristic features often persist after privatization, particularly if privatization was done hastily under 'mass privatization' or 'voucher' schemes. On the other hand, local private enterprises are, in most cases too small, financially unstable and not mature enough. Because of these reasons a 'go alone' strategy can be recommended in most cases. If local participation is required by the government (for example in such sectors as banking, insurance, telecommunications and so on), MNCs should nearly always seek management control.

MNCs need to decide whether to *give significant autonomy* to subsidiaries. This decision depends on such factors as the likely business experience and sophistication of local managers. Transition leaders (Hungary, Poland, the Czech Republic and Slovenia) score five stars, the same as Germany, others four or three stars. Autonomy is recommended, not only because of the cost of control and the need for quick response to market signals, but also because of the uniqueness of the C&EE business environment, which is difficult to grasp by global headquarters of MNCs.

MNCs need to decide whether to use *global management processes* or use local ones. This decision depends on the experience and sophistication of local managers, but also on whether there are strong local practices. For obvious reasons, such practices could not be developed under communism. Global management practices can be installed relatively easily, though, in such areas as marketing, finance, production, logistics, and development. Culturally-bounded human-resource management practices should be implemented with more flexibility. Some differences can also result from specificity of legal systems and inadequate physical infrastructure, communication systems, and so on.

Managers from C&EE can *participate in global management processes* provided they have sufficient communication skills and educational backgrounds compatible with Western management practices. Unprecedented efforts in management education, undertaken in the entire region with active Western participation, has considerably enhanced the chance of successful local participation in global management practices.[5]

Management education is closely linked to the extent to which MNCs can *use local managers*. The whole region scores relatively high, with leading transition economies, which have developed the most extensive network of management education, coming out slightly ahead of the others.

The complementary consideration is whether MNCs can use non-local managers. Here, the main issues are the local acceptance of foreigners, combined with the distinctiveness of local management approaches. In this regard, the whole region can be characterized by relatively high acceptance levels, provided that foreign managers demonstrate minimum sensitivity to local issues, respect for the local culture, and the like. People are willing to accept foreigners because they feel they can learn from them. For the same reasons, MNCs can relatively easily *instil global culture* into their subsidiaries in the region. In the local environment, however, some elements of this culture might be dysfunctional and eventually will undergo some form of 'spontaneous adjustment'. Corporate bureaucracies, for example, share some common characteristics with the old communist bureaucracy (over-formalization and over-regulation), and people will instinctively use old defensive strategies of subversion of rules through extensive networks of informal relations, exchange of services, information distortion and so on.[6]

MNCs may also wish to *source global and regional managers* from particular C&EE countries. Educational background, language skills, experience, adaptability and willingness to travel all come into play. The relatively low incomes of Central and Eastern European managers combined with high educational levels and ambition, and willingness to succeed in the West to overcome the 'underdog syndrome', make many of them willing and qualified to adopt expatriate careers. Some MNCs have already started promoting local managers worldwide but it is too early yet to evaluate the results.

Conclusion

Readers should recognize that our evaluations and recommendations concern the current situation, with allowance for future changes. For example, Ukraine, Bulgaria and Romania now rate relatively poorly on most items as compared with the transition leaders (countries which have recently become NATO members and are included in the first group of candidates for the accession to the EU). On the other hand, companies entering such less-advanced and lower-scoring countries will have the potential to reap first mover advantage in the future, provided that these economies enter the path of sustainable economic growth and consistent market reforms.

In contrast, for the more mature economies, such as Hungary, the Czech Republic and Poland, benefits are immediate but will improve less. We particularly recommend MNCs to take a portfolio approach both to the whole region of C&EE and within it. First, the region is far too important to not have a role in the global strategies of MNCs, especially if Russia continues to be a high risk, unstable economy, and that C&EE will almost certainly become the highest growth region of the EU. Second, the region offers many diverse economies that can play differing roles within a regional strategy. In conclusion, managers who develop strategies and organization approaches that exploit the specific globalization potentials of individual economies will build successful global strategies for the region of C&EE. For such managers, the sun will really rise in the East of Europe.

Notes

1. Kozminski, A.K. (1995), 'Lessons from the restructuring of post-communist enterprises', in: D.P. Cushman and S.S.King (eds.) *Communicating Organizational Change: A Management Perspective* (Albany, NY: State University of New York Press) pp. 311–28.
2. Lavigne, M. (1995), *The Economics of Transition. From Socialist Economy to Market Economy* (New York: St.Martin's Press).
3. Yip, G.S. and Madsen, T.L. (1996), 'Global account management: the new frontier in relationship management'. Special issue of *International Marketing Review*, 13: 3, pp. 24-42.
4. For a star-rating evaluation of Asian economies, see George S. Yip 2000, *Asian Advantage: Key Strategies for Winning in the Asia-Pacific Region*, updated edition, (Cambridge, MA: Perseus Books).
5. Kozminski, A.K. (1996), 'Management education in Central and Eastern Europe', in. M. Warner (ed.), *International Encyclopedia of Business and Management* (London and New York: Routledge), pp. 2775–81.
6. Kozminski, A.K. and Tropea, J.L. (1982), 'Negotiation and command: managing in the public domain'. *Human Systems Management*, 3: 1, pp. 21–31.

Appendix: Globalization Measures

Country Globalization Drivers

We use four sets of country globalization drivers, diagnosing the current and potential situation of each:

Market globalization drivers

Common customer needs

- Extent to which customer needs are common with the rest of the region and the world.
- Extent to which economic growth and social changes have moved the country toward the consumption patterns of developed Western economies.
- Examples, with statistics, of the consumption of key lead product and service categories, for example, automobiles, telephones, TV sets, personal computers, fast food.

Global customers

- Extent to which corporations who behave as global customers have regional headquarters or other significant operations in the country.
- Extent to which corporations who behave as global customers purchase/source from the country.

Global channels

- Extent to which global channels of distribution have regional headquarters or other significant operations in the country.
- Extent to which corporations who behave as global channels of distribution purchase/source from the country.

Regional customers

- Extent to which corporations who behave as regional customers have regional headquarters or other significant operations in the country.
- Extent to which corporations who behave as regional customers purchase/source from the country

Regional channels

- Extent to which regional channels of distribution have regional headquarters or other significant operations in the country.
- Extent to which corporations who behave as regional channels of distribution purchase/source from the country.

Transferable marketing

- Extent to which marketing approaches need to be adapted in this country in these three (and other) industries.
- Extent to which the country is generally amenable to marketing strategies and approaches used in other countries, for example, acceptance of foreign brand names, packaging and advertising; availability of media.

Lead countries

- Extent to which the country accounts for major product or process innovations
- Summary of patent filing statistics.

Cost globalization drivers

Global and regional scale economies

- Extent to which the country has markets that can contribute sales volume to MNCs needing to achieve global or regional scale economies (is the local market large enough to support a minimum efficient scale plant? If not, are there sufficient exports to support a minimum efficient scale plant?).

Sourcing efficiencies

- Extent to which the country can provide critical factors of production in efficient volumes.

Favorable logistics

- Extent to which the country has favorable logistics (transporting goods and services to and within the country) as either a production site or a market or both.

Good infrastructure

- Financial (banking system, credit, insurance, and so on).
- Physical (roads, power, communications, and so on).
- Business (advisory services, investor support, and so on).

Favorable country costs (including exchange rates)

- Extent to which the country offers low production costs, taking into account not just labor wage costs, but also overhead costs and productivity.
- Summary statistics on country's labor rates, hourly and with benefits/social charges.

Technology role

- Extent to which the country can be used as a base for developing technology.

Government globalization drivers

Favorable trade policies

- Extent to which the country has favorable trade policies, including both tariff and non-tariff barriers and how these are changing.

Favorable foreign direct investment rules

- Extent to which the country has rules that favor foreign direct investment, including currency regulations, repatriation of capital and foreign ownership.

Role of regional trade blocs

- How the country's participation in trade blocs affects the opportunities for multinational companies.

Freedom from government intervention

- Sensitivity of the country's government to foreign dominance of key industries, and response via intervention.

Absence of state-owned competitors

- Extent to which industries of interest to foreign MNCs are dominated by government-owned competitors or customers.

Reliable legal protection of contracts, trademarks, and intellectual property

- Extent to which contracts, trademarks and intellectual property are protected from imitation.

Compatible technical standards

- Extent to which the country uses global technical standards.

Common marketing regulations

- Extent to which the country has similar marketing regulations to the rest of the world, (for example, rules on television advertising).

Competitive globalization drivers

Global/regional strategic importance

- Extent to which this country has global or regional strategic importance in these three and other industries, and the role it should play in the market portfolio of a multi-national company. Global or regional strategic importance is defined in terms of:
  - large source of revenues or profits.
  - home market of global or regional customers.
  - home market of global or regional competitors.
  - significant market of global or regional competitors.
  - major source of industry innovation.

Presence of foreign competitors

- Extent to which foreign competitors participate in business activity.
- Concerns about competitors based in other countries in the region.

Interdependence via MNC value chains

- Extent to which MNCs create strategic interdependence between this country and others through sharing of activities such as factories or other parts of the value chain.

Internationalized domestic competitors

- Extent to which the largest local companies in the country are themselves globalized in terms of international revenues.
- Statistics of largest export and import sectors.

Global and Regional Strategy Levers

To analyze companies we focus on five dimensions of global and regional strategy:

Market participation strategies

- Extent to which MNCs do, and should participate in the country's markets.

Product strategies

- Extent to which MNCs do, and should, market in the country globally standardized products and services, and the extent to which they do, and should, make local adaptations.

Activity location strategies

- Actual and potential use of this country as a location for different value-chain activities, raw material sourcing, and manufacturing.
  - research
  - development
  - procurement office
  - production operations
  - regional marketing team
  - regional sales force
  - regional distribution center
  - regional customer service
  - regional headquarters
- Significance of this country's activities in the overall regional and global value chains of the MNCs.
- Actual and potential evolution of value chain role.

Marketing strategies

- Extent to which MNCs do, or should adapt their marketing in the country, evaluated by each element of the marketing mix: positioning, brand names, packaging, labeling, advertising, promotion, distribution and selling methods, sales representatives, and service personnel.
- Extent to which price levels in this country differs from the rest of the region and the world.
- Special aspects of marketing success factors needed in this country.

Competitive move strategies

- Extent to which MNCs should include the country when they make global or regional competitive moves, as opposed to making competitive moves in this country independent of moves in other countries.

Global Organization and Management

Can go alone

- Extent to which a foreign MNC does not need to have local partners.

Give significant autonomy

- Way in which MNCs typically organize operations in the country, for example, as autonomous subsidiaries or as tightly-controlled extensions of units outside the country.

Use global management processes

- Extent to which MNCs can use global management processes or need to adapt management processes for the country: particularly strategic planning, budgeting, motivation, performance review and compensation, human resource management (including career planning and employment terms, for example, lifetime employment) and information systems.

Participate in global processes

- Extent to which MNCs can expect subsidiaries/partners in the country to participate in global management processes.

Can use local managers

- Extent to which local managers are qualified to work in MNCs.

Can use non-local managers

- Extent to which foreign or expatriate managers are acceptable in the country;

Source global and regional managers

- Potential of local managers for transfer to other countries;

Instil global culture

- Extent to which global strategies of MNCs need to take account of local culture and the extent to which they can instil their global corporate culture.

Table A1 GDP (% change)

	1990	1991	1992	1993	1994	1995	1996	1997	1998
Hungary	−3.5	−11.9	−3.1	−0.6	2.9	1.5	1.3	4.6	5.1
Poland	−11.6	−7.0	2.6	3.8	5.2	7.0	6.1	6.8	4.8
Czech Rep.	−1.2	−11.5	−3.3	0.6	3.2	6.4	3.9	1.0	−2.7
Slovak Rep.	−2.5	−14.6	−6.5	−3.7	4.9	6.9	6.6	6.5	4.4
Slovenia	−4.7	−8.9	−5.5	2.8	5.3	4.1	3.1	3.8	4.0
Latvia	−3.5	−10.4	−34.9	−14.9	0.6	−0.8	3.3	8.6	3.6
Lithuania	−6.9	−5.7	−21.3	−16.2	−9.8	3.3	4.7	7.3	5.1
Estonia	−8.1	−13.6	−14.2	−9.0	−2.0	4.3	4.0	11.6	4.0
Romania	−5.6	−12.9	−8.8	1.5	3.9	7.1	3.9	−6.6	−7.3
Bulgaria	−9.1	−11.7	−7.3	−1.5	1.8	2.1	−10.9	−6.9	3.5
Ukraine	−2.6	−11.6	−13.7	−14.2	−23.0	−12.2	−10.0	−3.2	−1.7

Source: Business Central Europe Magazine, The Economist Group, Historical data collected from WIIW, EBRD, Reuters, national statistics. http://www.bcemag.com/

Table A2 Industrial Production (% change)

	1990	1991	1992	1993	1994	1995	1996	1997	1998
Hungary	−10.2	−18.3	−9.7	4.0	9.6	4.6	3.3	11.0	12.8
Poland	−24.2	−8.0	2.8	6.4	12.1	9.7	8.3	11.5	4.8
Czech Rep.	−3.3	−22.3	−7.9	−5.3	2.1	8.7	1.8	4.5	2.2
Slovak Rep.	−4.0	−19.4	−9.2	−3.8	4.9	8.3	2.5	2.7	5.0
Slovenia	−10.5	−11.6	−12.6	−2.5	6.6	2.3	1.2	4.7	3.7
Latvia	n.a.	−2.1	−46.2	−38.1	−9.5	−6.3	1.4	6.1	−0.7
Lithuania	n.a.	−26.4	−28.5	−34.4	−26.5	5.2	1.3	0.7	4.7
Estonia	n.a.	−9.5	−38.7	−19.0	−3.0	2.0	3.4	13.4	3.5
Romania	−19.0	−22.8	−21.9	1.2	3.3	9.4	9.9	−5.9	−17.3
Bulgaria	−16.7	−21.0	−6.4	−6.2	6.0	−5.4	−8.3	−7.0	−9.4
Ukraine	−0.1	−4.8	−6.4	−8.0	−27.3	−12.0	−5.7	−1.8	−1.5

Source: Business Central Europe Magazine, The Economist Group, Historical data collected from WIIW, EBRD, Reuters, national statistics. http://www.bcemag.com/

Table A3 *Budget Balance (% of GDP)*

	1990	1991	1992	1993	1994	1995	1996	1997	1998
Hungary	0.4	–2.9	–6.8	–5.5	–8.4	–6.7	–3.1	–4.1	–4.7
Poland	3.1	–6.7	–6.7	–3.1	–3.1	–2.8	–3.3	–1.6	–2.5
Czech Rep.	n.a.	–1.9	–3.1	0.5	–1.2	–1.8	–1.2	–1.0	–1.6
Slovak Rep.	n.a.	n.a.	n.a.	–7.0	–1.3	0.2	–1.9	–5.7	–5.5
Slovenia	–0.3	2.6	0.2	0.3	–0.2	0.0	0.3	–1.2	–0.3
Latvia	n.a.	n.a.	–0.8	0.6	–4.1	–3.5	–1.4	1.2	0.7
Lithuania	–5.4	2.7	0.5	–3.3	–5.5	–4.5	–4.5	–1.2	–1.6
Estonia	n.a.	5.2	–0.3	–0.7	1.3	–1.2	–1.9	2.0	–0.3
Romania	1.0	3.3	–4.6	–0.4	–1.9	–2.6	–5.1	–3.5	–2.6
Bulgaria	–4.9	–3.7	–5.2	–10.9	–5.8	–5.6	–11.5	–2.9	1.3
Ukraine	n.a.	n.a.	–25.4	–16.2	–9.1	–7.1	–3.2	–5.6	–2.5

Source: Business Central Europe Magazine, The Economist Group, Historical data collected from WIIW, EBRD, Reuters, national statistics. http://www.bcemag.com/

Table A4 *Unemployment (%)*

	1990	1991	1992	1993	1994	1995	1996	1997	1998
Hungary	1.9	7.4	12.3	12.1	10.4	10.4	10.5	10.4	9.1
Poland	6.3	11.8	13.6	16.4	16.0	14.9	13.2	10.3	10.4
Czech Rep.	0.8	4.1	2.6	3.5	3.2	2.9	3.5	5.2	7.5
Slovak Rep.	0.8	0.0	4.8	12.2	13.7	13.1	12.8	12.5	15.6
Slovenia	4.7	8.2	11.5	14.4	14.4	13.9	13.9	14.8	14.5
Latvia	n.a.	n.a.	2.3	5.8	6.5	6.6	7.2	7.0	8.8
Lithuania	n.a.	0.3	1.3	4.4	3.8	6.2	7.1	5.9	n.a.
Estonia	n.a.	n.a.	n.a.	6.5	7.6	9.7	10.0	10.5	n.a.
Romania	0.4	3.0	8.2	10.4	10.9	9.5	6.6	8.8	10.3
Bulgaria	1.7	11.1	15.3	16.4	12.8	11.1	12.5	13.7	12.2
Ukraine	n.a.	0.0	0.3	0.4	0.4	0.5	1.5	2.8	4.6

Source: Business Central Europe Magazine, The Economist Group, Historical data collected from WIIW, EBRD, Reuters, national statistics. http://www.bcemag.com/

Table A5 *Inflation (%)*

	1990	1991	1992	1993	1994	1995	1996	1997	1998
Hungary	28.9	35.0	23.0	22.5	18.8	28.2	23.6	18.3	14.3
Poland	585.8	70.3	43.0	35.3	32.2	27.8	19.9	14.9	11.8
Czech Rep.	9.7	56.6	11.1	20.8	10.0	9.1	8.8	8.5	10.7
Slovak Rep.	10.6	61.2	10.1	23.2	13.4	9.9	5.8	6.1	6.7
Slovenia	549.7	117.7	207.3	32.9	21.0	13.5	9.9	8.4	7.9
Latvia	10.5	172.0	951.0	108.0	36.0	25.0	17.6	8.4	4.7
Lithuania	8.4	225.0	1021.0	410.0	72.1	39.5	24.7	8.9	5.1
Estonia	17.2	211.0	1076.0	90.0	48.0	29.0	23.0	11.0	10.6
Romania	5.1	161.1	210.4	256.1	136.7	32.3	38.8	154.8	59.1
Bulgaria	n.a.	333.5	82.0	73.0	96.3	62.0	123.0	1082.3	22.3
Ukraine	4.8	91.0	1210.0	4735.0	891.0	376.0	80.0	16.0	10.6

Source: Business Central Europe Magazine, The Economist Group, Historical data collected from WIIW, EBRD, Reuters, national statistics. http://www.bcemag.com/

Table A6 *Foreign debt (US$bn)*

	1990	1991	1992	1993	1994	1995	1996	1997	1998
Hungary	21.3	22.7	21.4	24.6	28.5	31.7	27.6	22.1	24.5
Poland	49.4	48.0	47.6	47.2	42.2	43.9	40.4	38.0	33.1
Czech Rep.	6.4	6.7	7.1	8.5	10.7	16.5	20.8	21.6	24.4
Slovak Rep.	2.0	2.7	3.0	3.4	4.7	5.7	7.7	10.7	11.9
Slovenia	2.0	1.9	1.7	1.9	2.3	3.0	4.0	4.8	4.9
Latvia	n.a.	n.a.	0.0	0.2	0.4	1.4	2.0	0.4	0.4
Lithuania	n.a.	n.a.	0.1	0.3	0.5	0.8	2.3	1.4	1.7
Estonia	n.a.	n.a.	0.1	0.2	0.2	0.3	1.5	0.4	0.4
Romania	1.2	2.1	3.2	4.2	5.6	6.6	8.3	8.2	9.2
Bulgaria	10.9	11.8	12.5	13.9	11.4	10.2	9.7	9.7	10.1
Ukraine	n.a.	n.a.	0.5	3.7	7.7	8.1	9.2	11.8	15.5

Source: Business Central Europe Magazine, The Economist Group, Historical data collected from WIIW, EBRD, Reuters, national statistics. http://www.bcemag.com/

Table A7 Foreign Exchange Reserves (US$bn)

	1990	1991	1992	1993	1994	1995	1996	1997	1998
Hungary	1.1	3.9	4.3	6.7	6.7	12.0	9.7	8.4	9.3
Poland	4.5	3.6	4.1	4.1	5.8	14.8	17.8	24.4	28.0
Czech Rep.	1.1	0.7	0.8	3.9	6.2	14.0	12.4	9.8	12.6
Slovak Rep.	1.1	3.2	0.4	0.5	1.8	3.4	3.5	3.3	2.9
Slovenia	n.a.	0.1	0.7	0.8	1.5	1.8	2.3	3.3	3.6
Latvia	n.a.	n.a.	n.a.	0.4	0.5	0.5	0.7	0.7	1.0
Lithuania	n.a.	n.a.	0.0	0.4	0.5	0.8	0.8	1.0	1.5
Estonia	n.a.	n.a.	0.2	0.4	0.4	0.6	0.6	0.8	0.8
Romania	0.4	0.7	0.9	1.0	0.6	0.3	0.6	2.5	1.6
Bulgaria	n.a.	0.3	0.9	0.7	1.0	1.2	0.5	2.5	2.8
Ukraine	n.a.	n.a.	0.5	0.2	0.5	1.1	2.0	2.4	1.0

Source: Business Central Europe Magazine, The Economist Group, Historical data collected from WIIW, EBRD, Reuters, national statistics. http://www.bcemag.com/

Table A8 Exchange Rate (US$)

	1990	1991	1992	1993	1994	1995	1996	1997	1998
Hungary (forint)	63.2	74.8	79.0	92.0	105.1	125.7	152.6	186.8	214.3
Poland (zloty)	1.0	1.1	1.4	1.8	2.3	2.4	2.7	3.3	3.5
Czech Rep (koruna)	18.0	29.5	28.3	29.2	28.8	26.6	27.1	31.7	32.3
Slovak Rep (koruna)	18.0	29.5	28.3	30.8	32.0	29.7	30.7	33.6	37.5
Slovenia (tolar)	11.3	27.6	81.29	113.24	128.81	118.52	135.37	159.7	165.8
Latvia (lat)	n.a.	n.a.	0.7	0.7	0.6	0.5	0.6	0.6	0.6
Lithuania (litas)	n.a.	n.a.	1.8	4.3	4.0	4.0	4.0	4.0	4.0
Estonia (kroon)	n.a.	n.a.	12.7	13.2	13.0	11.5	12.0	13.9	13.4
Romania (leu)	24	76	308	760	1655	2033	3083	7168	11233
Bulgaria (lev)	0.8	16.7	23.3	27.7	54.3	67.2	175.8	1677.0	1754.4
Ukraine (hryvnia)	n.a.	n.a.	n.a.	0.1	0.3	1.5	1.8	1.9	3.4

Source: Business Central Europe Magazine, The Economist Group, Historical data collected from WIIW, EBRD, Reuters, national statistics. http://www.bcemag.com/

Table A9 Trade Balance (US$bn)

	1990	1991	1992	1993	1994	1995	1996	1997	1998
Hungary	0.5	0.2	0.0	−3.2	−3.6	−2.4	−2.6	−1.8	−2.1
Poland	3.6	0.1	0.5	−2.3	−0.8	−1.8	−8.2	−11.3	−13.6
Czech Rep.	−0.7	−0.5	−1.9	−0.3	−0.9	−3.7	−5.9	−4.4	−2.5
Slovak Rep.	−0.7	−0.5	−0.6	−0.9	0.1	−0.2	−2.3	−1.9	−2.3
Slovenia	−0.6	−0.3	0.8	−0.2	−0.3	−1.0	−0.9	−1.0	−1.0
Latvia	n.a.	n.a.	−0.2	0.0	−0.3	−0.6	−0.8	−1.0	−1.3
Lithuania	n.a.	n.a.	0.1	−0.2	−0.2	−0.7	−0.9	−1.7	−2.1
Estonia	n.a.	n.a.	−0.1	−0.1	−0.4	−0.7	−1.0	−1.2	−1.5
Romania	−1.7	−1.3	−1.4	−1.1	−0.5	−1.6	−2.5	−2.9	−3.5
Bulgaria	−0.8	0.4	−0.2	−0.9	0.0	0.1	0.2	0.0	−0.7
Ukraine	n.a.	−3.4	−0.6	−2.5	−2.6	−2.7	−4.3	−4.2	−1.7

Source: Business Central Europe Magazine, The Economist Group, Historical data collected from WIIW, EBRD, Reuters, National Statistics. http://www.bcemag.com/

Table A10 Current Account Balance (US$bn)

	1990	1991	1992	1993	1994	1995	1996	1997	1998
Hungary	0.4	0.3	0.3	−3.5	−3.9	−2.5	−1.7	−0.9	−2.3
Poland	3.1	−2.0	0.9	−0.6	2.3	5.5	−1.3	−4.3	−6.9
Czech Rep.	−1.0	0.3	−0.3	0.1	0.0	−1.4	−4.3	−3.2	−1.0
Slovak Rep.	n.a.	n.a.	n.a.	−0.6	0.7	0.4	−2.1	−1.3	−2.1
Slovenia	1.1	0.1	0.9	0.2	0.6	0.0	0.0	0.0	0.0
Latvia	n.a.	n.a.	0.0	0.3	0.0	−0.2	−0.2	−0.5	−0.7
Lithuania	n.a.	n.a.	0.2	−0.1	−0.1	−0.6	−0.7	−1.0	−1.5
Estonia	n.a.	n.a.	0.0	0.0	−0.2	−0.2	−0.4	−0.6	−0.4
Romania	−1.7	−1.3	−1.5	−1.2	−0.5	−1.7	−2.6	−2.1	−3.0
Bulgaria	−1.2	−0.4	−0.8	−1.4	−0.2	−0.1	0.1	0.4	0.0
Ukraine	n.a.	−2.9	−0.6	−0.8	−1.2	−1.2	−1.2	−1.3	−1.3

Source: Business Central Europe Magazine, The Economist Group, Historical data collected from WIIW, EBRD, Reuters, national statistics. http://www.bcemag.com/

Index

activity location strategies 26–7, 297–9, 308
  Baltic States 199
  Bulgaria 257
  Czech Republic 111
  Hungary 51–2
  Poland 78
  Romania 227–8
  Slovenia 169–70
  Ukraine 277–8
adaptation, product *see* product adaptation
Akmaya 215
Alcatel 138–9, 255
Algida 77
Anti-Monopoly Office, Slovakia 134, 135
AOI NV Group 187
ArchiCAD 42
Arthur Andersen 259
Asea Brown Boveri (ABB) 10, 25
  in Baltic states 195, 197, 198–9
  in Romania 216, 228
  Service Ostrava 111
  in Ukraine 277–8
Asian Investments Ltd 201
Asko 246
attitudes, business 12
Audi 49
autonomy decisions, operational 29–30
AvtoZAZ 279

Baltic Beverages Holdings 183
Baltic States 177, 203–204, 285–6
  activity location strategies 199
  advertising 200
  background 177
  company ownership 203
  competitive globalization drivers 194
  consulting companies 198
  consumer market segments 196–7
  cost globalization drivers 186–8
  domestic companies 181, 182, 183
  domestic consumers 200
  economic prospects 179–80
  effects of Russian crisis 204
  foreign investment 181–3
  Free Economic Zones 204
  global competitive strategies 201
  global marketing 199–200
  government globalization drivers 188–93
  infrastructure 201
  labour force 201
  liberalization 193
  managers 201–202
  market globalization drivers 184–6
  market participation strategies 194–6
  MNC regional headquarters 202
  new technologies 188
  organization and management approaches 201–203
  power supplies 188
  privatization 193
  product and service strategies 196–8
  statistics 205–207
  technical standards 193
  telecommunications systems 187–8
  transition process 6, 7
  transportation routes 186–7
  wage/salary rates 188
Banco Portugues de Investimento 212
Bank Austria 169

BAT 281
BATA 125
BAZ (Bratislava Automobile Company) 142
BBAG 227
Belarus, transition process 6
Benckiser 10, 27, 68, 73, 78
BNP-Dresdner 243
BorsodChem 42
Bratislava Stock Exchange 124, 131
Briegl & Bergmeister 173
Bristol-Myers Squibb 51
British Aerospace 69
Budapest Bank 50
Budejovicky Budvar 90
budget balance, table of values by country 311
Bulgaria 240, 260–62
  activity location strategies 257
  advertising 248, 258
  business services 255
  communications 250
  competitive globalization drivers 253–4
  competitive move strategies 258–9
  cost globalization drivers 249–51
  distribution networks 258
  domestic companies 246
  education 249
  favorable country costs 249
  foreign direct investment 243–5, 252
  foreign involvement 245–6
  foreign trade policies 251
  global customers 247–8
  government globalization drivers 251–3
  government intervention 252–3
  history 241
  human resources management 259–60
  industrial market 255
  infrastructure 249–50
  legal protection 252
  market globalization drivers 247–8
  market participation strategies 254–5
  marketing strategies 257–8
  organization and management approaches 259–60
  pricing 258
  privatization 252–3
  product promotion 258
  product and service strategies 256–7
  state-owned companies 252
  technology market 250–51
  trade blocs 251
  transferable marketing 248
  transition process 7, 241–3
Bulgarian International Business Association 261
Burger King 26, 255
business practices 12

Cable & Wireless 245
Calver Decken 183, 201
Canon 255
Carlsberg 183
Centertel 67–8
Central and Eastern Europe (C&EE)
  definition 3–4
  foreign direct investment growth 14, 20, 21
  key statistics 1, 2
  map 5
  MNC operating levels 3
Central European Free Trade Association (CEFTA) 10, 13, 20, 105–106

Hungary 47
Poland 62
Romania 224
Slovenia 153, 165
Ukraine associate membership 274
Chinoin 48
CIS
　country-market 25
　grouping 20
　and Ukraine 274
Coca-Cola
　Amatil 49, 50, 51, 52
　in Baltic states 196, 201
　in Bulgaria 245, 256
　in Romania 228, 229
　in Ukraine 267, 271, 276, 277, 278–9, 281
Cokoladovny 109
Colgate-Palmolive 110, 216–17, 219, 227, 230, 233
COMECON (Council for Mutual Economic Help) 9,13, 25, 121, 151, 285
Comliet 195
COMPA Sibiu 219
competition
　foreign 24
　internationalized domestic 24
　state-owned 21
competitive globalization drivers 23–4, 294–6, 307
　Baltic States 194
　Bulgaria 253–4
　Czech Republic 106–107
　Hungary 48–9
　Poland 72–3
　Romania 225–6
　Slovakia 137–8
　Slovenia 167
　Ukraine 275–6
competitive move strategies 28, 309
　Bulgaria 258–9
　Czech Republic 113–14
　Hungary 53
　Poland 79–80
　Romania 230–31
　Slovenia 171–2
Computerland 27
CONEL 221
CONNEX GSM 229–30, 236
cost globalization drivers 17–19, 289–92, 305–306
　Baltic States 186–8
　Bulgaria 249–51
　Czech Republic 100–103
　Hungary 44–5
　Poland 68–9
　Romania 220–22
　Slovakia 131–4
　Slovenia 162–4
　Ukraine 271–3
Council for Mutual Economic Assistance (CMEA) 9
country globalization drivers 304
　Ukraine 269
country ratings 286–7
country strategy, development 3
country-markets, selection 25–6
CPC Foods 75
Crédit Lyonnais 272
cross-investment 13
cultural characteristics
　common 11
　differences 31
current account balance, table of values by country 314
Cussons 73, 79
customer needs, global/regional 15–16
Cyfra+ 80
Czech Republic 86, 115–17, 286
　activity location strategies 111
　advent of MNCs 24
　advertising strategies 112
　airport facilities 102
　automotive sector 96
　banking sector 97
　competitive globalization drivers 106–107
　competitive move strategies 113–14
　cost globalization drivers 100–103

country-market 25
cultural attitudes 115
customer needs 99, 112
domestic companies 90–92
entry strategies 108–109
external trade 89–92, 90–92
foreign investment 92–7
foreign MNCs 97–8
global channels 100
government globalization drivers 103–106
government regulation 105
history 86–8
infrastrucuture 102–103
investment incentives 103–104
investment protection 104
joint ventures 93, 95, 109
labour resources 101–102
leasing advantages 113
legal protection 104
market globalization drivers 99–100
market participation strategies 108–109
marketing strategies 112–13
natural resources 102
organization and management approaches 114–15
pricing policies 113
privatization 87
product and service strategies 110
Russian financial crisis effects 89–90
technical standards 104–105
technology capabilities 103
trade blocs 105–106
transferable marketing 99–100
transition process 6, 89
CzechInvest 106

Daewoo Corporation
　in Bulgaria 255, 257
　in Czech Republic 96, 100, 110, 111
　in Romania 216, 219, 227, 230, 231, 234–5
　in Slovenia 168
　in Ukraine 268, 279
Danone 109, 246, 255, 256, 257
Danubiana 235
DERO Ploiesti 230
distribution networks 29
Droga 172
Dunafferr ironworks 41
Dunkin Doughnuts 254

Eastman Kodak 79
economic performance 14
Eight S Formula (C&EE market benefits) 284–5
Elcoteq Tallinn AS 198
Electrolux 169
Elektrim 69
Elinta 197
English, as business language 12
Ericsson 48, 69, 194, 198, 229, 245, 255
Estonia *see* Baltic States
European Bank for Reconstruction and Development (EBRD) 50, 236, 268
European Union (EU) 25
　Baltic States associate membership 189
　Bulgaria associate membership 251
　car imports to Ukraine 279
　Czech Republic 106
　Free Trade Agreement 189
　Harmonized Duty System 135
　Hungary associate membership 35, 47, 48, 55
　new members 6
　Poland 62, 69, 74
　priority transport routes 286
　Slovenia associate membership 153, 165, 166
　technical standards 72
European Venture Capital Association (EVCA) 129
exchange rates, table of values by country 313
expatriate managers 31
Exxon 255

factors of production, availability 17–18
Farm Frites 70
FARMEC Cluj 216

Index

Festo 197
Fiat 26, 233
FIBV *see* International Federation of Stock Exchanges (FIBV)
First International Computer (FIC) 100, 108
Ford 49
foreign debt, table of values by country 312
foreign direct investment (FDI) 14, 20, 21
   Baltic States 181–3
   Bulgaria 243–5
   Czech Republic 92–7
   Hungary 47
   Poland 62–4
   Romania 213–17, 223–4
   Slovakia 127–31
   Slovenia 157, 159
   Ukraine 268, 274
foreign exchange reserves, table of values by country 313
Fotex 42
Free Economic Zones (FEZ), Baltic States 187, 204
Frigorex 228

Gazprom 24, 255
GDP, table of values by country 310
General Electric (GE) 28, 41, 44, 50, 51, 52, 95, 111, 212
Generalized System of Preferences (GSP) 189
GEROCOSSEN Bucharest 216
global retailers/distribution channels 16
Global Ukraine 273
Goodyear 168, 169, 173
government globalization drivers 19–23, 292–4, 306–307
   Baltic States 188–93
   Bulgaria 251–3
   Czech Republic 103–106
   Hungary 45–9
   Poland 69–72
   Romania 222–5
   Slovakia 134–7
   Slovenia 165–7
   Ukraine 273–5
governments
   differing forms 13
   level of intervention 21
Graphisoft 27, 42–3, 51
GSM Bite 195
GTV 197

Hellenic Telecommunication Organization (OTE) 212
Hellmann 27, 167
Henkel 28, 73, 141–2, 169
Hewlett-Packard 254
Hill International 259
Hitech 197
Hortex 65–6
human resources, using local/non-local 30–31
Hungarian National Bank 47
Hungary 33, 54–5
   activity location strategies 51–2
   advent of MNCs 24
   automobile industry 51–2
   banking industry 46, 50
   competitive globalization drivers 48–9
   competitive move strategies 53
   cost globalization drivers 44–5
   country-market 25
   domestic companies 41–3, 51
   economic reforms 35–6
   established MNCs 39
   EU associate membership 35, 47, 48, 51
   exports 40–41
   foreign direct investment 47
   foreign investment 36–41, 47
   foreign shareholdings 46
   foreign trade policies 46–7
   global strategies 49
   government globalization drivers 45–9
   history 33–5
   International Management Center, Budapest 53
   investor motivation 45
   legal protection 48
   management practices 53–4
   market globalization drivers 43–4
   market participation strategies 49–50

marketing strategies 52
NATO membership 35
New Economic Mechanism 35
privatization 39, 45–6
product and service strategies 51
shopping malls 52–3
skilled workers 44, 54
technical standards 48
transition process 6
Hyundai 111, 219

IBM 49, 140–41, 199, 245, 254, 257, 259, 260
IFC (World Bank subsidiary) 96
Ikarus 42
IKEA 25
industrial production, table of values by country 310
inflation, table of values by country 312
infrastructure 14–15
   regional 18
ING Bank 243
institutional arrangements, changing 13
integrating factors, business 9–15
inter-regional trade 9
Interbrew 227
International Federation of Stock Exchanges (FIBV) 61
International Management Center, Budapest 53
International Monetary Fund (IMF) 20
   Bulgarian agreement 243
intra-regional trade 9–10
investment rules 20
Itochu Trading House 48

Johnson & Johnson 168, 169, 170, 255

Kentucky Fried Chicken 26
KFC 254
KJS, in Ukraine 277
Korean companies 26
Koyo Seiko 235
KPMG 256
Kraft Jacobs Suchard (KJS)
   in Baltic States 185, 186, 195, 197, 198, 199, 202
   in Bulgaria 255, 257
   in Romania 219, 227
   in Ukraine 277
KRKA (pharmaceutical company) 75, 158
Krupp-Bilstein 219
Kyocera 96

Lafarge 234, 236
languages, business 12
Latvia *see* Baltic States
lead markets 16
legal protection 21
Lek 158
LG Electronics 219
LG Group 111
Lithuania, *see* Baltic States
local managers 31
local partners, selection 29
logistics, advantages 18
Lucent Technologies 273
LuckyNet 273
LUKoil 24, 196, 215, 236–7

McDonald's 26, 228–9, 231, 254, 256, 276, 278
   in Ukraine 280, 281
management processes, global 30, 309
market globalization drivers 15–24, 287–9, 304–305
   Baltic States 184–6
   Bulgaria 247–8
   Czech Republic 99–100
   Hungary 43–4
   Poland 66–8
   Romania 218–19
   Slovakia 131
   Slovenia 160–62
market participation strategies 25–6, 308
   Baltic States 194–6
   Bulgaria 254–5
   Czech Republic 108–109
   Hungary 49–50

Poland 74–6
Romania 226
Slovenia 168–9
Ukraine 276–81
marketing regulation 23
marketing strategies 3, 27–8, 308
  Baltic States 199–200
  Bulgaria 257–8
  Czech Republic 112–13
  Hungary 52
  Poland 78–9
  Romania 228–30
  Slovenia 170–71
marketing transferability 16
Mars advertising 100
MATAV (Hungarian state telephone company) 46
Matsushita 96, 100, 108
Mecatim Timisoara 235
Medicor 41
Mercedes 219
Metsa-Serla 202
Michelin 41
Microsoft 69
Minolta 255
MIRAJ Bucharest 216
MIRALON Bucharest 216
MobiFon 229–30, 231
Mobil Rom 231
mobile phone systems, Poland 68
MOL (Hungarian oil and gas company) 41, 42, 46
MOM 41
Motorola 69, 78, 107, 194, 273
multinational companies (MNCs)
  C&EE market benefits 284–5
  C&EE regional strategy 3, 296–7
  headquarters location 16
  investment examples 13–14
  operating levels 3
  regional scale economies 17
  value chains 24
Multiservis 95

National Bank of Slovakia (NBS) 121
National Bank of Ukraine (NBU) 272
NATO 7, 20, 62, 302
Neste Oy 194, 196, 203
Nestlé 70, 109, 167, 169, 248, 255, 257, 280
New Holland 233
Nokia 48, 51
Nomura International 235
NORVEA Brasov 216, 219
Novartis 78
Nutella advertising 100

Omnitel 194
OMV (oil company) 139–40, 158
Opel 49
Optimus 73
organization approaches, global 29–30, 299–302, 309
Organization for Economic Cooperation and Development (OECD) 47
OTELINOX Targoviste 216

Pannonplast 41, 42
Papirnica Vevce 173
PepsiCo 49, 50, 79, 80, 196, 201, 245, 256
Petrol (fuel distribution company) 171–2
Petromidia Cnstanta 215
Pfleiderer 170
Pharmavit 51
Philip Morris 195, 267, 277, 281
Philips 28, 49, 51, 183
PipeLife International 230
pirating activities 21
Pizza Hut 254
Podravka 75
Polam Pila 28
Poland 57, 82, 285
  activity location strategies 78
  advent of MNCs 24
  advertising sector 77
  Anti-Monopoly Office (AMO) 71

automotive market 67
banking services 76
base for MNC regional offices 67
Branch Office role 81
capital market 61–2
car markets 75
common customer needs 66–7
company forms 81
competitive globalization drivers 72–3
competitive moves 79–80
consumer goods 75
cost globalization drivers 68–9
country-market 25
digital television services 80
domestic companies 64–6, 73
economic performance 60–62
Fiat agreement 26
financial sector 75
foreign investment 62–4, 71, 73, 76
global retailers 75
government globalization drivers 69–72
government intervention 70
history 57–8
ice-cream market 77
infrastructure 69
labor force 68
legal protection 71, 72
management consultancy 77
market globalization drivers 66–8
market oriented reforms 59
market participation strategies 74–6
marketing strategies 78–9
Mass Privatization Program 60–61
mobile phone sector 67–8
non-financial sector credit 76
organization and management approaches 80–82
pension schemes 61–2
petrol stations 74
pricing strategies 80
privatization 70–71
product and service strategies 76–7
research and development facilities 68–9
Russian investment 75
Special Economic Zones (SEZs) 70
strategic importance 72–3
supermarkets 74
Supervisory Office role 80
Technical Information Office role 81
technical standards 71–2
telecommunication services 74–5
telecommunications 69
tourist industry 63–4
transition process 6, 58–9
WIG Index 61
Povazsko Kysucky Enterprise Fund (PKPF) 130
Procter & Gamble (P & G) 73, 78, 169, 227–8, 255, 267, 275, 280
product adaptation 26, 27
product and service strategies 26, 308
  Baltic States 196–8
  Bulgaria 256–7
  Czech Republic 110
  Hungary 51
  Poland 76–7
  Romania 226–7
  Slovakia 131
  Slovenia 168–9
production costs 18
PZL Mielec 69

quality awards, Hungary 48

R J Reynolds 281
Recognita 42
Reemtsma 277, 281
Regie Autonoma (RA) 212
regional headquarters, location 29
regional strategy development 3
Renault 158, 168, 169, 170, 199
Richter 41, 42
RJR Tobacco 277
RM System Slovakia 124, 131

Index

Romania 209, 237
  activity location strategies 227–8
  banking sector 212
  common market regulations 225
  competitive globalization drivers 225–6
  competitive move strategies 230–31
  consumer needs 218–19
  cosmetics industry 216–17
  cost globalization drivers 220–22
  domestic companies 217
  electrotechnical industry 216
  favorable country costs 221
  foreign direct investment 213–17, 223–4
  Galati Free Zone 223
  government globalization drivers 222–5
  government intervention 225
  history 209–10
  infrastructure 220
  labor costs 221–2
  legal protection 225
  liberalization process 212
  market globalization drivers 218–19
  market participation strategies 226
  marketing strategies 228–30
  metallurgical industry 216
  oil and gas sector 215
  organization and management approaches 231–7
  privatization 212
  product and service strategies 226–7
  resources 220–21
  retailing services 221
  SME sector 212–13
  technical standards 225
  technology level 222
  trade blocs 224–5
  trade policies 223
  transferable marketing 219
  transition process 7, 210–13
  TV sets market 219
Romcim 234
Romgaz 236
Roti Auto Dragasani 234
Rover 256
Royal Dutch Shell 236, 245
Rubic's Cube 51
Rulmenti Alexandria 235
Rulmenti Grei 233
Russia
  1998 financial crisis 14, 89–90, 204, 286
  advent of MNCs 24
  investment in Poland 75
  role in C&EE development 8
  transition process 6, 8

S C Johnson 277, 280
Samsung 111, 183, 219, 273
Sanofi 48
Sava 158, 168, 169, 173
scale economies, global/regional 17
Schöller 77
Schweppes 245
scientific resources 19
Seed Capital Company 130
Semanatoarea 233
Shell 194, 196, 203, 215, 236, 245, 255, 256
Showa Aluminium Corporation 108
Siemens 199, 255
Sigmanta 199
Skoda Auto 10, 96, 98, 105, 115
Slovak American Enterprise Fund (SAEF) 130
Slovak National Bank 131
Slovak Venture Capital Association (SLOVCA) 130
Slovakia 119, 142–4, 286
  arbitration rights 136
  automobile industry 142
  banking system 131
  bankruptcy law 136
  capital markets 130–31
  civil engineering industry 138
  communications industry 138–9
  competition law 134
  competitive globalization drivers 137–8
  computing and IT industry 140–41
  cost globalization drivers 131–4
  customs duty system 135
  detergent industry 141–2
  domestic companies 125, 125–6
  energy sector 133
  environment 137
  favorable country costs 132
  fiscal environment 124
  foreign direct investment 127–31
  foreign trade 124–5
  government globalization drivers 134–7
  Gross Domestic Product 121, 123
  history 119–20
  human resources 144
  inflation 123
  infrastructure and logistics 132
  Labor Code 145
  labor laws and regulations 144–7
  labour resources 132–3
  legal environment 123
  liberalization process 134
  macro-economic indicators 122
  market globalization drivers 131
  MNCs role 127, 129, 138–42
  oil and gas industry 139–40
  packaging industry 140
  political organization 120–21
  population 120
  product and service strategies 131
  property rights protection 136
  social benefits 144
  sourcing efficiencies 133
  tax treaties 135
  technical skills 133
  technical standards 136
  technology role 133–4
  trade liberalization 124–5
  transition process 121–5
  unemployment 146–7
  venture capital business 123–4, 129–30
  wage policy 144, 146
Slovasfalt 138
Slovenia 150, 174, 175
  activity location strategies 169–70
  automobile industry 158
  business infrastructure 164
  competitive globalization drivers 167
  competitive move strategies 171–2
  cost globalization drivers 162–4
  currency regime 153
  customer needs 160
  domestic companies 155, 156
  education 163–4
  employment levels 163
  favorable country costs 163
  financial institutions 153–4
  foreign investment 157, 159
  foreign retailers 158
  GDP 152
  global marketing strategies 170–71
  global strategies 167–8
  government globalization drivers 165–7
  history 150–51
  infrastructure 162–3
  insurance companies 154
  legal protection 166
  liberalization policies 167
  market globalization drivers 160–62
  market participation strategies 168–9
  marketing practices 165
  organization and management approaches 172–4
  pension systems 154
  price liberalization 153
  privatization 152
  product and service strategies 168–9
  research and development activities 170
  scale economies 162
  securities markets 154
  tax legislation 154
  technical standards 166
  technology role 164

trade blocs 165
trade policies 165–6
transferable marketing 161–2
transition process 6, 7, 151–5
Wages Agreement 154
Société Générale 212
SOLIDARITY movement 58–9
Solvay 257
SPT Telecom 107
Stambouli 246
standards, technical *see* technical standards
Statoil 194, 196, 203
STEAUE Sibiu 216
STELA Bucharest 219
strategy development
  activity location 26–7
  competitive move 28
  global/regional 23–4
  market participation 25–6
  marketing 27–8
  product and service 26
Suzuki 48–9
Svitoch 280

Tatra Raiffeisen Capital (TRC) 130
Taurus Rubber Works 41
technical standards 22–3
  Baltic States 193
  Czech Republic 104–105
  Hungary 48
  Poland 71–2
  Romania 225
  Slovakia 136
  Slovenia 166
  Ukraine 275
technology/scientific resources 19
Telia 194
Telsource 107
Tesla Sezam 107
TetraPak 140, 277
Time-Warner 219
Timken 233
Tofan Group 234, 235–6
Tolaram Group 183, 201
Toray Industries 95, 100
Toyota 82
TP SA (telecommunications services) 74
trade balance, table of values by country 314
trade blocs 20
trade policies, favorable 20
transition process 4–7
Transparency International (global corruption watch group), ranking report 21
Tungsram 28, 41, 44, 50
Tyn Group of Funds 130

Ukraine 264, 283
  activity location strategies 277–8
  advertising 270–71
  automotive industry 279
  capital market 271–2
  common market regulations 275
  competitive globalization drivers 275–6
  cost globalization drivers 271–3
  country globalization drivers 269
  customer needs 270
  domestic companies 267, 276
  economy 266–7
  energy resources 272
  financial system 272
  foreign competition 276
  foreign direct investment 268, 274
  foreign trade 267–8
  government globalization drivers 273–5
  government intervention 274
  GSM licenses 273
  history 264–5
  infrastructure 272
  labor resources 271
  land resources 271
  legal protection 274–5
  logistics 272
  market participation strategies 276–81
  marketing strategies 281
  organization and management approaches 282–3
  politics 265–6
  product and service strategies 277
  regions 270
  social climate 268–9
  technical standards 275
  technology resources 273
  telecommunications 272–3
  trade blocs 274
  trade policies 273–4
  transferable marketing 270–71
  transition process 7
unemployment, table of values by country 311
Unilever
  in Czech Republic 115
  in Hungary 50, 51, 52
  in Poland 73, 76, 77, 79
  in Romania 227, 230–31, 233, 238
Union Miniere 257
United Technologies Automobile Hungary (UTAH) 50, 51
Utel 273

Vienna Convention 251
Vilniaus Vingis 183
Volkswagen (VW) 10, 96, 98, 110, 142

wage rates, by country 18–19
Warsaw Stock Exchange (WSE) 61–2
Wassenar Agreement 251
Wedel 79, 80
Weinerberger Baustoffindustrie 115
Westel 900 Mobile Telecommunications Company 50
Williams International 194
Winiary 70
Wizja TV 80
World Bank 20
  see also IFC (World Bank subsidiary)
World Intellectual Property Organization (WIPO) 136
World Trade Organization (WTO)
  Hungary 47
  Poland 62, 69
  Romania 224
  Slovakia 136
  Slovenia 165

Xerox 248, 255, 257, 258

Zavarovalnica Triglav 172
Zwack Unicum 42